COMPUTATIONAL
NUMERICAL METHODS

ELLIS HORWOOD SERIES IN COMPUTERS AND THEIR APPLICATIONS
Series Editor: Brian Meek, Computer Centre, King's College, University of London

Atherton, R.	Structured Programming with COMAL
Bailey, R.	Functional Programming with HOPE
Barrett, R., Ramsay, A. & Sloman, A.	POP-11: A Practical Language for Artificial Intelligence
Berry, R.E.	Programming Language Translation
Bishop, P.	Fifth Generation Computers: Concepts, Implementations and Uses
Brailsford, D.F. & Walker, A.N.	Introductory ALGOL 68 Programming
Bull, G.M.	The Dartmouth Time Sharing System
Burns, A.	New Information Technology
Burns, A.	The Microchip: Appropriate or Inappropriate Technology
Carberry, J.C.	COBOL
Chivers, I.D.	A Practical Introduction to Standard PASCAL
Chivers, I.D. & Clark, M.W.	Interactive FORTRAN 77: A Hands-on Approach
Clark, M.W.	pc-Portable FORTRAN
Cope, T.	Computing using BASIC: An Interactive Approach
Dahlstrand, I.	Software Portability and Standards
Davie, A.J.T. & Morrison, R.	Recursive Descent Compiling
de Saram, H.	Programming in micro-PROLOG
Deasington, R.J.	A Practical Guide to Computer Communications and Networking, 2nd Edition
Deasington, R.J.	X.25 Explained: Protocols for Packet Switching Networks
Eisenbach, S.	Functional Programming: Languages, Tools and Architectures
Ellis, D.	Computers in Medical Practice
Ennals, J.R.	Artificial Intelligence: Applications to Logical Reasoning and Historical Research
Fernando, C.	Cluster Analysis on Microcomputers
Fossum, E. *et al.*	Computerization of Working Life
Gray, P.M.D.	Logic, Algebra and Databases
Harland, D.M.	Concurrency and Programming Languages
Harland, D.M.	Polymorphic Programming Languages
Hill, I.D. & Meek, B.L.	Programming Language Standardisation
Hogger, C.J. & Hogger, E.I.	PROLOG for Scientists and Engineers
Hutchison, D.	Fundamentals of Computer Logic
Hutchison, D. & Silvester, P.	Computer Logic: Principles and Technology
Hutchins, W.J.	Machine Tranlsation: Past, Present, Future
Lester, C.	Advanced Programming in PASCAL
McKenzie, J., Elton, L. & Lewis, R.	Interactive Computer Graphics in Science Teaching
Matthews, J.	FORTH
Meek, B.L., Heath, P. & Rushby, N.	Guide to Good Programming Practice, 2nd Edition
Millington, D.	Systems Analysis and Design for Computer Application
Moore, L.	Foundations of Programming with PASCAL
Moseley, L., Sharp, J. & Salenieks, P.	PASCAL in Practice
Moylan, P.	Assembly Language for Engineers
Narayanan, A. & Sharkey, N.E.	An Introduction to LISP
Paterson, A.	Office Systems: Planning, Procurement and Implementation
Pemberton, S. & Daniels, M.C.	PASCAL Implementation
Pesaran, M.H. & Slater, L.J.	Dynamic Regression: Theory and Algorithms
Peter, R.	Recursive Functions in Computer Theory
Phillips, C. & Cornelius, B.J.	Introduction to Computational Numerical Methods
Ramsden, E.	Microcomputers in Education 2
Ramsay, A. & Barrett, R.	POP-11 in Practice: Examples in Artificial Intelligence
Schirmer, G.	Programming in C
Sharp, J.A.	Data Flow Computing
Smith, I.C.H.	Microcomputers in Education
Späth, H.	Cluster Analysis Algorithms
Späth, H.	Cluster Dissection and Analysis
Stratford-Collins, M.J.	ADA: A Programmer's Conversion Course
Teskey, F.N.	Principles of Text Processing
Tizzard, K.	C for Professional Programmers
Turner, S.J.	An Introduction to Compiler Design
Whiddett, R.J., *et al.*	UNIX: A Practical Introduction for Users
Whiddett, R.J.	Concurrent Programming: Techniques and Implementations
Young, S.J.	An Introduction to ADA, 2nd (Revised) Edition
Young, S.J.	Real Time Languages

COMPUTATIONAL NUMERICAL METHODS

CHRIS PHILLIPS
Computing Laboratory
University of Newcastle upon Tyne
and
BARRY CORNELIUS
Department of Computer Science
University of Durham

ELLIS HORWOOD LIMITED
Publishers · Chichester

Halsted Press: a division of
JOHN WILEY & SONS
New York · Chichester · Brisbane · Toronto

First published in 1986 by
ELLIS HORWOOD LIMITED
Market Cross House, Cooper Street, Chichester, West Sussex, PO19 1EB, England

The publisher's colophon is reproduced from James Gillison's drawing of the ancient Market Cross, Chichester.

Distributors:

Australia and New Zealand:
Jacaranda-Wiley Ltd., Jacaranda Press,
JOHN WILEY & SONS INC.
GPO Box 859, Brisbane, Queensland 4001, Australia

Canada:
JOHN WILEY & SONS CANADA LIMITED
22 Worcester Road, Rexdale, Ontario, Canada

Europe and Africa:
JOHN WILEY & SONS LIMITED
Baffins Lane, Chichester, West Sussex, England

North and South America and the rest of the world:
Halsted Press: a division of
JOHN WILEY & SONS
605 Third Avenue, New York, NY 10158, USA

© 1986 C. Phillips and B. Cornelius/Ellis Horwood Limited

British Library in Cataloguing in Publication Data
Phillips, C.
Computational numerical methods. —
(Ellis Horwood series in computers and their applications)
1. Numerical analysis — Data processing
I. Title   II. Cornelius, B.
519.4'028'5   QA297

Library of Congress Card No. 86-7385

ISBN 0–85312–495–7 (Ellis Horwood Limited)
ISBN 0–470–20336–6 (Halsted Press)

Phototypset in Times by Ellis Horwood Limited
Printed in Great Britain by R. J. Acford, Chichester

COPYRIGHT NOTICE
All Rights Reserved. No part of this publication may be reproduced, stored in a retrieval system, or transmitted, in any form or by any means, electronic, mechanical, photocopying, recording or otherwise, without the permission of Ellis Horwood Limited, Market Cross House, Cooper Street, Chichester, West Sussex, England.

Contents

Preface 11

Acknowledgements 13

1 INTRODUCTION
    1.1 Numerical Analysis. 15
    1.2 Mathematical Formulae 17
    1.3 The Idea of Iteration 19
    1.4 Exploratory Work 21
    1.5 Computing Resources 22
    1.6 Numbers: Fixed and Floating Formats 23
    1.7 Errors 24
        1.7.1 Sources of error 24
        1.7.2 Rounding errors. 25
    1.8 Numerical Instability and Ill-conditioning 27
    1.9 Computer Programs 29
    1.10 Some Important Mathematical Results. 31
        Exercises 34

2 NON-LINEAR ALGEBRAIC EQUATIONS
    2.1 Introductory Remarks 37
    2.2 The Method of Bisection 38
    2.3 Functional Iteration 40
        2.3.1 Introduction 40
        2.3.2 A convergence theorem. 42
        2.3.3 Discussion of the convergence theorem 43
        2.3.4 A program for functional iteration 47
    2.4 Aitken Acceleration 49
    2.5 Newton–Raphson Iteration 51

2.5.1 Derivation 51
2.5.2 Discussion of Newton–Raphson iteration 51
2.5.3 Termination criteria 54
2.5.4 A program for Newton–Raphson iteration 55
2.6 The Order of Convergence of Functional Iteration 58
2.6.1 Introductory remarks and definition 58
2.6.2 Practical considerations 62
2.7 Interpolation Methods 62
2.7.1 Introduction 62
2.7.2 The rule of false position 63
2.7.3 The Illinois and Pegasus Methods 64
2.7.4 The secant method 65
2.7.5 Other methods 66
2.8 Polynomial Equations 66
2.8.1 Introduction 66
2.8.2 Evaluation of a polynomial and its derivative 67
2.8.3 A program for polynomial evaluation 70
2.8.4 Synthetic division of a polynomial by a quadratic 72
2.8.5 Conditioning 73
2.9 Non-linear simultaneous equations 74
2.9.1 Introduction 74
2.9.2 Functional iteration 75
2.9.3 Newton's method 76
2.9.4 Bairstow iteration 80
2.10 The Implementation of Iterative Processes 81
Exercises 82

3 LINEAR SIMULTANEOUS EQUATIONS
3.1 Introduction 84
3.2 Notation 86
3.3 Gauss Elimination and Back-substitution 88
3.3.1 Discussion of the basic method 88
3.3.2 A program for Gauss elimination 92
3.3.3 Several right-hand sides 96
3.3.4 Pivoting strategies and scaling 97
3.3.5 An improved version of the program for Gauss elimination 99
3.4 Conditioning and Error Analysis 105
3.4.1 Conditioning 105
3.4.2 Iterative refinement of an approximate solution 106
3.4.3 Error analysis 107
3.5 Iteration 108
3.5.1 Derivation of the Jacobi and Gauss–Seidel schemes . . . 108
3.5.2 A convergence theorem 110
3.5.3 Termination criteria 112
3.5.4 A program for Gauss–Seidel iteration 113
3.5.5 Computational efficiency 117

    3.5.6 A further discussion of iterative methods 118
  3.6 Factorisation Methods. 120
    3.6.1 Introduction 120
    3.6.2 The methods of Doolittle, Crout and Choleski 125
    3.6.3 Factorisation of a tridiagonal matrix 128
    3.6.4 A program for tridiagonal coefficient matrices 131
  3.7 Matrix Inversion. 137
  3.8 Concluding Remarks 138
    Exercises 139

4 APPROXIMATION OF CONTINUOUS FUNCTIONS
  4.1 Introduction. 143
  4.2 Taylor Series Approximation 143
  4.3 Least Squares Approximation. 147
    4.3.1 Introduction to best approximation 147
    4.3.2 Least squares 149
    4.3.3 A program for least squares approximation. 153
    4.3.4 Ill-conditioning of the normal equations 159
  4.4 Use of Orthogonal Polynomials 160
    4.4.1 Legendre polynomials. 160
    4.4.2 Polynomial orthogonalisation 164
    4.4.3 A program for polynomial evaluation. 169
  4.5 Generalised Least Squares Methods. 171
    4.5.1 Weighted least squares 171
    4.5.2 Gram–Schmidt orthogonalisation process. 174
    4.5.3 Chebyshev polynomials 176
    4.5.4 Choosing a basis 178
  4.6 Approximation on Non-finite Intervals 178
  4.7 Alternative Criteria for Best Approximation 180
    4.7.1 L_p approximation 180
    4.7.2 Minimax approximation. 181
    4.7.3 Chebyshev series approximation 184
    4.7.4 A program for Chebyshev series approximation 187
    4.7.5 Least first power approximation 191
  4.8 Further Comments 193
    Exercises 194

5 APPROXIMATION OF NUMERICALLY DEFINED FUNCTIONS
  5.1 Introduction. 197
  5.2 Lagrange Interpolation 198
    5.2.1 Interpolation of function values. 198
    5.2.2 Function minimisation using LIP's 205
    5.2.3 A program for function minimisation based on quadratic
         interpolation 207
    5.2.4 The case of equally spaced points 211
  5.3 Numerical Differentiation and the Approximation of
     Derivatives 213
  5.4 Hermite Interpolation. 215

      5.4.1 Interpolation of function and derivative values 215
      5.4.2 Function minimisation using HIP's 218
      5.4.3 A program for function minimisation based on cubic
           interpolation 219
  5.5 Spline Approximation 223
      5.5.1 Approximation using piecewise polynomials 223
      5.5.2 B-splines 227
  5.6 Discrete Approximation 229
      5.6.1 Introduction to discrete best approximation 229
      5.6.2 Discrete least squares approximation 230
      5.6.3 Discrete Gram–Schmidt orthogonalisation 233
      5.6.4 Discrete minimax and least first power approximation . . 236
      5.6.5 Non-uniqueness of l_p approximations 239
      5.6.6 A program for least first power straight line approximation 241
  5.7 Available Algorithms 244
      Exercises 244

6 NUMERICAL INTEGRATION
  6.1 Introductory Remarks 247
  6.2 Newton–Cotes Quadrature Rules 251
      6.2.1 Closed and open rules 251
      6.2.2 Composite Newton–Cotes rules 257
      6.2.3 A program for trapezium rule approximation 261
  6.3 The Romberg Quadrature Scheme 266
      6.3.1 The Romberg T-table 266
      6.3.2 A program for Romberg integration 269
  6.4 Gauss Quadrature 273
      6.4.1 Gauss Legendre quadrature 273
      6.4.2 A program for Gauss Legendre integration 277
  6.5 Other Gauss Quadrature Rules 280
  6.6 Patterson Quadrature Rules 282
  6.7 Clenshaw–Curtis Quadrature 283
  6.8 Adaptive Quadrature 284
      6.8.1 Adaptive Simpson's rule 284
      6.8.2 A program for adaptive quadrature 290
  6.9 The Computational Error 296
  6.10 The Quadrature Chapter of the NAG Algol 68 Library . . . 298
      Exercises 298

7 ORDINARY DIFFERENTIAL EQUATIONS
  7.1 Introduction 301
  7.2 Euler's Method 303
      7.2.1 Derivation of Euler's method 303
      7.2.2 A program implementing Euler's methods 305
      7.2.3 Use of the test equation 307
      7.2.4 Estimating the truncation error 308

7.3	The Use of Taylor Series	311
7.4	Trapezoidal Method	313
	7.4.1 Deriving the trapezoidal method	313
	7.4.2 A program implementing the trapezoidal method	316
7.5	An Introduction to Predictor-Corrector Method	316
	7.5.1 A simple predictor-corrector pair	316
	7.5.2 The local truncation error	322
7.6	Stability	324
7.7	Further Predictor–Corrector Methods	326
7.8	Runge–Kutta Methods	328
7.9	Second Order Equations	331
	7.9.1 Introduction	331
	7.9.2 A pair of simultaneous first order equations	333
	7.9.3 A program for a system of first order ordinary differential equations	335
	7.9.4 Stiffness	340
	7.9.5 The second order boundary value problem	341
7.10	Available Algorithms	344
	Exercises	345

Solutions to Exercises 348

References 369

Index 372

Preface

When preparing early drafts of this book, we attempted to keep four principal aims in mind. The first was to outline the derivation of results which form the basis of a wide range of topics in numerical analysis, but within a framework which is not too demanding of the reader's mathematical ability. The topics covered in this book, root-finding, linear systems of equations, approximation, quadrature and ordinary differential equations, are those which are traditionally regarded as core components of an undergraduate course in numerical analysis. The second aim was to illustrate the performance of numerical schemes obtained from these fundamental results and to indicate the conditions which dictate whether a particular method will, or will not, work successfully.

It is occasionally possible to obtain a wide range of numerical evidence using little more than a pocket calculator. However, even a simple algorithm may involve a considerable amount of computation and the third aim was, therefore, to encourage the reader to program the numerical schemes in order to gain further insight. The text of this book is interspersed with a number of computer codes and it is hoped that the reader will make use of these programs, possibly in an amended form, to obtain and interpret results. Two versions of each program are given, one in Algol 68 and the other in Pascal. Whilst we would argue that Algol 68 is better equipped for serious numerical work, it cannot be denied that Pascal compilers are more widely available. All the programs in this book are available in machine-readable form on application to the authors. The reader whose main programming language is other than Algol 68 or Pascal should be able to translate many of the programs with ease. It should be emphasised that this book assumes a working knowledge of one of the languages and that, rather than being a text on either language, the book illustrates the way the constructs of the two languages may be used. Further, the programs provide examples of some of the objectives which must be borne in mind when developing numerical software. The routines are not meant to form the basis of a mathematical software library; rather, frequent references are made to the facilities available in the library of the Numerical Algorithms Group (NAG). This is in line with the fourth aim, which was to indicate how the simple schemes outlined in this work may be extended, and to provide pointers for further study.

Chris Phillips
Barry Cornelius

Acknowledgements

First, and foremost, we wish to acknowledge the considerable assistance we have received during the preparation of this book from Geoff Cook. We are indebted to Geoff for permission to use notes which form the basis of portions of this book, for constructive criticism of early drafts, and for encouragement to complete the task in hand. We would also like to acknowledge the help given by colleagues at the University of Hull, particularly Tony Fisher and John Griffiths. In addition, Helen Lythgoe deserves special mention for her excellent typing of the final manuscript. We are also grateful to Mike Delves for advice and encouragement throughout the past few years, and to Michael Horwood who has shown remarkable patience and understanding. Finally, we wish to put on record our appreciation of the support given by our wives Margaret and Carol.

1

Introduction

1.1 NUMERICAL ANALYSIS

The subject of numerical analysis is concerned with the *derivation*, *analysis* and *implementation* of methods for obtaining *reliable* numerical answers to mathematical problems. We use the adjective 'reliable' to indicate that it is essential to have confidence in any answers produced and an assessment of reliability can form part of the analysis of a method. In subsequent chapters we shall apply numerical methods to a variety of problems including finding some or all of the roots of an algebraic equation, solving a set of linear simultaneous equations and calculating the value of a definite integral. We can at this stage make a number of preliminary observations on the words 'derivation', 'analysis' and 'implementation' which appear at the beginning of this paragraph.

The 'derivation' stage is concerned with deriving and describing the sequence of numerical steps which it is expected will eventually lead to the required numerical answers. The complete description of these steps, perhaps written in some pseudo-programming language, is called an *algorithm*. This stage may or may not be easy and often intuition and experience will play an important role. As a simple example of a derivation, we may cite the so-called trapezium rule in which we try to estimate the value of the definite integral

$$I = \int_a^b f(x)\,dx \quad b > a. \tag{1.1}$$

In Fig. 1.1(a) we suppose the curve $y = f(x)$ is plotted. Then, remembering the interpretation of a definite integral, if A is the point $(a, f(a))$ and B is the point $(b, f(b))$, we can say that an approximation to I is given by the shaded area. This corresponds to the area of the trapezium with base $b - a$ and vertical sides $f(a)$ and $f(b)$; that is,

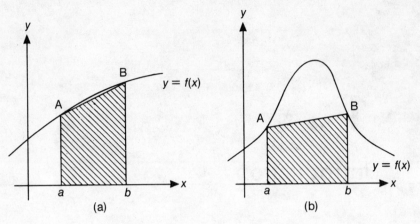

Fig. 1.1.

$$I \simeq \frac{b-a}{2} \{f(a) + f(b)\}. \tag{1.2}$$

We ought now to ask immediately about the accuracy of the approximation (1.2). Trying to find an answer to this question forms a part of the 'analysis' stage to which we have referred. For the present we observe that the error is represented by the unshaded part of the area under the curve. It may be 'small' in the sense that the trapezium rule approximation is sufficiently accurate for the purpose in hand, or it may be unacceptably large as illustrated in Fig. 1.1(b). It is important to realise that, in general, we shall not be able to find this error. What we can do, however, is to try to find a bound for it; that is, if E denotes the error, we try to find a positive number M for which we can assert that $|E| \leq M$.

Intuitively, we would expect that an improvement in accuracy can be obtained by dividing the interval of integration into a number of sub-intervals and then applying the trapezium rule to each sub-interval in turn. In section 6.8 we describe a method which employs interval subdivision to estimate a definite integral correct to some specified tolerance.

In producing a computer program which implements a numerical method, due attention must be paid to *efficiency*; by this we mean that time and storage requirements must not be excessive. Suppose that we wish to estimate (1.1) using the trapezium rule with $f(x) = 2x^3 - 3x^2 + 4x + 1$. Clearly, we need to evaluate $2a^3 - 3a^2 + 4a + 1$ (and, of course, $2b^3 - 3b^2 + 4b + 1$) and this would appear to involve six multiplications, one subtraction and two additions. However, by expressing $f(a)$ as $((2a - 3)a + 4)a + 1$ we can reduce the number of multiplications to just three. Although this may seem a fairly small reduction in the number of operations (and hence the time taken to evaluate the expression) for high degree polynomials the savings achievable are very large indeed.

1.2 MATHEMATICAL FORMULAE

There are occasions when a numerical answer can be obtained satisfactorily using a known mathematical result. A simple example of this is furnished by the calculation of

$$I = \int_0^\alpha \cos(x) \, dx$$

where α is a given number. The reader will easily verify that $I = \sin(\alpha)$ and, using a pocket calculator, say, I can be computed for any given value of α (to an accuracy which will depend on the particular calculator used).

However, there are also occasions when a valid mathematical result is not always useful for *computational* purposes or is even useless. A simple example of the first situation is furnished by the problem of finding the roots of the quadratic equation

$$ax^2 + bx + c = 0 \quad a \neq 0.$$

We can write these down formally as

$$x_1 = \frac{-b + \sqrt{b^2 - 4ac}}{2a}$$

and

$$x_2 = \frac{-b - \sqrt{b^2 - 4ac}}{2a}.$$

To show that these results may not be satisfactory for computational purposes, we suppose a, b and c are such that b^2 is very much greater than $4ac$. Then the calculation of x_1 (if $b > 0$) or x_2 (if $b < 0$) involves forming the difference of two nearly equal numbers (since $\sqrt{(b^2 - 4ac)} \simeq b$) with the consequence that there will be a loss of significant figures. (The reader who is unfamiliar with the notion of significant figures should return to this point after reading the material of section 1.7.) As an example, the equation $x^2 - 200x + 5 = 0$ has roots $100 \pm \sqrt{9995}$ which, working to six significant figures, gives 100.000 ± 99.9750, that is, 199.975 or 0.0250 and the second root is comparatively inaccurate. (This root may be obtained more accurately by observing that the product of the roots is 5 so that the second root can be estimated as $5/199.975$.)

A second example shows that a known result can be useless for computational purposes. Consider the integral

$$I_n = \int_0^1 \frac{x^n}{x + \alpha} \, dx \qquad (1.3)$$

where α is a given positive number and n is a positive integer. Suppose we wish to evaluate this integral for $n = 0, 1, 2, \ldots$. An apparently satisfactory way of proceeding is to observe that for $n \geq 1$

$$I_n + \alpha I_{n-1} = \int_0^1 \frac{x^n + \alpha x^{n-1}}{x + \alpha} \, dx = \int_0^1 x^{n-1} \, dx = \frac{1}{n}$$

so that successive members of the sequence of integrals (1.3) satisfy the *recurrence relation*

$$I_n = \frac{1}{n} - \alpha I_{n-1}. \tag{1.4}$$

Now

$$I_0 = \int_0^1 \frac{dx}{x + \alpha} = \ln(\alpha + 1) - \ln(\alpha) = \ln\left(\frac{\alpha + 1}{\alpha}\right)$$

(where ln denotes the natural logarithm) and so we can easily evaluate I_0. Equation (1.4) can then be used to find I_n for $n = 1, 2, \ldots$; I_1 is found as $I_1 = 1 - \alpha I_0$ and, having evaluated I_1, we find I_2 using $I_2 = \frac{1}{2} - \alpha I_1$, and so on.

We now investigate what happens when $\alpha = 0.5$ and $\alpha = 5$. The results are summarised in Table 1.1 and are quoted correct to five decimal places.

Table 1.1

	I_n	
n	$\alpha = 0.5$	$\alpha = 5$
0	1.098 61 = ln(3)	0.182 32 = ln(1.2)
1	0.450 70	0.088 40
2	0.274 65	0.058 00
3	0.196 01	0.043 33
4	0.152 00	0.033 35
5	0.124 00	0.033 25
6	0.104 67	0.000 42
7	0.090 52	0.140 76
8	0.079 74	−0.578 80

Each individual arithmetic operation (addition, subtraction, multiplication or division) was performed to seven decimal place accuracy on a desk calculator and the result then rounded. What we have, in fact, done here is to work to five decimal place accuracy throughout with each individual operation being performed in slightly greater precision before rounding. In this way we have, to some extent, attempted to simulate computer arithmetic (although we have worked to a fixed number of decimal places rather than significant figures). We return to this in section 1.7. From (1.3) it is not too difficult to see that I_n should decrease with n and for $\alpha = 0.5$ the results given in Table 1.1 behave in the expected manner. This is an important observation: we should always examine the results of a numerical computation and, if possible, verify that they behave as anticipated. Confirmation that this is the case does not necessarily imply that the results are correct but it does, at least, encourage some confidence. For $\alpha = 5$ the results behave as expected initially; however, we observe that $I_7 > I_6$ and that I_8 is negative. This result for I_8 is clearly nonsense since the integrand in (1.3) is positive at every point in the interval of integration $0 \leq x \leq 1$ and hence I_n should be positive. We can have no confidence in the results for $\alpha = 5$. By increasing the accuracy we might expect to obtain a more realistic estimate of I_8. However, this merely serves to delay matters. If we work to a finite number of decimal places a point will eventually be reached at which the results become meaningless. The method based on the recurrence relation (1.4) fails for computational purposes and it is now necessary to explain this behaviour (as a part of the analysis of the method). We return to the point in section 1.8.

1.3 THE IDEA OF ITERATION

Iteration is an important numerical technique and we introduce the idea by deriving in a simple way a possible method for the estimation of the mth root of a given positive number a. Consider $m = 2$, so that the square root is required, and suppose that an initial approximation, x_0, to $\sqrt{a}$ is available. (For example, if $a = 50$ we could take $x_0 = 7$.) Since x_0 is an approximation to $\sqrt{a}$ then so is a/x_0 and, moreover, if $x_0 > \sqrt{a}$ then $a/x_0 < \sqrt{a}$ and conversely. This suggests that a better approximation to $\sqrt{a}$ can be calculated as

$$x_1 = \tfrac{1}{2}\left(x_0 + \frac{a}{x_0}\right).$$

We can repeat the argument with x_1 in place of x_0 and obtain a new approximation as

$$x_2 = \tfrac{1}{2}\left(x_1 + \frac{a}{x_1}\right).$$

Proceeding in this way we can summarise the method as

$$x_n = \tfrac{1}{2}\left(x_{n-1} + \frac{a}{x_{n-1}}\right) \quad n = 1, 2, \ldots$$

so that from x_0 we generate a sequence of numbers $x_1, x_2, \ldots$ which can be conveniently written as $\{x_n\}$. This is another example of the derivation of a numerical method. The method is called *iterative* (that is, repetitive) and, among other things, such methods always require an initial approximation (in some methods, more than one) to the quantity we wish to find, in this case $\sqrt{a}$.

The idea can now be extended to the estimation of $\sqrt[m]{a}$. Suppose x_0 is an initial approximation. Then, observing that a/x_0^{m-1} is also an approximation to $\sqrt[m]{a}$ and following the same arguments as before, we suggest the iterative method

$$x_n = \tfrac{1}{2}\left(x_{n-1} + \frac{a}{x_{n-1}^{m-1}}\right) \quad n = 1, 2, \ldots . \tag{1.5}$$

Let us now see how (1.5) performs when $a = 10$ and $m = 2, 3, 4, 5$. The results are summarised in Table 1.2 where in each case a 'good' initial

Table 1.2

	$m = 2$	$m = 3$	$m = 4$	$m = 5$
x_0	3.200 0	2.200 0	1.800 0	1.600 0
x_1	3.162 5	2.133 1	1.757 4	1.563 0
x_2	3.162 3	2.165 5	1.799 9	1.619 3
x_3		2.149 0	1.757 5	1.536 9
x_4		2.157 2	1.799 8	1.664 6
x_5		2.153 1	1.757 5	1.483 6
x_6		2.155 1		1.773 9
x_7		2.154 1		1.391 9
x_8		2.154 6		2.028 0
x_9		2.154 4		1.309 6
True value	3.162 3	2.154 4	1.778 3	1.584 9

approximation has been taken and the working is to four places of decimals throughout.

We make the following observations on the results of Table 1.2. For $m = 2$ an estimate of $\sqrt{10}$ to four places has been obtained quickly. For $m = 3$ an estimate of $\sqrt[3]{10}$ to four places has been obtained but after a substantially larger number of iterations. For $m = 4$ and $m = 5$, however, the method clearly fails to work. In the former case a very small amount of progress is made initially but then the estimates oscillate indefinitely between the values 1.7998 and 1.7575. In the latter case it can be seen that as the process progresses the results obtained get further and further away from 1.5849. These results must be explained and in order to proceed we need some analysis of iterative methods for the estimation of the roots of an algebraic equation (in this case $x^m - a = 0$). This analysis forms much of the material of Chapter 2 and the present results are discussed further in section 2.3 and section 2.6.

1.4 EXPLORATORY WORK

In many practical situations it is often possible to make some progress using elementary mathematical ideas and simple tools such as pencil, paper and a pocket calculator. To give the flavour of this idea we consider the problem of determining the roots of the equation $e^x \sin(x) - 1 = 0$. There is much that we can say about the nature and approximate position of these roots without carrying out any serious computation.

We choose to express the equation as $e^{-x} - \sin(x) = 0$ and observe that the roots of this, and hence the original, equation correspond to the intersections of the graphs $y = e^{-x}$ and $y = \sin(x)$. The reader should be able to sketch these functions without difficulty and obtain Fig. 1.2.

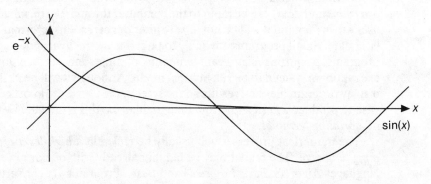

Fig. 1.2.

We can now make the following assertions.

(i) There are no negative roots of the equation since for $x < 0$, $e^{-x} > 1$ and $|\sin(x)| \leq 1$ and there can be no intersections.
(ii) There are an infinite number of positive roots since for $x > 0$, $0 < e^{-x} < 1$ and $\sin(x)$ oscillates indefinitely between -1 and 1.
(iii) Since e^{-x} decreases very rapidly ($e^{-\pi} \simeq 0.043$, $e^{-2\pi} \simeq 0.0019$) the roots (apart from the first) get closer and closer to the zeros of $\sin(x)$; that is, $x = n\pi : n = 1, 2, \ldots$. Moreover, those for odd n are just less than $n\pi$ and those for even n just greater than $n\pi$.

In Chapter 2 we shall look at a number of iterative methods for finding the roots of a given equation. Each method will require the provision of one, or more, initial estimates to start the process off and the exploratory work can be of crucial importance. By studying the problem carefully it may be possible to obtain starting values which will ensure that the iterative process will not only yield the root we require but hopefully will do it using only a small number of iterations.

1.5 COMPUTING RESOURCES

In the previous section we gave an example of the way in which modest resources can be useful in exploratory work. Any serious computation will require more sophisticated resources and we now review briefly what these are likely to be. The reader may have access to a number of computing facilities which range from the large *mainframe* system (which, among other things, is designed to handle large-scale computations) through the *mini* system (which, despite its name, may well be used to carry out computations of some size and complexity) to the personal *microcomputer* (which is rapidly becoming a resource of importance in many types of computation). Access to these systems is usually through a *terminal* consisting of a *keyboard* to which a *video screen* is attached. Information (such as *computer programs* and *data*) can be input to the computer through the terminal to the user's own *files* and it will be possible to alter or correct this information at a later date. Small programs are likely to be run in *interactive* mode but larger programs, requiring significant amounts of time and storage for their execution, may need to be run in *batch* mode. Although it will be possible to display program files and results on the screen, the user will almost certainly wish to produce a *hard copy* from time to time and this is effected by the use of a *printing terminal*.

The numerical processes will usually be coded in a *high-level programming language*. The reader may be familiar already with one or more of the languages *Algol 68*, *Pascal*, *Fortran* and *Basic*. Programs are converted into instructions which the computer can obey directly via an appropriate *compiler*. There may already exist an efficient, general purpose routine which implements the method to be used. However, in order to make good use of such routines, it is often necessary to have a basic understanding of the underlying algorithm and hence their existence adds weight to, rather than detracts from, the need to study numerical analysis.

Many routines have been collected into packages or libraries. It is essential that library programs are easy to use and well documented. Further, they must be robust; by this we mean that they should be capable of detecting data which is unacceptable and, in addition, they should return some indication of the accuracy of the results produced. One of the most widely used libraries of mathematical software is that marketed by the

Numerical Algorithms Group (*NAG*) (see the Foreword to the NAG Library Manual written by Fox and Wilkinson for a discussion on the philosophy behind the library). Routines are available in Algol 68, Algol 60 and Fortran and a Fortran subset, designed to run on small machines, is available. A Pascal subset is also in preparation and, in addition, it is often possible to call a NAG Fortran routine directly from a Pascal program. Note that in the full library the facilities available in the Fortran and Algol 60 versions are the same. However the Algol 68 version has, to a large extent, been developed independently. The NAG library is divided into a number of chapters, each chapter covering a particular branch of computation (quadrature, approximation, etc.). For each chapter, documentation is available which describes the purpose of the routines of that chapter and gives advice on which of the (possibly many) codes available the user should employ for a particular problem. Individual routines also have documentation associated with them and it should be noted that this conforms to a common style. The routine documentation always contains an example of a call of the routine together with corresponding output.

In subsequent chapters we shall make reference to the NAG library and so, at this stage, the reader is encouraged to investigate whether the library is mounted on any of the machines available to him. If it is, the mechanism for accessing the library should be determined and the NAG documentation studied so that insight may be gained into how the library is used.

1.6 NUMBERS: FIXED AND FLOATING FORMATS

We recall first the use of the familiar *decimal* system to represent numbers. Each digit in a decimal number will be one of the digits $0, 1, \ldots, 9$ and the contribution of a digit p to the magnitude of a number is $p \times 10^{k-1}$ if it is k places to the left of the decimal point and $p \times 10^{-k}$ if it is k places to the right. Thus the sequence of digits 721.5438 means

$$7 \times 10^2 + 2 \times 10^1 + 1 \times 10^0 + 5 \times 10^{-1} + 4 \times 10^{-2} + 3 \times 10^{-3} + 8 \times 10^{-4}.$$

We can attach a sign to the number, although it is usual to omit the sign if the number is positive.

In a decimal number we say that the *base* (or *radix*) is 10. We can, however, use other bases and 2 (*binary*), 8 (*octal*) and 16 (*hexadecimal*) are cases of particular interest. In general, if the base is r (an integer ≥ 2), the digits in the number will be one of the digits $0, 1, \ldots, r-1$ and a positive number

$$d_1 d_2 \ldots d_n . d_{n+1} d_{n+2} \ldots d_{n+m}$$

means

$$d_1 \times r^{n-1} + d_2 \times r^{n-2} + \ldots + d_{n-1} \times r^1 + d_n \times r^0 + d_{n+1} \times r^{-1} + d_{n+2} \times r^{-2} + \ldots + d_{n+m} \times r^{-m}.$$

We refer to the digits to the left of the point as the *integer part* of the number and those to the right as the *fractional part*. The point itself is called the *binary point* ($r = 2$), the *octal point* ($r = 8$) and so on. We note here that the digits 0 and 1 occurring in a binary number are called *bits* (binary dig*its*). Again, we attach a sign to the number if it is negative.

The number 721.5438 is called a *fixed point* decimal number. Other examples are 123.456, −83.249 and 0.00000014. In order to avoid confusion we sometimes attach a suffix to a number to indicate its base. Thus 1101.01_2 is an example of a fixed point binary number (and the reader should verify that this is equivalent to the fixed point decimal number 13.25).

Decimal numbers may also be expressed in *floating point* format. By the floating point decimal representation of a number A ($\neq 0$), we mean the representation

$$A = p \times 10^q$$

where q is an integer and p is a fixed point decimal number. If p is in the range $0.1 \leq p < 1$ the representation is said to be *normalised*. Thus the normalised floating point decimal representations of 721.5438 and 0.00000014 are 0.7215438×10^3 and 0.14×10^{-6} respectively. We call p the *mantissa* and q the *exponent* of A. If the number is negative, the mantissa is negative so that −123.456 has floating representation -0.123456×10^3. If $A = 0$, it is conventional to take $p = q = 0$.

We can extend the definition of floating point format to numbers expressed in other bases. The normalised floating point binary representation of the number A takes the form

$$A = p \times 2^q$$

where q is an integer and p is a fixed point binary number such that $\frac{1}{2} \leq p < 1$. Thus the normalised floating binary representation of 1101.01_2 is 0.110101×2^4.

1.7 ERRORS

1.7.1 Sources of error

We note the following sources of error, most if not all of which will be present in a computation of any size. (We exclude from detailed discussion the blunder due to human error in programming or implementation and the malfunction due to a computer error through hardware or through some failure in a software system that is supposed to be reliable.)

(i) We have said that we are trying to obtain numerical answers to a mathematical problem. The mathematical problem will very probably have arisen as a part of modelling some situation in the physical, life or social sciences. The numerical answers may be reliable but how far they contribute to an understanding of the situation being modelled is a matter for the modeller. This is not a part of numerical analysis but the reader should understand that the provision of reliable answers is a matter of crucial importance so that the model can be altered or refined.

(ii) Many numerical processes involve a *truncation error*. Examples of this situation include the use of a finite number of terms from an infinite series or the use of an approximation to estimate the value of a given definite integral (and the unshaded area under the curve in Fig. 1.1 can be regarded as the truncation error in applying the trapezium rule).

(iii) Some or all of the initial data of the problem may be subject to uncertainty. The measurement of length, weight, etc. will produce data whose accuracy is limited by the calibration of the equipment being used.

(iv) The numbers we work with will be subject to rounding and we now spend some time examining some elementary ideas concerned with rounding, the representation of numbers in a binary computer, and the way in which a computer performs the basic arithmetic operations of addition, subtraction, multiplication and division.

1.7.2 Rounding errors

In section 1.6 we described the positional notation for the representation of numbers in any base. Thinking of the decimal case for the time being, some rational numbers require an infinite number of digits after the decimal point. Examples are 5/3 and 22/7 where we have the fixed point representations 1.6666... and 3.1428571428571.... In the second case the reader will verify that the sequence of digits 142857 repeats indefinitely. In practice, we can work only with numbers containing a finite number of digits and we can approximate 22/7 by, say, the number 3.142857 with error 0.00000014.... We say that 3.142857 is 22/7 *correct* (*rounded*) to six *decimal place accuracy* and 0.00000014... is known as the *rounding error*. In contrast, 1.666667 is 5/3 rounded to six decimal places; here we have rounded up rather than rounded down because 1.666667 is closer to 5/3 than 1.666666. The process of rounding should be compared with that of chopping. In the latter we simply remove all digits after a specified number of them. Another way of describing the accuracy of 3.142857 is to say that it is 22/7 correct to seven *significant figures*. If we express a number in its normalised floating point format, the number of digits after the point gives the number of significant figures. Thus, 0.00000014 ($= 0.14 \times 10^{-6}$) is $0.000000142857...$ correct to eight decimal places but only two significant figures. We note that rounding errors may be incurred even when it is possible to express a decimal number using a finite number of digits. Thus, an exact number in which we are interested might be 187.452963 but, because of limitations imposed by computing machinery, suppose we can work only with six digits available for the mantissa. The approximation to this number is 0.187453×10^3 and a rounding error of 0.37×10^{-4} has been incurred.

The ideas introduced in the previous paragraph may be extended to numbers expressed in other bases. We continue the discussion in the context of normalised floating point binary numbers. Recall that in normalised binary floating point format a (positive) number A is expressed as

$$A = p \times 2^q$$

where q is an integer and p is such that $\frac{1}{2} \leq p < 1$. In a computer, the number of bits available for storing p and q is limited and so only a finite set of numbers (called the *machine numbers*) can be represented. A particular case is discussed in section 1.9 in the context of the PR1ME 550 system; here the number of bits for q (expressed as a binary integer) is 8 and so the machine numbers are contained in the interval $\frac{1}{2} 2^{-2^7} \leq A < 2^{2^7 - 1}$ (that is, $2^{-129} \leq A < 2^{127}$) (plus zero). The number A will be represented as

$$\text{fl}(A) = \bar{p} \times 2^q$$

where $\bar{p}$ is p appropriately rounded. The *absolute error* in the representation of A is

$$|A - \text{fl}(A)| = |p - \bar{p}| \times 2^q.$$

If t bits are available for the representation of p, then

$$|p - \bar{p}| \leq \tfrac{1}{2} 2^{-t} = 2^{-t-1}.$$

The *relative error* in the representation of A is

$$\left| \frac{A - \text{fl}(A)}{A} \right| = \left| \frac{p - \bar{p}}{p} \right|$$

so, since $|p| \geq \frac{1}{2}$,

$$\left| \frac{A - \text{fl}(A)}{A} \right| \leq \frac{2^{-t-1}}{2^{-1}} = 2^{-t}$$

and we refer to 2^{-t} as the *machine round-off unit*.

We now consider the effect of the four basic arithmetic operations. Suppose $\text{fl}(A_1) = \bar{p}_1 2^{q_1}$ and $\text{fl}(A_2) = \bar{p}_2 2^{q_2}$. Then formally we have

$$\text{fl}(A_1) \times \text{fl}(A_2) = \bar{p}_1 \bar{p}_2 \, 2^{q_1 + q_2}$$

which, if $\bar{p}_1 \bar{p}_2 < \frac{1}{2}$, must be expressed in normalised form by multiplying $\bar{p}_1 \bar{p}_2$ by 2 and subtracting 1 from the exponent. In many computers, the result $\text{fl}(\text{fl}(A_1) \times \text{fl}(A_2))$ is equal to the rounded value of the exact result of multiplying $\text{fl}(A_1)$ by $\text{fl}(A_2)$. For addition we have

$$\mathrm{fl}(A_1) + \mathrm{fl}(A_2) = \bar{p}_1 \times 2^{q_1} + \bar{p}_2 \times 2^{q_2}$$

and if $q_1 \geq q_2$

$$\mathrm{fl}(A_1) + \mathrm{fl}(A_2) = (\bar{p}_1 + \bar{p}_2 \times 2^{q_2-q_1}) \times 2^{q_1}$$

which must, if necessary, be brought to normalised form by adjusting the mantissa and exponent. Again, it is usual to be able to assert that $\mathrm{fl}(\mathrm{fl}(A_1) + \mathrm{fl}(A_2))$ is equal to the rounded value of the exact operation with operands $\mathrm{fl}(A_1)$ and $\mathrm{fl}(A_2)$. The same is usually true of subtraction and division and we therefore have

$$\left| \frac{\mathrm{fl}(\mathrm{fl}(A_1) o\ \mathrm{fl}(A_2)) - \mathrm{fl}(A_1) o\ \mathrm{fl}(A_2)}{\mathrm{fl}(A_1) o\ \mathrm{fl}(A_2)} \right| \leq 2^{-t} \qquad (1.6)$$

where o denotes one of the operations $+, -, *, /$.

1.8 NUMERICAL INSTABILITY AND ILL-CONDITIONING

Our mathematical problem will usually have *input data* such as the coefficients of successive powers of x in a polynomial equation or the elements of the matrix of coefficients and of the right-hand side vector in a set of linear simultaneous equations. The calculation will produce output data; in the case of a polynomial equation we output one or more of the roots and, for a set of linear simultaneous equations, we output the elements of the solution vector.

It sometimes happens that there is poor accuracy in the *output data* and, in extreme cases, that they are completely valueless. An example of this situation was encountered in using the recurrence relation (1.4) with $\alpha = 5$ and we now examine the reason for this numerical disaster. We have mathematically that $I_0 = \ln((\alpha + 1)/\alpha)$ which for $\alpha = 5$ gives $I_0 = \ln(1.2)$. We approximated $\ln(1.2)$ by 0.18232, the value correct to five decimal places, and worked to five decimal places throughout. Thus, in calculating I_1 we used $I_0 + \varepsilon$, where ε is the rounding error in I_0 and so there is an error of $-\alpha\varepsilon$ in I_1. If we follow the further progress of this error, it is easy to see that there will be an error of $(-\alpha)^n \varepsilon$ in I_n. (There are other errors introduced as the computation proceeds; for example, it is not possible to represent $1/n$ exactly as a five decimal place number for $n = 3, 6$, and 7.) When $\alpha \leq 1$ the initial error is not magnified. However, when $\alpha > 1$ it is magnified to $(-\alpha)^n \varepsilon$ (to 390625ε, when $n = 8$ and $\alpha = 5$). This build-up of error, accompanied by a reduction in the values which we are estimating, gives rise to results which are quite meaningless. We refer in this case to the problem of *numerical instability*. This phenomenon arises from the growth of errors in the initial data and of errors introduced by rounding as the computation proceeds.

If these errors grow at a faster rate than the values we are estimating a point is eventually reached beyond which they completely dominate the computed results.

The process (1.4) can also be written as

$$I_{n-1} = \frac{1}{\alpha}\left(\frac{1}{n} - I_n\right). \tag{1.7}$$

For n sufficiently large (>20) we can take $I_n = 0$ and use (1.7) to recurse backwards. Now, if $\alpha = 5$ the error in I_n will be multiplied by 1/5 when forming I_{n-1}. Since, at the same time, the terms are increasing in size as n decreases, the accumulation of error is unlikely to be significant. We conclude that it is sometimes possible to reformulate an algorithm in such a way that the difficulties associated with numerical instability are avoided.

Poor accuracy in the output data can also depend on the problem itself in the sense that, independently of the choice of numerical method, the output data is sensitive to small changes in the input data. If this is the case, we refer to the phenomenon of *ill-conditioning*. Two examples will illustrate this situation.

(i) The quadratic $x^2 - 2x + 1 = 0$ has a pair of coincident roots at $x = 1$. However, $x^2 - 2x + 0.99 = 0$ has roots $x = 1.1$ and $x = 0.9$ and $x^2 - 2x + 1.01$ has roots $x = 1 \pm 0.1i$. Thus a 1 per cent change in one of the coefficients in the original polynomial equation has produced comparatively large changes in the roots.

(ii) The set of four linear simultaneous equations

$$10x_1 + 7x_2 + 8x_3 + 7x_4 = 32$$
$$7x_1 + 5x_2 + 6x_3 + 5x_4 = 23$$
$$8x_1 + 6x_2 + 10x_3 + 9x_4 = 33$$
$$7x_1 + 5x_2 + 9x_3 + 10x_4 = 31$$

has the solution $x_1 = x_2 = x_3 = x_4 = 1$. If the right-hand sides are replaced by 32.1, 22.9, 32.9 and 31.1 respectively, that is, altered by about one-third of 1 per cent, the solution is $x_1 = 6$, $x_2 = -7.2$, $x_3 = 2.9$, $x_4 = -0.1$.

We repeat that the phenomenon of ill-conditioning is a property of the problem itself, not the numerical method used for its solution. The 'small changes in the input data' could simply be the result of rounding polynomial coefficients or the components of the right-hand side vector in a system of linear simultaneous equations. We shall see in Chapter 4 that it is sometimes possible to reformulate the problem in order to avoid ill-conditioning.

1.9 COMPUTER PROGRAMS

Processes such as the iteration (1.4) can be very tedious when performed by hand (and it is very easy to make mistakes). More complex processes will prove impossible to carry out by hand. By developing software implementing the numerical methods we propose, the reader will be able to assess the performance of such methods on a wide variety of problems. Each chapter of this book contains sample programs which the reader is encouraged to copy and run. These programs are written in Algol 68 and Pascal. Both are *structured* programming languages and are in many ways similar. Indeed they are both a development of Algol 60. However, Algol 68 possesses a number of features (for example, *dynamic arrays*) which make its use particularly relevant to the development of numerical software. We highlight various constructs of Algol 68 and Pascal and refer the reader to Colin (1978) and Findlay and Watt (1985) for introductions to these languages.

Case Study 1.1 gives a simple example of the way programs are presented in this book. There are three distinct components:

(i) The Algol 68 and Pascal programs.
(ii) A sample set of data values.
(iii) The output produced by the Algol 68 program when supplied with the sample data. (The output produced by the Pascal program will be the same apart from possible small changes in the numerical values.)

All Algol 68 programs listed in this book conform to the specification given in the Revised Report for Algol 68 (van Wijngaarden *et al.*, 1976). They were developed on a PR1ME 550 computer running PRIMOS using a compiler written by Fisher (1983). The PR1ME has a 16 bit *word length*. Integer numbers are stored, in binary, in a single word which means that an Algol 68 *INT* can take any value in the range $-2^{15} \leq INT \leq 2^{15} - 1$, that is, $-32768 \leq INT \leq 32767$. Real numbers are stored (again in binary) in floating point format in two words. As a consequence an Algol 68 *REAL* can take values in the range $(\pm) 2^{-129} (\simeq 10^{-39}) \leq REAL < 2^{127} (\simeq 10^{38})$ plus zero (see subsection 1.7.2). In addition, in real arithmetic we have approximately seven significant base 10 figure accuracy.

CASE STUDY 1.1

Algol 68 Program

```
BEGIN
   COMMENT
   a program to find the roots of a quadratic
   COMMENT
   REAL a , b , c ;
   read ( ( a, b, c ) ) ;
   print ( ( newline , "the roots of the quadratic" ,
              newline , a , "  x↑2  +" , b , "  x  +" , c , newline ,
              "are" ) ) ;
   REAL discrim = b ↑ 2 - 4.0 * a * c ;
   IF discrim < 0.0
   THEN
      COMMENT
      the roots are complex
      COMMENT
      print ( ( " complex" , newline ) )
   ELSE
      COMMENT
      the roots are real
      COMMENT
      REAL twoa = a + a ;
      REAL bovertwoa = b / twoa ;
      REAL rootovertwoa = sqrt ( discrim ) / twoa ;
      REAL root1 = - bovertwoa + rootovertwoa ;
      REAL root2 = - bovertwoa - rootovertwoa ;
      print ( ( " real and equal to" , newline , root1 , " and   " , root2 ,
               newline ,
               "and when substituted back into the quadratic these values give" ,
               newline , a * root1 ↑ 2 + b * root1 + c , " and    " ,
               a * root2 ↑ 2 + b * root2 + c , newline , "respectively" ,
               newline ) )
   FI
END
```

Pascal Program

```
PROGRAM quadratic(input, output);
   { a program to find the roots of a quadratic  }
   VAR
       a, b, c:real;
       discrim, twoa, bovertwoa, rootovertwoa:real;
       root1, root2:real;
BEGIN
   read(a, b, c);
   writeln;
   writeln('the roots of the quadratic');
   writeln(a, '  x↑2  +', b, '  x  +', c);
   write('are');
   discrim:= b*b - 4.0*a*c;
   IF discrim<0.0
   THEN
       { the roots are complex  }
       writeln(' complex')
   ELSE
       { the roots are real  }
       BEGIN
          twoa:= a + a;
          bovertwoa:= b/twoa;
          rootovertwoa:= sqrt(discrim)/twoa;
          root1:= -bovertwoa + rootovertwoa;
          root2:= -bovertwoa - rootovertwoa;
          writeln(' real and equal to');
          writeln(root1, '  and    ', root2);
          writeln('and when substituted back into the quadratic',
                  ' these values give');
          writeln(a*root1*root1 + b*root1 + c, '  and   ',
                  a*root2*root2 + b*root2 + c);
          writeln('respectively')
       END
END { quadratic } .
```

Sample Data

    1.2   10.5   4.6

Sample Output

```
the roots of the quadratic
   1.20000e  0  x↑2   +  1.05000e  1  x   +  4.60000e  0
are real and equal to
 -4.62546e -1  and     -8.28745e  0
and when substituted back into the quadratic these values give
   1.60933e -6  and     -1.52588e -5
respectively
```

All Pascal programs listed in this book

(i) Comply with Level 0 of the Pascal standard (see BSI, 1982);
(ii) Have been written so that they can be compiled by compilers that regard only the first 8 characters of an identifier as being significant;
(iii) Have been tested on an IBM PC/XT microcomputer running PC-DOS using the ProPascal compiler from Prospero Software. (The programs have also been tested on the same machine with the Pascal MT+ compiler from Digital Research.)

Programs compiled using the ProPascal compiler use 4 bytes of storage for integer variables; integer values must, therefore, lie in the range $-2147483647 \ldots 2147483647$. However, variables which are declared to be a subrange of the integer type are stored in 1, 2 or 4 bytes depending on the actual subrange. For real values, ProPascal uses a representation which is similar to that described above for Algol 68.

In either Algol 68 or Pascal, operations on integers will be exact unless *overflow* occurs (giving a result which is outside the stated limits). (If larger integer values are required it is possible, in Algol 68, to use *multiple length precision*. A *LONG INT*, for example, occupies two words. Standard Pascal has no such facility.) Any attempt to use a real number which is outside the specified limits will result in overflow (if it is too large) or *underflow* (if it is too small). Underflow is not usually a matter for concern. (In some implementations of Algol 68 it is possible to increase precision by using a *LONG REAL*, a *LONG LONG REAL*, etc. Again, such facilities are not available in Standard Pascal.)

1.10 SOME IMPORTANT MATHEMATICAL RESULTS

In this book we will use $[a,b]$ to denote a *closed interval* consisting of all points between a and b inclusive; $]a,b[$ will be used to denote an *open interval* which excludes the end points a and b; and $[a,b[$ will be used to refer to all points between a and b including the point a but excluding the point b. Note that some authors use (a,b) instead of $]a,b[$ and $[a,b)$ instead of $[a,b[$.

We conclude this introductory chapter by giving a number of mathematical results which will be used in subsequent chapters. (Proofs of these results may be found in Flett (1966).) The first result we state forms the basis of many of the numerical schemes we outline in this book.

THEOREM 1.1 (Taylor)
Let n be a non-negative integer and let $f(x)$ be a real-valued function with the properties that $f(x)$ and its first n derivatives are continuous everywhere in $[a,b]$ and $f^{(n+1)}(x)$ is defined everywhere in $]a,b[$. Then there exists at least one value $\xi \in \,]a,b[$ such that

$$f(b) = f(a) + \frac{b-a}{1!} f'(a) + \frac{(b-a)^2}{2!} f''(a) + \ldots + \frac{(b-a)^n}{n!} f^{(n)}(a)$$
$$+ \frac{(b-a)^{n+1}}{(n+1)!} f^{(n+1)}(\xi).$$

Theorem 1.1 applies to functions of one real variable only. For multivariable functions we can apply the result to each dimension in turn. The following result is required in subsection 2.9.3.

THEOREM 1.2 (Taylor in two dimensions)
Let $\mathbf{a}^T = (a_1, a_2)$ and $\mathbf{b}^T = (b_1, b_2)$ be two coordinates in real two-dimensional space and let $f(\mathbf{x})$ (with $\mathbf{x}^T = (x_1, x_2)$) be a real-valued function of two real variables with the property that $f(\mathbf{x})$ and its first two derivatives are continuous on the line joining $\mathbf{a}$ and $\mathbf{b}$. Then

$$f(\mathbf{b}) = f(\mathbf{a}) + (b_1 - a_1) \frac{\partial f}{\partial x_1}(\mathbf{a}) + (b_2 - a_2) \frac{\partial f}{\partial x_2}(\mathbf{a}) + R$$

where the remainder term, R, may be expressed in terms of second order partial derivatives of $f(\mathbf{x})$.

For a full statement of Taylor's theorem in p dimensions $(p > 1)$ see Flett (1966).

Now consider Fig. 1.3 in which we have drawn a function $f(x)$ which is such that $f(a) = f(b)$. Clearly, for any function satisfying this condition there must be at least one point between a and b at which the tangent to $f(x)$ is parallel to the x-axis (and in Fig. 1.3 there are three such points, ξ_1, ξ_2 and ξ_3). This observation is summarised by the following theorem.

THEOREM 1.3 (Rolle)
Let $f(x)$ be a real-valued function with the properties that $f(x)$ is continuous on $[a,b]$, $f(a) = f(b)$ and $f'(x)$ is defined everywhere in $]a,b[$. Then there exists at least one $\xi \in \,]a,b[$ such that $f'(\xi) = 0$.

We now relax the constraint that $f(a) = f(b)$. From Fig. 1.4 it is clear that

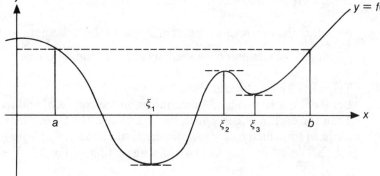

Fig. 1.3

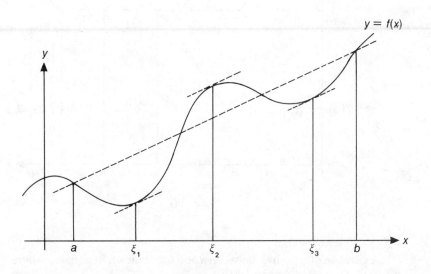

Fig. 1.4

for any function $f(x)$ there must be at least one point between a and b at which the tangent to $f(x)$ is parallel to the line joining $(a, f(a))$ and $(b, f(b))$ (which has slope $(f(b) - f(a))/(b - a)$). The following is a generalisation of Theorem 1.3. (The only difference is now the slope is not necessarily zero.)

THEOREM 1.4 (Mean Value)
Let $f(x)$ be a real-valued function with the properties that $f(x)$ is continuous on $[a,b]$ and $f'(x)$ is defined everywhere in $]a,b[$. Then there exists at least one $\xi \in]a,b[$ such that

$$f(b) - f(a) = (b-a) f'(\xi).$$

Note that the mean value theorem is just Taylor's theorem with $n=0$.

Finally, in Chapter 6 we shall make use of the following.

THEOREM 1.5 (Intermediate Value)
Let $f(x)$ be a real-valued function continuous on $[a,b]$ and let $f(a) < f(b)$. Then for each number γ such that $f(a) < \gamma < f(b)$ there exists at least one $\lambda \in [a,b]$ for which $f(\lambda) = \gamma$. A similar result holds if $f(a) > f(b)$.

This result is clear from a close inspection of Fig 1.5. We require the

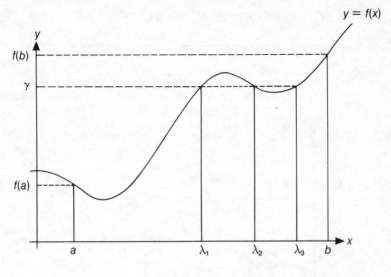

Fig. 1.5

existence of a point λ between a and b at which $f(\lambda)$ equals γ where γ is some value between $f(a)$ and $f(b)$.

EXERCISES

1.1 Let

$$I_n = \int_0^{\pi/2} \sin^n(x) \, dx.$$

Write I_n as $\int_0^{\pi/2} \sin(x) \sin^{n-1}(x) \, dx$ and use integration by parts to show that

$$I_n = \frac{n-1}{n} I_{n-2}.$$

Starting with $I_0 = \int_0^{\pi/2} dx = \pi/2$ generate I_2, I_4 and I_6, working to five significant figure accuracy throughout. Similarly, starting with

$$I_1 = \int_0^{\pi/2} \sin(x)\, dx = [-\cos(x)]_0^{\pi/2} = 1,$$

generate I_3, I_5 and I_7.

1.2 Use the identity

$$\cos(A+B) + \cos(A-B) = 2\cos(A)\cos(B)$$

to show that

$$\cos(nx) + \cos((n-2)x) = 2\cos(x)\cos((n-1)x)$$

and deduce that if $y_n = \cos(nx)$ then

$$y_n = 2\cos(x)\, y_{n-1} - y_{n-2}.$$

Choose $x = \pi/6$ and use the values $y_0 = 1$, $y_1 = \cos(\pi/6) = \sqrt{3}/2$ to generate y_2, y_3, y_4, y_5 and y_6 from this recurrence relation. (We shall meet it again in subsection 4.5.3.)

1.3 Let $x_n = 1/3^{n-1}$. Show that

$$x_n = (3001 x_{n-1} - 1000 x_{n-2})/3.$$

Working to six decimal place accuracy, generate x_2, x_3, x_4 and x_5 using this recurrence relation and the starting values $x_0 = 3$, $x_1 = 1$. Why are the results you obtain so poor?

1.4 The cubic polynomial equation $x^3 - 3x + 2 = 0$ has roots $x = 1$ (a double root) and $x = -2$. Starting with $x_0 = 0.9$, find $x_1, x_2, \ldots, x_7$ using the iteration

$$x_n = x_{n-1} - \frac{1}{3} \frac{(x_{n-1} - 1)(x_{n-1} + 2)}{(x_{n-1} + 1)}.$$

Show that the iteration

$$x_n = x_{n-1} - \frac{2}{3} \frac{(x_{n-1} - 1)(x_{n-1} + 2)}{(x_{n-1} + 1)}$$

converges to the root $x = 1$ at a much faster rate by starting with $x_0 = 0.9$ and forming x_1 and x_2. (The reasons for this improvement are given in section 2.6.)

1.5 Using Taylor's theorem (Theorem 1.1) with $a = 0$, $b = 1$ and $n = 8$ we have

$$e = 1 + 1 + \frac{1}{2} + \frac{1}{3!} + \frac{1}{4!} + \ldots + \frac{1}{8!} + \frac{1}{9!} e^{\xi}$$

where $0 < \xi < 1$. Obtain an estimate of e from this series expansion by ignoring the term $1/(9!)\, e^{\xi}$, working to five decimal place accuracy throughout. What errors have you incurred in doing this? Why do you think the estimate you obtain is so accurate?

1.6 Find out how integer and real values are stored in the computer on which you run your programs. What are the largest permitted integer and real values?

1.7 Stirling's formula, $f(n) = \sqrt{2\pi}\, n^{n+\frac{1}{2}} e^{-n}$, may be used as an approximation to $n!$. Write a program which evaluates $n!$, $f(n)$, the relative error $(n! - f(n))/n!$ and $1/(12n)$ for $n = 1, 2, \ldots, 10$. Deduce the relationship between the last two quantities. (You may experience problems with integer overflow when running the program for large values of n.)

1.8 Amend the program of Case Study 1.1 so that it finds the two real roots of the quadratic equation $ax^2 + bx + c = 0$ as $x_1 = (-b + \sqrt{b^2 - 4ac})/2a$ if $b < 0$ or $x_1 = (-b - \sqrt{b^2 - 4ac})$ if $b > 0$ and $x_2 = c/x_1$. Use your program to find the roots of $x^2 - 200x + 5$.

1.9 If you have access to the NAG library, use a NAG routine to solve the system of equations given in section 1.8. Make the given small changes to the right-hand sides and call the routine again to show the effect of ill-conditioning.

2

Non-linear Algebraic Equations

2.1 INTRODUCTORY REMARKS

In this chapter we are concerned with the problems of finding one or more of the roots of the algebraic equation

$$f(x) = 0 \qquad (2.1)$$

where $f(x)$ is some given real-valued function. We shall assume that $f(x)$ possesses all the necessary analytical properties (particularly with regard to continuity) for the methods we develop and analyse to be mathematically valid. We recall that a root of (2.1) is a number α such that $f(\alpha) = 0$; in practice we shall aim to calculate an estimate of α correct to some prescribed accuracy.

EXAMPLE 2.1
(i) $f(x) \equiv e^{-x} - \sin(x) = 0$
(ii) $f(x) \equiv x^3 - 3x + 1 = 0$

The first of these equations was considered in detail in section 1.4. The second is a (cubic) polynomial equation since $f(x)$ is a polynomial of degree three.

Although most of this chapter is concerned with finding the roots of a single function of one real variable, we shall also consider the more general case of finding the roots of a non-linear system of equations. Further, although we are principally concerned with finding the real roots, or zeros, of a given equation, we shall also have something to say about determining the complex roots of a polynomial equation. In discussing real roots, we shall concentrate principally on *simple, isolated* roots. By 'simple' we mean that $f(\alpha) = 0$ but $f'(\alpha) \neq 0$ (that is, $f(x)$ actually crosses the x-axis) and, by 'isolated', we mean that the root is well separated in some sense from all other roots of (2.1). These ideas are shown diagrammatically in Fig. 2.1.

Most of the methods we discuss are iterative in nature. The idea of iteration was introduced in section 1.3 and, as presented there, it depends on the availability of an initial estimate of the value of the root. Some methods, however, will require the provision of two, or more, starting values. Using the initial information, we generate a sequence of values which hopefully will converge to the root we seek.

Before implementing an iterative method, the following four items need to be investigated.

(i) We must establish the conditions under which the sequence of iterates generated by the method will converge to the root.
(ii) If the iteration is convergent we need to know how and when to stop the iterative process. This, of course, depends on the accuracy we are demanding and we have to formulate a *termination criterion* which can be easily implemented.
(iii) There may be several iterative schemes which are equally acceptable in that they are all expected to converge to the root in question starting from the same initial approximation. We need to determine whether one method is preferable to another and are led to the notion of trying to formulate some measure of *computational efficiency*.
(iv) If the chosen method is to be programmed, *good programming practice* must be observed to ensure that the storage and time requirements are not excessive. In addition, we must take account of as many eventualities as possible when designing a piece of code to ensure that it does not halt due to a run-time failure.

2.2 THE METHOD OF BISECTION

Let $f(x)$ be a given function whose root(s) we wish to find and suppose that we have been able to find (probably by tabulation of $f(x)$) two points, a_0 and b_0 (with $b_0 > a_0$), which are such that $f(a_0) f(b_0) < 0$. Now, for this condition to be satisfied, $f(a_0)$ and $f(b_0)$ must be of opposite sign and consequently (since $f(x)$ is continuous) we can say that there is at least one root of (2.1) lying in the interval $]a_0, b_0[$. Suppose there is only one root. The method of bisection aims to generate a sequence of successively smaller intervals in which the root is known to lie. The first step is to take the mid-point $M_0 = (a_0 + b_0)/2$ of the initial interval, evaluate $f(M_0)$ and form $f(a_0) f(M_0)$. If this quantity is negative, the root lies in $]a_0, M_0[$; and if it is positive the root lies in $]M_0, b_0[$. Let $]a_1, b_1[$ be the new interval containing the root. We then repeat the process. In this manner we generate a sequence of intervals $\{]a_n, b_n[\}$ of decreasing size containing the root and the method may be summarised as follows.

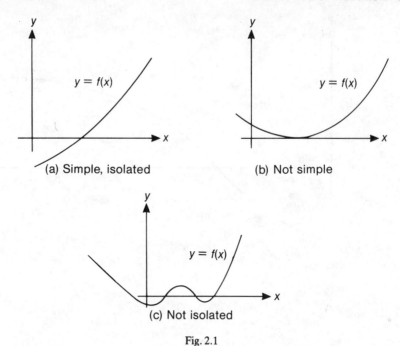

Fig. 2.1

form $M_n = (a_n + b_n)/2$;
evaluate $f(M_n)$ and form $f(M_n) f(a_n) = p_n$, say;
if $p_n < 0$
then take $a_{n+1} = a_n$ and $b_{n+1} = M_n$
else take $a_{n+1} = M_n$ and $b_{n+1} = b_n$
fi

Note that if $f(M_n) = 0$ we have found a root of (2.1) and need go no further. We illustrate the performance of this method in Fig. 2.2.

After k steps of the above method, we know that the root lies within the interval $]a_k, b_k[$ which has length

$$b_k - a_k = (b_{k-1} - a_{k-1})/2 = \ldots \quad = (b_0 - a_0)/2^k.$$

Hence, if we take M_k as an estimate of α, the root to which we are working, the absolute difference between M_k and α is at most $(b_0 - a_0)/2^{k+1}$. The rate of convergence of the method is, admittedly, slow but at least convergence is guaranteed. The method makes use only of the sign of $f(x)$ at the iteration points and it should be possible to obtain an improved scheme by employing the magnitude of $f(x)$ as well. We return to this in section 2.7.

The method of bisection is of interest and importance in its own right. However, we observe that it may also be of use in providing an initial approximation for some other method which is expected to give more rapid convergence.

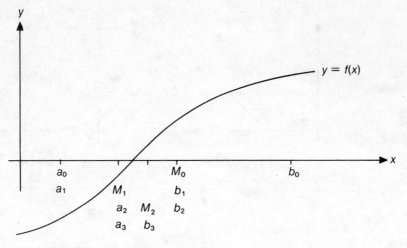

Fig. 2.2

2.3 FUNCTIONAL ITERATION
2.3.1 Introduction
The starting point here is to rewrite (2.1) in the form

$$x = \phi(x) \tag{2.2}$$

so that if α is a root of (2.1) then $\alpha = \phi(\alpha)$. There will usually be several obvious ways in which the given equation can be rewritten in the required form and, indeed, if we are prepared to modify slightly the problem, there will be an infinity of ways.

EXAMPLE 2.2

(i) $x^3 - 3x + 1 = 0$ can be rewritten as $x = (x^3 + 1)/3$ or as $x = 1/(3 - x^2)$.
(ii) $x + \ln(x) = 0$ can be rewritten as $x = -\ln(x)$ or as $x = e^{-x}$.
(iii) $x^3 - 3x + 1 = 0$ can be rewritten as
$x = (1 - k)x + k(x^3 + 1)/3$, where k ($\neq 0$) is some chosen value.

Suppose we have an initial approximation, x_0, to the root α whose value we wish to determine. We can generate a sequence of iterates $\{x_n\}$ using the iterative process

$$x_n = \phi(x_{n-1}) \quad n = 1, 2, \ldots$$

Sec. 2.3] FUNCTIONAL ITERATION 41

but before using it, we must try to establish the conditions under which the iterates converge to α.

Some insight into the question of convergence can be gained by graphical arguments (see Fig. 2.3). Using the same axes, we draw the graphs of $y = x$

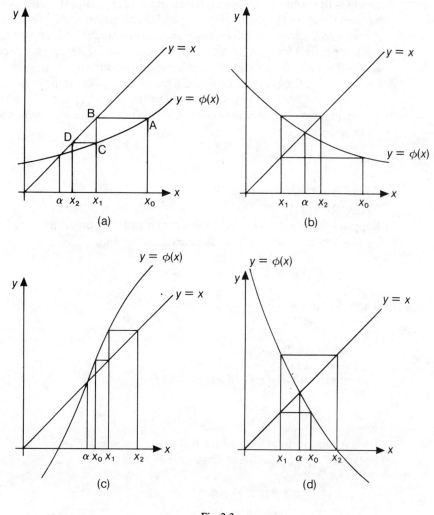

Fig. 2.3

and $y = \phi(x)$. The points of intersection of these two curves give the roots of (2.1). We describe the progress of the iteration for Fig. 2.3(a) and encourage the reader to verify the results for the three other cases illustrated by using the same constructions.

Let x_0 be the chosen starting value. Then A is the point on the curve $y = \phi(x)$ with coordinates $(x_0, \phi(x_0))$ and B is the point on the straight line

$y = x$ obtained by drawing a line through A parallel to the x-axis. The y coordinate of B is therefore $\phi(x_0)$ and, since B lies on $y = x$, the x coordinate of B is also $\phi(x_0)$. However, $x_1 = \phi(x_0)$ and so we are able to mark the position of x_1 by drawing a line through B parallel to the y-axis. We now repeat the argument. C is the point on $y = \phi(x)$ with coordinates $(x_1, \phi(x_1))$ and D is the point on $y = x$ with coordinates $(\phi(x_1), \phi(x_1))$. Since $x_2 = \phi(x_1)$ we are able to mark the position of x_2 and the argument can be repeated once more. The diagram we have drawn suggests the iteration is converging and the same is true for case (b) although we note that the iterates are oscillating about the root. In (c) and (d) the iteration is diverging, that is, successive iterates are getting further away from the root. The key thing to note here is that, near the root, $|\phi'(x)| < 1$ in (a) and (b) but in (c) and (d) $|\phi'(x)| > 1$. These qualitative observations can be formalised in a theorem which we now state and prove.

2.3.2 A convergence theorem

THEOREM 2.1

Suppose the equation $x = \phi(x)$ has a root α and that, on the interval I defined by $\alpha - a \leq x \leq \alpha + a$, $\phi'(x)$ satisfies the condition

$$|\phi'(x)| \leq L < 1.$$

Then, for any $x_0 \in I$

(i) $x_n \in I \quad n = 1, 2, \ldots$;
(ii) $\lim_{n \to \infty} x_n = \alpha$; and
(iii) α is the unique root of $x = \phi(x)$ in I.

Proof

(i) We use induction to prove this part. Suppose $x_{n-1} \in I$.
Now $x_n = \phi(x_{n-1})$ and $\alpha = \phi(\alpha)$ so

$$\begin{aligned} x_n - \alpha &= \phi(x_{n-1}) - \phi(\alpha) \\ &= (x_{n-1} - \alpha) \phi'(\xi_{n-1}) \end{aligned} \quad (2.3)$$

(using Theorem 1.4) where ξ_{n-1} lies between α and x_{n-1}. Since $x_{n-1} \in I$ it follows that $\xi_{n-1} \in I$ and hence that

$$|x_n - \alpha| = |x_{n-1} - \alpha| |\phi'(\xi_{n-1})| \leq L|x_{n-1} - \alpha|. \quad (2.4)$$

Since $L < 1$, it follows that $x_n \in I$. By hypothesis, $x_0 \in I$ and so part (i) of the theorem is proved.

(ii) Here, we use the result (2.4) repeatedly. We have

$$|x_n - \alpha| \leq L|x_{n-1} - \alpha|$$
$$\leq L^2|x_{n-2} - \alpha|$$
$$\leq \ldots \leq L^n|x_0 - \alpha|.$$

Since $L < 1$, $\lim_{n \to \infty} L^n = 0$ which implies that $\lim_{n \to \infty} |x_n - \alpha| = 0$ and hence that $\lim_{n \to \infty} x_n = \alpha$.

(iii) Assume to the contrary and suppose the equation $x = \phi(x)$ has another root β ($\neq \alpha$) lying in I. Then

$$\alpha - \beta = \phi(\alpha) - \phi(\beta)$$
$$= (\alpha - \beta)\phi'(\eta)$$

(using Theorem 1.4 again), where η lies between α and β and hence lies in I. We know that $|\phi'(\eta)| < 1$ from the conditions of the theorem and so

$$|\alpha - \beta| = |\alpha - \beta||\phi'(\eta)| < |\alpha - \beta|$$

which gives a contradiction.

In proving the above theorem we have made use of three important tools of mathematical analysis, namely, induction, the repeated application of a known result, and contradiction. These techniques are frequently used in the derivation of important results in numerical analysis.

2.3.3 Discussion of the convergence theorem

The theorem is important theoretically but to verify that its hypotheses are valid in a particular instance may not be easy. What we can do, however, is attempt to establish whether it is possible to find an interval centred on the root such that the magnitude of the first derivative of the iteration function, $\phi(x)$, is less than one for all points of the interval. The following (easy) examples give some indication of the procedure.

EXAMPLE 2.3

(i) The equation $x + \ln(x) = 0$ has a root lying between 0.5 and 0.6. If we rewrite the equation as $x = -\ln(x)$, the iteration function is $\phi(x) = -\ln(x)$ so $\phi'(x) = -1/x$. Hence, $|\phi'(x)| \geq 1$ for $x \in [0,1]$ (and, in particular, for $x \in [0.5, 0.6]$). It is therefore not possible to find an interval so that the conditions of Theorem 2.1 are satisfied. This means (strictly) only that we have failed to prove that this functional iteration will converge for any interval centred on the root. In fact, this functional iteration will diverge for any starting value $x_0 \in [0,1]$.

Now, the given equation may also be expressed as $x = e^{-x}$. Here, $\phi(x) = e^{-x}$ so $\phi'(x) = -e^{-x}$ and $|\phi'(x)| < 1$ for all $x > 0$. It is therefore possible to satisfy the conditions of the theorem and, in fact, the functional iteration $x_n = e^{-x_{n-1}}$ will converge for any starting value $x_0 > 0$.

In the first and second columns of Table 2.1 we list some numerical

Table 2.1

	$\phi(x) = -\ln(x)$	$\phi(x) = e^{-x}$	$\phi(x) = (x^3+1)/3$	$\phi(x) = (x^3+1)/3$
x_0	0.500 000	0.500 000	0.500 000	0.100 000
x_1	0.693 147	0.606 530	0.375 000	0.333 667
x_2	0.366 513	0.545 239	0.350 911	0.345 716
x_3	1.003 72	0.579 703	0.347 737	0.347 107
x_4	−0.003 715 04	0.560 064	0.347 349	0.347 273
x_5		0.571 172	0.347 303	0.347 293
x_6		0.564 863		
x_7		0.568 438		
x_8		0.566 409		
x_9		0.567 559		
x_{10}		0.566 907		
x_{11}		0.567 277		
x_{12}		0.567 067		
x_{13}		0.567 186		
x_{14}		0.567 119		
x_{15}		0.567 157		

results which support the above claims. These six significant figure numbers were obtained using the program of subsection 2.3.4 and a termination criterion of 0.00005; that is, the process was continued (if possible) until the difference between successive iterates was less than 5×10^{-5} in magnitude and this means that if we quote the final result to four decimal places we are at most 1 out in the final digit (but see below). We note that when $\phi(x) = -\ln(x)$ the iterates do, indeed, diverge from the root. (In fact, a situation is soon reached in which the latest iterate is negative. It is not possible to continue beyond this point since we cannot take the natural logarithm of a negative number.) In contrast, when $\phi(x) = e^{-x}$ the iterates converge, albeit slowly, to the value 0.5672.

(ii) It can be easily verified that the equation $x^3 - 3x + 1 = 0$ has roots lying in the intervals $[-2, -1]$, $[0,1]$ and $[1,2]$. This equation can be rewritten as $x = (x^3+1)/3$ so $\phi(x) = (x^3+1)/3$ and $\phi'(x) = x^2$. Thus,

$\phi'(x) \geq 1$ for $|x| \geq 1$ and the conditions of the theorem can be satisfied only for the middle root (see the last two columns of Table 2.1). If x_0 is in $[-2, -1]$ or $[1,2]$ the process will either diverge or converge to the root in $[0,1]$. The reader is encouraged to demonstrate that this is the case by experimenting with starting values within these ranges.

(iii) In section 1.3 we derived an iterative method for estimating $\sqrt[m]{a}$ ($a > 0$) of the form $x_n = \phi(x_{n-1})$ in which $\phi(x) = (x + a/x^{m-1})/2$. Hence, $\phi'(x) = (1 + (1-m) a/x^m)/2$ so that at the root $\phi'(\sqrt[m]{a}) = 1 - m/2$ (and this is independent of a, the number whose root we wish to find). Thus, the absolute value of the derivative of the iteration function at the root is

(a) 0 when $m = 2$
(b) $\frac{1}{2}$ when $m = 3$
(c) ≥ 1 when $m \geq 4$.

The convergent behaviour when $m = 2$ and 3 is therefore explained and it is clear why the method fails to work for larger values of m. However, we have not yet explained why the rate of convergence is faster for $m = 2$ than for $m = 3$ and we shall examine this in detail in section 2.6.

Equation (2.4) enables us to find a bound for $|x_n - \alpha|$, the magnitude of the error in the nth iterate of a convergent process. We have

$$x_{n-1} - \alpha = (x_n - \alpha) + (x_{n-1} - x_n)$$

so that

$$|x_{n-1} - \alpha| \leq |x_n - \alpha| + |x_{n-1} - x_n|$$

(using the triangle inequality) and, multiplying throughout by the positive number L defined in the statement of Theorem 2.1, we obtain

$$L|x_{n-1} - \alpha| \leq L|x_n - \alpha| + L|x_{n-1} - x_n|.$$

From equation (2.4)

$$|x_n - \alpha| \leq L|x_n - \alpha| + L|x_{n-1} - x_n|$$

so that

$$|x_n - \alpha| \leq \frac{L}{1-L}|x_{n-1} - x_n|. \tag{2.5}$$

This tells us that the distance of the latest iterate from the root is at most $L/(1-L)$ times the distance between it and the previous estimate of α. As such, this is not a particularly useful result since it depends on the availability of a realistic value for L. It is instructive, however, to pursue the consequences of (2.5) for the case $L < \frac{1}{2}$ and the case $L \geq \frac{1}{2}$. If $L < \frac{1}{2}$, $L/(1 - L) < 1$ so

$$|x_n - \alpha| < |x_{n-1} - x_n|$$

that is, the distance of x_n from the root is less than the magnitude of the difference between the two most recent iterates. For $L \geq \frac{1}{2}$, however,

$$|x_n - \alpha| \leq C|x_{n-1} - x_n|$$

where $C = L/(1-L) \geq 1$ and we cannot make the same assertion about the magnitude of $x_n - \alpha$. These observations are of significance when we come to formulate a termination criterion since the condition $|x_n - x_{n-1}| < \varepsilon$ may not guarantee that $|x_n - \alpha| < \varepsilon$.

We have observed already that, for a convergent iteration, on some occasions, iterates oscillate about the root and on others they are all on one side of the root. From equation (2.3) we can see that if $0 < \phi'(x) < 1$ and $x_0 - \alpha > 0$ ($x_0 - \alpha < 0$) then $x_n - \alpha > 0$ ($x_n - \alpha < 0$) and so we get *monotonic* convergence. However, if $-1 < \phi'(x) < 0$, $x_n - \alpha$ is alternately positive and negative and we get *oscillatory* convergence.

EXAMPLE 2.4

Consider Table 2.1 again. In the final two columns, $\phi(x) = (x^3 + 1)/3$ so that $\phi'(x) = x^2$ which is everywhere positive in $[0,1]$. Hence if we start with an initial value which is to the left or right of the root, then all subsequent iterates are on the same side. However, when $\phi(x) = e^{-x}$, $\phi'(x) = -e^{-x}$ which is negative for all x and so successive iterates are on alternate sides of the root. The reader is encouraged to compute the errors in the iterates in each case in order to gain further insight. (The equations $x + \ln(x) = 0$ and $x^3 - 3x + 1 = 0$ have roots at 0.567143 and 0.347296 respectively to six significant figures.)

Theorem 2.1 provides us with the conditions under which functional iteration will converge to a root of (2.2). If the iteration is convergent we are, of course, also interested in how rapidly it will converge. This is a subject that we treat in detail in section 2.6 and, for the moment, we infer what we can from equation (2.4). Now $|x_n - \alpha| = |x_{n-1} - \alpha||\phi'(\xi_{n-1})|$, where ξ_{n-1} lies between α and x_{n-1} and so the magnitude of the error in x_n is just $|\phi'(\xi_{n-1})|$ times the magnitude of the error in x_{n-1}. We do not know ξ_{n-1} but if we are close to the root and assume that $\phi'(x)$ is not varying rapidly, the nth iteration decreases the magnitude of the error by a factor of approximately $|\phi'(x_n)|$. If this factor is close to unity the convergence will obviously be slow; if, on the other hand, it is small, the convergence will be much more rapid.

EXAMPLE 2.5
For the last two columns of Table 2.1 the derivative of the iteration function

is x^2 so that when x is close to the root, say $x = 0.35$, $\phi'(x) = 0.1225$ and the magnitude of the error is reduced by a factor of about $\frac{1}{8}$th at each iteration. For the iteration function $\phi(x) = e^{-x}$, $\phi'(x) = -e^{-x}$ and so when x is near the root, say $x = 0.567$, $|\phi'(x)|$ is about the same value and hence the error is reduced only by a factor which is slightly greater than $\frac{1}{2}$ at each iteration. This explains the convergence rate of each iteration in Table 2.1.

It is possible that $\phi'(\alpha) = 0$ and, in this case, a rather different sort of convergence behaviour is encountered. (We have already observed that Example 2.3 (iii) falls into this category when $m = 2$.) This case is considered, in detail, in section 2.6.

2.3.4 A program for functional iteration

The programs given in Case Study 2.1 estimate the root of the equation $x - e^{-x} = 0$ using the iteration $x_n = e^{-x_{n-1}}$. The reader is recommended to study carefully the constructions used, particularly the looping and the conditional constructions, and also the use of a Boolean (*BOOL* in Algol 68) variable. The programs require three numbers as data: the first two are real numbers and the last one is an integer. These are used to provide the starting value, an accuracy criterion, and an upper bound on the total number of iterations.

It should be noted that the iterates x_n are not stored as array elements $x[n]$. Only two variables are required, one corresponding to the previous iterate x_{n-1} (called *xnminus1*) and the other to the current iterate x_n (called *xn*).

The number of iterations is counted using the integer variable *iter* which is initially set to zero. This variable is incremented each time the loop is executed. (The Algol 68 program uses the $+ :=$ operator in order to do this. Remember that $l + := r$ is equivalent to $l := l + r$; so, $iter + := 1$ is equivalent to $iter := iter + 1$.)

Each time the loop is executed the value of *iter* is compared with the value of *itermax* and the loop terminates if $iter = itermax$. This device should always be included in a program which implements an iterative process. It is intended to guard against some unexpected behaviour in the iterates with the possibility of a gross waste of computer time if the iteration is allowed to continue.

The loop will also terminate as soon as the difference between successive iterates is less than the specified accuracy, *tol*, in absolute value. Note that, in the Algol 68 program, the variable *converged* has to be given the value *FALSE* initially in order to ensure that the loop is executed at least once; in the Pascal program this is guaranteed by using a *REPEAT* statement (which tests at the end of the loop) rather than a *WHILE* statement (which tests at the beginning).

CASE STUDY 2.1

Algol 68 Program

```
BEGIN
  COMMENT
  a program to find the root of the equation x-exp(-x)=0 using
  functional iteration
  COMMENT
  REAL xn , xnminus1 , tol ;
  INT itermax ;
  read ( ( xn , tol , itermax ) ) ;
  print ( ( newline , "starting value                      " , xn , newline ,
            "accuracy tolerance                 " , tol , newline ,
            "maximum number of iterations allowed " , itermax ) ) ;
  BOOL converged := FALSE ;
  INT iter := 0 ;
  WHILE NOT converged AND iter < itermax
  DO
    iter +:= 1 ;
    xnminus1 := xn ;
    COMMENT
    compute the next iterate
    COMMENT
    xn := exp ( - xnminus1 ) ;
    COMMENT
    test for convergence
    COMMENT
    converged := ABS ( xn - xnminus1 ) < tol
  OD ;
  IF converged
  THEN
    print ( ( newline , "convergence to specified tolerance after" , iter ,
              " iterations to" , xn , newline ) )
  ELSE
    print ( ( newline , "no convergence to specified tolerance after" ,
              itermax , " iterations" , newline ) )
  FI
END
```

Pascal Program

```
PROGRAM exampleoffunctionaliteration(input, output);
    { a program to find the root of the equation x-exp(-x)=0 using
      functional iteration  }
    VAR
       xn, xnminus1, tol: real;
       iter, itermax: integer;
       converged:boolean;
BEGIN
   read(xn, tol, itermax);
   writeln;
   writeln('starting value                      ', xn);
   writeln('accuracy tolerance                  ', tol);
   writeln('maximum number of iterations allowed ', itermax);
   iter:= 0;
   REPEAT
      iter:= iter + 1;
      xnminus1:= xn;
      { compute the next iterate }
      xn:= exp(-xnminus1);
      { test for convergence }
      converged:= abs( xn - xnminus1 ) < tol
   UNTIL converged OR (iter=itermax);
   IF converged
```

```
        THEN
             writeln('convergence to specified tolerance after', iter,
                    '  iterations to', xn)
        ELSE
             writeln('no convergence to specified tolerance after',
                    itermax, ' iterations')
   END { exampleoffunctionaliteration } .
```

Sample Data

   0.5   0.00005   20

Sample Output

```
starting value                            5.00000e -1
accuracy tolerance                        5.00000e -5
maximum number of iterations allowed           20
convergence to specified tolerance after       15  iterations to  5.67157e -1
```

2.4 AITKEN ACCELERATION

We now describe an important and useful device for accelerating convergence in a convergent iteration. Let x_n, x_{n+1} and x_{n+2} be three successive iterates. Then from (2.3)

$$x_{n+2} - \alpha = (x_{n+1} - \alpha)\, \phi'(\xi_{n+1}) \qquad (2.6)$$
$$x_{n+1} - \alpha = (x_n - \alpha)\, \phi'(\xi_n). \qquad (2.7)$$

We do not, of course, know ξ_{n+1} or ξ_n but near the root we can suppose that $\phi'(x)$ does not vary too rapidly and hence that $\phi'(\xi_{n+1})$ is approximately equal to $\phi'(\xi_n)$. We can therefore expect the quantity x^*_{n+2} satisfying

$$\frac{x_{n+2} - x^*_{n+2}}{x_{n+1} - x^*_{n+2}} = \frac{x_{n+1} - x^*_{n+2}}{x_n - x^*_{n+2}}$$

(obtained by dividing (2.6) by (2.7) and setting $\alpha = x^*_{n+2}$) to be an improved estimate of α. Rearranging we have

$$x^*_{n+2} = \frac{x_n x_{n+2} - x_{n+1}^2}{x_{n+2} - 2x_{n+1} + x_n}$$

which gives x^*_{n+2} in terms of x_n, x_{n+1} and x_{n+2}. The right-hand side is, however, not in a form which is suitable for computation since both the numerator and denominator are tending to zero. We can manipulate the right-hand side into an acceptable form in which x_n is corrected. This is

$$x^*_{n+2} = x_n - \frac{(x_{n+1} - x_n)^2}{x_{n+2} - 2x_{n+1} + x_n}. \tag{2.8}$$

x^*_{n+2} can now be used to generate two further iterates using straightforward functional iteration and these values combined using (2.8) again. The procedure, known as *Aitken acceleration*, can be summarised as follows. From an initial approximation x_0 (which is not necessarily the x_0 used to start the iterative process), we generate $x_1 = \phi(x_0)$ and $x_2 = \phi(x_1)$ and use (2.8) with $n = 0$ to form x^*_2. We then generate $x_3 = \phi(x^*_2)$ and $x_4 = \phi(x_3)$ and use (2.8) with $n = 2$ and x^*_2 in place of x_2 to compute x^*_4, and so on.

EXAMPLE 2.6

The second column of figures listed in Table 2.1 indicate that the iterative process $x = e^{-x}$ is slowly convergent. Let us apply Aitken acceleration to the first three iterates. We find that $x^*_2 = 0.567623$ and the magnitude of the error in this approximation is less than that for x_8.

Having introduced the idea of Aitken acceleration we must make some further comments. The basis of the derivation of the method is equation (2.3) which is characteristic (see section 2.6) of so-called *linear* convergence. If an iterative method does not have linear convergence, Aitken acceleration should not be used; the new sequence of iterates is likely to converge even more slowly. As a second comment we note that in deriving the method we assumed that $\phi'(\xi_{n+1})$ and $\phi'(\xi_n)$ were approximately equal and this will usually be the case when we are near the root. However, in the early stages of a convergent iteration in which x_0 is a poor (but acceptable) initial value, it is possible that this assumption is not valid and consequently that the application of Aitken acceleration will not be beneficial. We therefore need some criterion which, ideally, would tell us when to start applying the acceleration. There is no easy answer to this problem but a simple strategy which is often satisfactory is to monitor the ratio $(x_n - x_{n-1})/(x_{n+1} - x_n)$ for $n = 1, 2, \ldots$ until it is approximately constant. It is then probable that the application of the acceleration procedure will be beneficial.

As a final observation, we note that the acceleration (2.8) can be expressed as

$$z_n = \Phi(z_{n-1}) \quad n = 1, 2, \ldots$$

where

$$\Phi(z) = z - \frac{(\phi(z) - z)^2}{\phi^2(z) - 2\phi(z) + z} \tag{2.9}$$

and $z_0 = x_0$ and $z_n = x^*_{2n}$. To see this, substitute $z = x_n$ into the right-hand side of (2.9) in order to derive the right-hand side of (2.8). We shall require this form when, in section 2.6, we discuss the relationship between Aitken acceleration and Steffenson iteration.

2.5 NEWTON–RAPHSON ITERATION

2.5.1 Derivation

Suppose we have an approximation x_{n-1} to a simple root $x = \alpha$ of the equation $f(x) = 0$. Let $\alpha = x_{n-1} + h$. Then, since $f(\alpha) = 0$ we have

$$0 = f(x_{n-1} + h) = f(x_{n-1}) + hf'(x_{n-1}) + \ldots$$

using Taylor's Theorem (Theorem 1.1). Ignoring the rest of the terms we see that an approximation to h can be obtained as $-f(x_{n-1})/f'(x_{n-1})$ and so we take the next approximation to the root as

$$x_n = x_{n-1} - \frac{f(x_{n-1})}{f'(x_{n-1})}.$$

Using this for $n = 1, 2, \ldots$ we have a process known as *Newton–Raphson* iteration. We notice that each iteration involves the evaluation of $f(x)$ and its first derivative at the previous iterate and, since we are now using more information about the function than before, we might hope to obtain some benefits.

2.5.2 Discussion of Newton–Raphson iteration

There is a simple geometrical interpretation of the iteration. Referring to Fig. 2.4, where a supposed x_{n-1} is shown, the tangent to the curve $y = f(x)$ is drawn at the point $x = x_{n-1}$. The point of intersection of this tangent with the x-axis gives the new estimate x_n. To justify this statement, we observe that $\tan(\theta) = -f(x_{n-1})/(x_n - x_{n-1})$. However, $\tan(\theta) = f'(x_{n-1})$ and, on equating the two right-hand sides of these equations, the result follows.

Newton–Raphson iteration is a functional iteration with iteration function

$$\phi(x) = x - \frac{f(x)}{f'(x)}.$$

We must now investigate whether the conditions of the convergence theorem (Theorem 2.1) can be satisfied which means we must examine the behaviour of $\phi'(x)$ near the root. Differentiating, we find

$$\phi'(x) = \frac{f(x)f''(x)}{(f'(x))^2}.$$

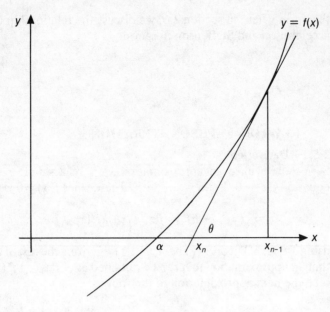

Fig. 2.4

so that, at the root, $\phi'(\alpha) = 0$. (Note that $f'(\alpha) \neq 0$ since we are considering simple roots only.) Assuming $\phi'(x)$ to be continuous, there must be an interval centred on $x = \alpha$ for which $|\phi'(x)| < 1$ and hence, for a suitably chosen x_0, Newton–Raphson iteration will always converge. The question of how close x_0 must be to the root is difficult to answer in general terms but the following theorem gives sufficient conditions for convergence.

THEOREM 2.2
Suppose there is an interval $[a, b]$ which is such that
(i) $f(a)f(b) < 0$;
(ii) $f'(x) \neq 0$, $x \in [a, b]$;
(iii) $f''(x) \geq 0$ or $f''(x) \leq 0$, for all $x \in [a, b]$; and

(iv) $\left|\dfrac{f(a)}{f'(a)}\right| < b - a$ and $\left|\dfrac{f(b)}{f'(b)}\right| < b - a$.

Then Newton–Raphson iteration will converge to the unique root of the equation $f(x) = 0$ lying in $]a, b[$ for any $x_0 \in [a, b]$.

A formal proof of Theorem 2.2 may be found in Dahlquist and Björck

(1974) and we consider here only the implications of the four conditions. Condition (i) shows that there is at least one root of $f(x) = 0$ lying in $]a,b[$ and (ii) ensures that it is unique, (iii) demands that $f(x)$ be convex or concave on $[a,b]$ whilst (iv) ensures the convergence of the iterates for any $x_0 \in [a,b]$ by requiring that the tangents to the curve $y = f(x)$ at the points $x = a$ and $x = b$ intersect the x-axis at points lying within $]a,b[$.

In Fig. 2.5 we illustrate what happens if conditions (i), (ii) and (iv) of

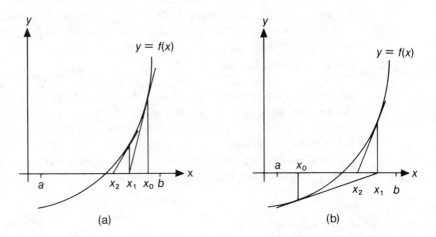

Fig. 2.5

Theorem 2.2 are satisfied and $f''(x) \geq 0$. In (a) $x_0 > \alpha$ and we have monotonic convergence of the iterates to the root from the right. In (b) $x_0 < \alpha$ and we find that $x_1 > \alpha$ and then, once again, we have monotonic convergence from the right. The reader should now draw diagrams corresponding to the case $f''(x) < 0$ and infer the behaviour of the iterates.

In practice it may not be too difficult to find an interval $[a,b]$ such that the conditions of Theorem 2.2 are satisfied as indicated by the following example.

EXAMPLE 2.7
Consider the equation $f(x) \equiv x^3 + 2x^2 + 3x - 1 = 0$ and the interval $[0,1]$. Then $f(0) = -1$ and $f(1) = 5$ so that condition (i) of Theorem 2.2 is satisfied. In addition $f'(x) = 3x^2 + 4x + 3$ has no root in $[0,1]$ and $f''(x) = 6x + 4$ is everywhere positive in this interval and hence conditions (ii) and (iii) are satisfied. Further $f'(0) = 3$ and $f'(1) = 10$ and so condition (iv) holds which means that convergence is guaranteed for any starting value $x_0 \in [0,1]$.

If the conditions of Theorem 2.2 are not satisfied we may be fortunate and still have convergence. It is likely, however, that we will converge to some other root of the equation or that the iteration will exhibit strange behaviour.

EXAMPLE 2.8
(i) The equation $f(x) \equiv 4x^3 - 13x + 6$ has roots in $]0,1[$ and $]1,2[$ since $f(0) = 6$, $f(1) = -3$ and $f(2) = 12$. But $f'(x) = 12x^2 - 13$ so that $f'(1) = -1$ and $|f(1)/f'(1)| = 3 > 1$. Hence, the conditions of Theorem 2.2 are not satisfied for the interval $[0,1]$. In fact, if we choose as an initial value $x_0 = 1$ we converge to the root lying in $]1,2[$ (we note that $f'(x)$ has a root at $x = \sqrt{13/12}$) and the reader should sketch the function and verify that this does indeed happen.
(ii) The equation $f(x) = e^x - 1 - \cos(\pi x)$ has a root lying in $]0,1[$. Applying Newton–Raphson iteration with $x_0 = 0$ and noting that $f'(x) = e^x + \pi \sin(\pi x)$ we have $x_1 = 0 - (-1)/1 = 1$ and $x_2 = 1 - e/e = 0$. Thus the iterates will oscillate indefinitely between 0 and 1. (This is a theoretical result only; it is possible that in practice convergence will be obtained due to the round-off errors incurred in evaluating e and $\cos(\pi)$ to some finite precision when calculating x_2.)

2.5.3 Termination criteria
In implementing Newton–Raphson iteration there are a number of important devices which should be included in a robust program. The termination criterion will usually monitor the absolute difference between successive iterates, $|x_n - x_{n-1}|$, and also perhaps $|f(x_n)|$, continuing until either or both of these are sufficiently small to ensure that a satisfactory estimate of the root is found. If the magnitude of the numbers is not known in advance it is best to use a relative criterion in which $|x_n - x_{n-1}|/|x_n|$ and $|f(x_n)/F|$ are monitored, where F is some estimate of the value of $f(x)$ near the root which can be obtained during the iteration or preset before the calculation begins.

As usual, in implementing an iterative process the number of iterations should be counted and the process stopped if this number exceeds some preassigned value.

It is necessary here to observe that a computer implementation of Newton–Raphson iteration will result in the values $f(x_{n-1})$ and $f'(x_{n-1})$ being obtained to limited precision only. The values used in the termination criteria must take account of this restriction so that an accuracy which is impossible to satisfy is not specified.

In Newton–Raphson iteration it is usual also to monitor the behaviour of $|f'(x)|$. If the value of $f'(x_{n-1})$ becomes small in some sense, the program user should be made aware of this fact through an appropriate message although the iteration may be allowed to continue. A small value of $f'(x_{n-1})$ may be symptomatic of the presence of a double root, or a root of even higher multiplicity, or of the existence of two or more roots which are close together, or of the absence of a root. These situations are illustrated in Fig. 2.6.

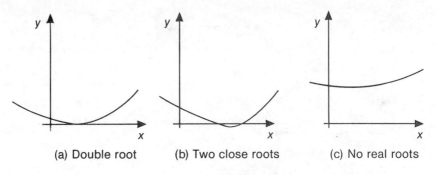

Fig. 2.6

2.5.4 A program for Newton–Raphson iteration

The programs of Case Study 2.2 use Newton–Raphson iteration to solve the equation $x - e^{-x} = 0$. The devices used in the programs of subsection 2.3.4 have once again been employed together with those discussed in subsection 2.5.3. In particular, the termination criterion based on $|f(x_n)/F|$ is implemented by asking the programmer to supply a value for *festimate*, an estimate of the value of $f(x)$ near the root. Further, a warning is given if the derivative of $f(x)$ becomes small, although the iteration is allowed to continue.

The user is able to indicate in the data whether he wishes individual iterates to be printed so that he can observe the rate of convergence. In Algol 68, on the PR1ME 550, the external denotation of a *BOOL* is Y (for *YES*) and N (for *NO*). It is possible that the reader will have to use T (for *TRUE*) and F (for *FALSE*) in order to run the Algol 68 program successfully. In Pascal, it is not possible to read a Boolean value. Instead, the Pascal program of Case Study 2.2 looks for the next non-space character, sees whether this is 'Y' or 'y' and sets the Boolean variable *monitoriterates* appropriately.

We observe that, using the same starting value (0.5) and the same accuracy tolerance (5×10^{-5}) as in Case Study 2.1, we achieve convergence with just 3 (as opposed to 15) iterations (and the final value happens to be more accurate). This represents a significant saving since the function $f(x)$ and its derivative $f'(x)$ are evaluated at a relatively small number of points only. It should be noted that these functions are called *f* and *fdash* in the programs. In the Algol 68 program they are *PROC*s of mode *PROC (REAL)REAL* and in the Pascal program they are *FUNCTION*s of type *real*. In both cases they are real-valued functions with one real parameter.

The reader should verify that it is permissible to use Newton–Raphson iteration for the equation $x - e^{-x} = 0$ on the interval [0,1], i.e., that the conditions of Theorem 2.2 are satisfied.

CASE STUDY 2.2

<u>Algol 68 Program</u>

```
BEGIN
  COMMENT
  a program to find a root of the equation x-exp(-x)=0 using
  newton-raphson iteration
  COMMENT
  PROC f = ( REAL x ) REAL :
    x - exp ( - x ) ;
  PROC fdash = ( REAL x ) REAL :
    1.0 + exp ( - x ) ;
  REAL xn , tol , festimate ;
  BOOL monitoriterates ;
  INT itermax ;
  read ( ( xn , tol , itermax , festimate , monitoriterates ) ) ;
  print ( ( newline ,
    "starting value                           " , xn , newline ,
    "accuracy tolerance                       " , tol , newline ,
    "maximum number of iterations allowed     " , itermax , newline ,
    "estimated value of the function at the root  " , festimate ) ) ;
  IF monitoriterates
  THEN
    print ( ( newline , newline , "      xn           f(xnminus1)" ) )
  FI;
  REAL twotol = tol + tol ;
  REAL xnminus1 , fxnminus1 , fdashxnminus1 , changeiniterates ;
  BOOL converged := FALSE ;
  BOOL smallfdash := FALSE ;
  INT iter := 0 ;
  WHILE NOT converged AND iter < itermax
  DO
    iter +:= 1;
    xnminus1 := xn ;
    fxnminus1 := f ( xnminus1 ) ;
    fdashxnminus1 := fdash ( xnminus1 ) ;
    xn := xnminus1 - fxnminus1 / fdashxnminus1 ;
    IF monitoriterates
    THEN
      print ( ( newline , xn , "    " , fxnminus1 ) )
    FI ;
    COMMENT
    changeiniterates is the absolute difference between the iterates if
    xn is regarded as being small;  otherwise it is the relative difference
    COMMENT
    changeiniterates := IF ABS xn < tol
                        THEN ABS ( xn -xnminus1 )
                        ELSE ABS ( ( xn - xnminus1 ) / xn )
                        FI ;
    converged := changeiniterates + ABS ( fxnminus1 / festimate )
                        < twotol ;
    IF NOT smallfdash
    THEN
      smallfdash := ABS ( fdashxnminus1 ) < tol
    FI
  OD ;
  IF smallfdash
  THEN
    print ( ( newline , newline ,
              "warning! the derivative of the function became less than" , tol ,
              "   at at least one iteration point" ) )
  FI;
  IF converged
  THEN
    print ( ( newline , newline ,
              "convergence to specified tolerance after " , iter ,
              "   iterations to " , xn , newline ) )
```

```
        ELSE
          print ( ( newline , newline ,
                  "no convergence to the specified tolerance after   " ,
                  itermax , "   iterations" , newline ) )
    FI
END
```

Pascal Program

```pascal
PROGRAM newtonraphson(input, output);

   {  a program to find a root of the equation x-exp(-x)=0 using
      newton-raphson iteration  }

     VAR
        xn, xnminus1, festimate, fxnminus1, fdashxnminus1,
                          tol, twotol, changeiniterates: real;
        monitoriterates, converged, smallfdash: boolean;
        character:char;
        iter, itermax: integer;

    FUNCTION f(x:real):real;
    BEGIN
        f:= x - exp(-x)
    END { f } ;

    FUNCTION fdash(x:real):real;
    BEGIN
        fdash:= 1.0 + exp(-x)
    END { fdash } ;

BEGIN
    read(xn, tol, itermax, festimate);
    REPEAT
        read(character)
    UNTIL character<>' ';
    monitoriterates:= (character='Y') OR (character='y');
    writeln;
    writeln('starting value                              ', xn);
    writeln('accuracy tolerance                          ', tol);
    writeln('maximum number of iterations allowed        ', itermax);
    writeln('estimated value of the function at the root ', festimate);
    writeln;
    smallfdash:= false;
    twotol:= tol + tol;
    IF monitoriterates
    THEN
        writeln('     xn            f(xnminus1)');
    iter:= 0;
    REPEAT
       iter:= iter + 1;
       xnminus1:= xn;
       fxnminus1:= f(xnminus1);
       fdashxnminus1:= fdash(xnminus1);
       xn:= xnminus1 - fxnminus1/fdashxnminus1;
       IF monitoriterates
       THEN
           writeln(xn, ' ':3, fxnminus1);
       { changeiniterates is the absolute difference between the iterates
         if xn is regarded as being small;  otherwise it is the absolute
         difference  }
       IF abs(xn)<tol
       THEN
           changeiniterates:= abs(xn - xnminus1)
       ELSE
           changeiniterates:= abs( (xn - xnminus1)/xn );
       converged:=  changeiniterates + abs(fxnminus1/festimate) < twotol;
       IF NOT smallfdash
       THEN
           smallfdash:= abs(fdashxnminus1) < tol
    UNTIL converged OR (iter=itermax);
```

```
writeln;
IF smallfdash
THEN
    writeln('warning! the derivative of the function became less than', tol,
    '      at at least one iteration point');
IF converged
THEN
    writeln('convergence to specified tolerance after  ', iter,
    '   iterations to ', xn)
ELSE
    writeln('no convergence to the specified tolerance after   ', itermax,
    '   iterations')
END { newtonraphson } .
```

Sample Data

    0.5   0.00005   20   1.0   Y

Sample Output

    starting value   5.00000e -1
    accuracy tolerance   5.00000e -5
    maximum number of iterations allowed   20
    estimated value of the function at the root   1.00000e 0

          xn   f(xnminus1)
    5.66311e -1   -1.06531e -1
    5.67143e -1   -1.30466e -3
    5.67143e -1   -2.55182e -7

    convergence to specified tolerance after   3 iterations to   5.67143e -1

2.6 THE ORDER OF CONVERGENCE OF FUNCTIONAL ITERATION

2.6.1 Introductory remarks and definition

In our discussions so far of the iteration $x_n = \phi(x_{n-1})$, we have been concerned with trying to establish the conditions under which the sequence of iterates $\{x_n\}$ is guaranteed to converge to a root $x = \alpha$ of the equation $x = \phi(x)$. We now turn to an important and practical point; if the iteration is convergent we would like to know how rapid the convergence is. The starting point is to relate the error $\varepsilon_n (= x_n - \alpha)$ in the nth iterate to the error ε_{n-1} in the previous iterate making use of Taylor's theorem.

Since $\alpha = \phi(\alpha)$ and $x_n = \phi(x_{n-1})$ we have

$$x_n - \alpha = \phi(x_{n-1}) - \phi(\alpha). \qquad (2.10)$$

Now, from Theorem 1.1

$$\phi(x_{n-1}) = \phi(\alpha) + (x_{n-1} - \alpha)\phi'(\alpha) + \ldots + \frac{(x_{n-1} - \alpha)^{p-1}}{(p-1)!}\phi^{(p-1)}(\alpha)$$

$$+ \frac{(x_{n-1} - \alpha)^p}{p!}\phi^{(p)}(\xi_{n-1})$$

Sec. 2.6] THE ORDER OF CONVERGENCE OF FUNCTIONAL ITERATION 59

where ξ_{n-1} lies somewhere between x_{n-1} and α. Using the definition of ε_n and (2.10) we have

$$\varepsilon_n = \varepsilon_{n-1}\phi'(\alpha) + \ldots + \frac{\varepsilon_{n-1}^{p-1}}{(p-1)!}\phi^{(p-1)}(\alpha) + \frac{\varepsilon_{n-1}^p}{p!}\phi^{(p)}(\xi_{n-1}).$$

Suppose the iteration function $\phi(x)$ is such that $\phi'(\alpha) = \phi''(\alpha) = \ldots = \phi^{(p-1)}(\alpha) = 0$ (that is, the first $p-1$ derivatives of $\phi(x)$ are all zero at the root) but that $\phi^{(p)}(\alpha) \neq 0$. Then

$$\varepsilon_n = \frac{\varepsilon_{n-1}^p}{p!}\phi^{(p)}(\xi_{n-1}).$$

This result shows that near the root ε_n is approximately proportional to ε_{n-1}^p and, the larger the value of p, the more rapid the convergence will be. Further, in a convergent iteration ξ_{n-1} will get closer and closer to α. Thus

$$\lim_{n \to \infty} \left| \frac{\varepsilon_n}{\varepsilon_{n-1}^p} \right| = \frac{1}{p!} |\phi^{(p)}(\alpha)|. \tag{2.11}$$

We now summarise these observations.

DEFINITION
If, in the convergent iteration $x_n = \phi(x_{n-1})$, the sequence of iterates converges to α and the iteration function $\phi(x)$ is such that

$$\phi'(\alpha) = \phi''(\alpha) = \ldots = \phi^{(p-1)}(\alpha) = 0$$

but $\phi^{(p)}(\alpha) \neq 0$, the iteration is said to have *pth order convergence* and the quantity on the right-hand side of (2.11) is called the *asymptotic error constant*.

When $p = 1, 2$ and 3 we refer to the order of convergence as being linear, quadratic and cubic respectively.

To find the order of convergence of a particular form of functional iteration, we have therefore to examine the successive derivatives of the iteration function at the root and find the first which is non-zero.

EXAMPLE 2.9
(i) In Example 2.3 (i) we observed that the equation $x = e^{-x}$ has a root lying between 0.5 and 0.6 and that the iteration $x_n = e^{-x_{n-1}}$ is convergent. Here $\phi(x) = e^{-x}$ and $\phi'(x) = -e^{-x}$ so that $\phi'(x) \neq 0$ at the root. The iteration therefore has first order (or linear) convergence only. (If, as suggested, in Example 2.4, the reader has computed the errors for the

results of Table 2.1, he may care to verify that the ratio of the errors in successive iterates is about 0.57. Since we have linear convergence here, this means that the asymptotic error constant is approximately equal to this value).

(ii) We return again to the iteration suggested in section 1.3 for estimating $^m\sqrt{a}$. In Example 2.3 (iii) we were able to show why the iteration is successful for $m = 2$ and $m = 3$ but fails to converge for $m \geq 4$. We can now show why the iteration for $m = 2$ is much more rapidly convergent than that for $m = 3$. We have $\phi'(^m\sqrt{a}) = 1 - (m/2)$ so, when $m = 2$, the first derivative of the iteration function is zero at the root. Hence, we would expect the iterative process to exhibit (at least) second order convergence for a square root ($m = 2$). However, $\phi'(^3\sqrt{a}) \neq 0$ and we expect only linear convergence for a cube root.

In Newton–Raphson iteration $\phi(x) = x - f(x)/f'(x)$ so that

$$\phi'(x) = \frac{f(x)f''(x)}{(f'(x))^2} \quad (2.12)$$

and

$$\phi''(x) = \frac{f''(x)}{f'(x)} + \frac{f(x)f'''(x)}{(f'(x))^2} - 2\frac{f(x)(f''(x))^2}{(f'(x))^3}. \quad (2.13)$$

Thus, at a simple root $x = \alpha$, $\phi'(\alpha) = 0$ (since $f(\alpha) = 0$) and $\phi''(\alpha) = f''(\alpha)/f'(\alpha) \neq 0$ unless exceptionally $f''(\alpha) = 0$. Newton–Raphson iteration therefore has second order convergence in general. There are exceptional cases, however.

If α is a double root of $f(x) = 0$, that is, $f(\alpha) = f'(\alpha) = 0$ we cannot immediately determine $\phi'(\alpha)$ because of the term $(f'(\alpha))^2$ in the denominator of (2.12). It is possible to show, although we do not do so, that $\phi'(\alpha) = \frac{1}{2}$ and so Newton–Raphson iteration applied to a double root has first order convergence only. However, it is also possible to show that the iteration

$$x_n = x_{n-1} - 2\frac{f(x_{n-1})}{f'(x_{n-1})}$$

is quadratically convergent to a double root. More generally if α has multiplicity m, that is, $f(\alpha) = f'(\alpha) = \ldots = f^{(m-1)}(\alpha) = 0$, Newton–Raphson iteration exhibits first order convergence but the iteration

$$x_n = x_{n-1} - m\frac{f(x_{n-1})}{f'(x_{n-1})}$$

is quadratically convergent. This result is not of much practical use since it is unlikely that the multiplicity of a root will be known in advance. However, a slowly convergent Newton–Raphson iteration can indicate the presence of a multiple root. It may be possible to estimate the multiplicity m and adjust the iteration accordingly.

If α is a simple root and $f''(\alpha) = 0$ then equations (2.12) and (2.13) show that $\phi'(\alpha) = \phi''(\alpha) = 0$ and Newton–Raphson iteration has at least third order convergence.

EXAMPLE 2.10

Newton–Raphson iteration applied directly to the equation $e^{2x} - 3 = 0$ converges quadratically to the root $\alpha = 0.549306$ (correct to six decimal places) provided that a suitable initial starting value is chosen. But

$$e^{2x} - 3 = 2e^x \left(\frac{e^x - e^{-x}}{2} - e^{-x} \right) = 2e^x (\sinh(x) - e^{-x}).$$

Therefore, α is a root of $f(x) \equiv e^{-x} - \sinh(x)$. But $f''(x) = e^{-x} - \sinh(x) = f(x)$ and Newton–Raphson iteration gives at least third order convergence.

It is possible to construct iterative methods of the form $x_n = \phi(x_{n-1})$ of arbitrarily high order. However, if the method has order p then at each iteration the evaluation of $f(x), f'(x), \ldots, f^{(p-1)}(x)$ at $x = x_{n-1}$ will usually be required. An example of a method which is generally of third order is *Richmond's iteration*

$$x_n = x_{n-1} - \frac{2f(x_{n-1})f'(x_{n-1})}{2(f'(x_{n-1}))^2 - f(x_{n-1})f''(x_{n-1})}.$$

Clearly this process requires knowledge of the first two derivatives of $f(x)$.

Steffensen iteration achieves second order convergence without requiring derivative values. It can be stated formally as

$$x_n = x_{n-1} - \frac{f(x_{n-1})}{F(x_{n-1})}$$

where

$$F(x) = \frac{f(x + f(x)) - f(x)}{f(x)}.$$

There is a close relationship between this process and Aitken acceleration. At the end of section 2.4 we observed that the generation of the sequence of accelerated iterates $\{z_n\}$ could be described as follows. With $z_0 = x_0$, form $z_n = \Phi(z_{n-1}) : n = 1, 2, \ldots$ where (equation (2.9))

$$\Phi(z) = z - \frac{(\phi(z) - z)^2}{\phi^2(z) - 2\phi(z) + z}.$$

If we apply Steffensen iteration to the equation $f(z) \equiv \phi(z) - z = 0$ we find that the sequence of iterates is the same. The extension of this method to simultaneous non-linear equations is of special interest (see Ortega and Rheinboldt, 1970).

2.6.2 Practical considerations

Although a high order method is theoretically desirable because of the expected rapid convergence near the root, there may be disadvantages. The successive derivatives of $f(x)$ have to be found analytically and then programmed; this is frequently a tedious and difficult operation and, in extreme cases, may be almost impossible to achieve. In addition, it must be remembered that at each iteration a number (equal to the order of the method) of function evaluations are required (where, by a 'function evaluation', we mean the evaluation of $f(x)$ or one of its derivatives at an iteration point) and this could prove costly in terms of computer time. The ultimate measure of the efficiency of a particular iterative method for a given algebraic equation is (apart from programming effort and storage requirements) the amount of computer time required. There is a complicated trade-off between a reduction in the number of iterations required and an increase in the number of function evaluations. It is arguable that Newton–Raphson iteration represents a reasonable compromise provided a good initial approximation is known. However, if $f'(x)$ is difficult and expensive to obtain, the more sophisticated of the methods to be discussed in the next section may well be more efficient.

2.7 INTERPOLATION METHODS

2.7.1 Introduction

In section 2.2 we described the method of bisection for a simple root and observed that we were making no use of the information that we were gathering about the function $f(x)$ (apart from its sign) and that it might be possible to derive more robust methods if we did so. The subject of interpolation is treated in detail in Chapter 5 but here we note that if values of $f(x)$ are available at two distinct points x_{n-1} and x_n, say, then we can join the coordinates $(x_{n-1}, f(x_{n-1}))$ and $(x_n, f(x_n))$ by a straight line. If x_{n-1} and x_n are suitably chosen, the point of intersection of this straight line with the x-axis can be taken as an approximation to the root of the equation $f(x) = 0$. The points x_{n-1} and x_n are then updated in some way, a new straight line obtained and hence a new approximation to the root found. There is no reason, in principle, why we should restrict ourselves to straight line (that is, linear) approximations; we could equally well consider successive quadratic approximations or, indeed, any of the approximating functions discussed in

Sec. 2.7]

INTERPOLATION METHODS 63

Chapter 5. Here we put forward some elementary ideas based on successive approximations of $f(x)$ by some simple function near the root. The simplest idea, using straight line approximation, is contained in the so-called *rule of false position* which we now discuss.

2.7.2 The rule of false position

The starting point here, as with the method of bisection, is the knowledge of two points x_0 and x_1 for which $f(x_0)f(x_1) < 0$ (see Fig. 2.7). The straight line

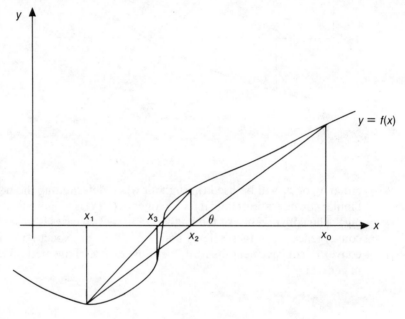

Fig. 2.7

joining the points $(x_0, f(x_0))$ and $(x_1, f(x_1))$ is formed and x_2, the next estimate of the root, is taken as the point of intersection of this line with the x-axis. Formally, observing that $\tan(\theta) = (f_0 - f_1)/(x_0 - x_1) = f_0/(x_0 - x_2)$, we have

$$x_2 = \frac{f_1}{f_1 - f_0} x_0 + \frac{f_0}{f_0 - f_1} x_1 \qquad (2.14)$$

where we have used the notation $f_i = f(x_i)$ $(i = 0, 1)$. The next stage is to evaluate $f(x_2)$ and to determine whether $f(x_2)f(x_0)$ or $f(x_2)f(x_1)$ is negative. This allows us to find x_3 and the process is repeated.

The rule of false position will always converge to a simple root lying between x_0 and x_1. We observe that if the function is convex (that is, the gradient is monotonically increasing) in the vicinity of the root (see Fig. 2.8),

64 NON-LINEAR ALGEBRAIC EQUATIONS [Ch. 2

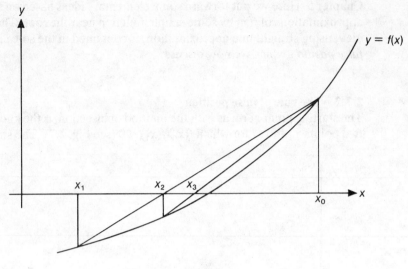

Fig. 2.8

either x_0 or x_1 will be used throughout when determining the next iterate. Similar considerations hold if $f(x)$ is concave ($f'(x)$ is monotonically decreasing). The convergence of the method in these two cases is slow (the rate of convergence is, in fact, linear) and, since we can expect $f(x)$ to be either convex or concave near the root, the convergence of the method will be slow in general.

2.7.3 The Illinois and Pegasus methods

These methods were designed to try to improve the slow rate of convergence of the method of false position. An examination of Fig. 2.8 indicates why convergence is so slow; all new iterates lie on the same side of the root. We would like to be able to introduce a new iterate on the same side of the root as x_0. Suppose x_n, x_{n+1} and x_{n+2} are three successive iterates which are such that $f_n f_{n+1} < 0$ and $f_{n+1} f_{n+2} > 0$ (which will be the case in applying false position to x_0 and x_1 to give x_2 for a convex function, although a node number re-ordering may be necessary). We now apply false position to x_n and x_{n+2} but with f_n replaced by $f^*_n = \alpha f_n$, for some $0 < \alpha < 1$, to give x_{n+3}. If then $f_{n+2} f_{n+3} < 0$ we apply false position to x_{n+2} and x_{n+3}; otherwise we carry out another modified iteration based on information at x_n and x_{n+3}. In the *Illinois* method (see Fig. 2.9) we choose $\alpha = \frac{1}{2}$, whilst for *Pegasus* $\alpha = f_0/(f_1 + f_0)$. It can be shown (see Ralston and Rabinowitz, 1978) that these methods have a greater-than-linear rate of convergence and that Pegasus is inherently faster than Illinois.

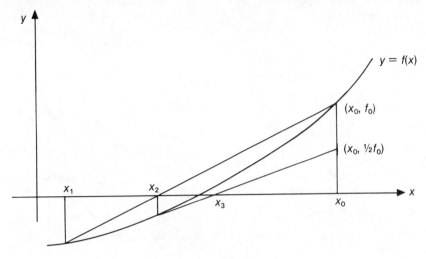

Fig. 2.9

2.7.4 The secant method

Here, two initial approximations, x_0 and x_1, to the root α are again required but we do not demand that $f(x_0)f(x_1)<0$. As with the rule of false position, the straight line (or *secant*) joining the coordinates (x_0,f_0) and (x_1,f_1) is drawn and x_2 taken as the point at which this line meets the x-axis. Now, of course, there is no guarantee that x_2 will lie between x_0 and x_1. x_0 is discarded and the process repeated with x_1 and x_2 (see Fig. 2.10). In general, if x_n and x_{n+1} are the two most recent approximations, the new estimate of the root is found as

$$x_{n+2} = x_{n+1} - \frac{x_{n+1} - x_n}{f_{n+1} - f_n} f_{n+1}.$$

This equation is just another way of writing the generalised form of (2.14) but it does give us some additional insight into the method. It suggests that the secant method is Newton–Raphson iteration with $f'(x_{n+1})$ approximated by $(f_{n+1}-f_n)/(x_{n+1}-x_n)$. The validity of this approximation is discussed in Chapter 5.

It can be shown (see Ralston and Rabinowitz, 1978) that the order of convergence of the secant method is approximately 1.62 (that is, $\varepsilon_n \sim \varepsilon_{n-1}^{1.62}$). This is considerably better than linear convergence but not as good as that for Newton–Raphson iteration. However, in Newton–Raphson iteration the evaluation of $f'(x)$ as well as $f(x)$ is required at each iteration. If the evaluation of the derivative of $f(x)$ is expensive in terms of computer time compared with the evaluation of $f(x)$ itself, the secant method is likely to prove to be more efficient than Newton–Raphson.

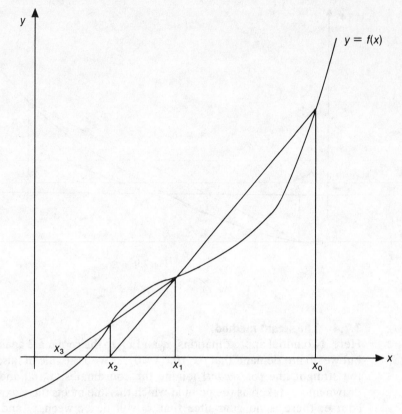

Fig. 2.10

2.7.5 Other methods
Although we shall not pursue the details, the ideas of the previous subsections, which are based on linear interpolation, can be extended in an obvious manner. For example, if we have three estimates, x_0, x_1 and x_2, of α and corresponding function values it is possible to fit a quadratic to the coordinates (x_0,f_0), (x_1,f_1) and (x_2,f_2). One of the roots of this quadratic will give an improved estimate of the root of $f(x) = 0$. We shall consider other types of approximating functions in detail in Chapter 5.

2.8 POLYNOMIAL EQUATIONS

2.8.1 Introduction
Here we are concerned with estimating some or all of the roots of the equation $f(x) = 0$ where $f(x)$ is a polynomial of degree m in x, that is,

$$f(x) = a_m x^m + a_{m-1} x^{m-1} + \ldots + a_1 x + a_0 = \sum_{i=0}^{m} a_i x^i. \quad (2.15)$$

We assume that the coefficients $a_0, a_1, \ldots, a_m$ are all real. The polynomial equation has m roots, some or all of which may be complex. We recall that if complex roots are present they occur in conjugate pairs, that is, if $\alpha + i\beta$ is a root then so is $\alpha - i\beta$.

Any of the methods which have been considered in this chapter may, in principle, be used to estimate the real roots of a polynomial equation and some of them may be adapted to deal with complex roots. Complex roots, however, are usually best determined using special techniques and we make some reference to these in subsection 2.9.4.

The subject of root-finding in the case of polynomial equations is very large and we shall not attempt to give an exhaustive survey. Rather we concentrate on a number of basic ideas concerning the evaluation of a polynomial and its first derivative, the division of a polynomial by a polynomial of degree 2 and the extremely important notion of the conditioning of a polynomial equation. The reader is referred to Ralston and Rabinowitz (1978) for a more comprehensive discussion of the subject.

2.8.2 Evaluation of a polynomial and its derivative

Suppose we wish to evaluate the polynomials $f(x)$ and $f'(x)$ for some real number λ. (This would be required, for example, if we were employing Newton–Raphson iteration to find the root of $f(x) = 0$.) The starting point is to use the *nested form* of a polynomial which, for (2.15) is

$$((\ldots((a_m x + a_{m-1})x + a_{m-2})x + \ldots)x + a_1)x + a_0. \tag{2.16}$$

Using (2.16) the evaluation of $f(x)$ at the point $x = \lambda$ can be described by the recursive algorithm

$$\begin{aligned} b_{m-1} &= a_m \\ b_i &= a_{i+1} + \lambda b_{i+1} \qquad i = m-2, m-3, \ldots, 0 \\ f(\lambda) &= a_0 + \lambda b_0 \end{aligned} \tag{2.17}$$

which defines a sequence of intermediate values $b_{m-1}, b_{m-2}, \ldots, b_0$.

EXAMPLE 2.11
If $m = 3$ we have

$$a_3 x^3 + a_2 x^2 + a_1 x + a_0 = ((a_3 x + a_2)x + a_1)x + a_0$$

and hence, from (2.17)

$$\begin{aligned} b_2 &= a_3 \\ b_1 &= a_2 + \lambda b_2 \quad (= a_2 + \lambda a_3) \end{aligned}$$

$$b_0 = a_1 + \lambda b_1 \quad (= a_1 + \lambda(a_2 + \lambda a_3))$$
$$f(\lambda) = a_0 + \lambda b_0 \quad (= a_0 + \lambda(a_1 + \lambda(a_2 + \lambda a_3))).$$

For the purposes of hand computation, we can set out the coefficients of successive powers of x in a row, supplying a zero entry if any power of x is missing. The b_is and $f(\lambda)$ are then formed in a straightforward way.

EXAMPLE 2.12
To evaluate $f(x) \equiv x^5 - 6x^3 + x^2 + 7x - 4$ at the point $x = 2$ we form the following table

	(x^5)	(x^4)	(x^3)	(x^2)	(x^1)	(x^0)
	1	0	-6	1	7	-4
$x=2$	1	$2\times 1 + 0$ $=2$	$2\times 2 - 6$ $=-2$	$2\times(-2)+1$ $=-3$	$2\times(-3)+7$ $=1$	$2\times 1 - 4$ $=-2$
	(b_4)	(b_3)	(b_2)	(b_1)	(b_0)	$f(2)$

and deduce that $f(2) = -2$.

We now show that the b_is defined by (2.17) have a significance which is of interest. The following theorem defines a process which is known as *synthetic division*.

THEOREM 2.3

$$\frac{f(x) - f(\lambda)}{x - \lambda} = \sum_{i=0}^{m-1} b_i x^i \tag{2.18}$$

where the b_is are defined by (2.17).

The proof of this theorem is straightforward and depends on multiplying the right-hand side of (2.18) by $x - \lambda$ and showing the result equals $f(x) - f(\lambda)$. Rearranging (2.18) further we have

$$f(x) = (x - \lambda) \sum_{i=0}^{m-1} b_i x^i + f(\lambda) \tag{2.19}$$

so that if $\lambda = \alpha$, a root of the equation $f(x) = 0$, the remaining roots may be found as the roots of the *deflated equation*

$$g(x) \equiv \sum_{i=0}^{m-1} b_i x^i = 0. \tag{2.20}$$

This suggests that we can determine the real roots of a polynomial equation one at a time using the appropriate deflated equation after each root is found. The process can be quite unsatisfactory in practice, however. Suppose we have found the first root correct to some precision. Call this value $\bar{\alpha}$. Since $f(\bar{\alpha})$ will not be exactly zero, the b_is in the deflated equation will not be exact and the root of this equation will also be calculated correct only to the required precision. The cumulative effect of the errors introduced at each deflation can be disastrous and this superficially attractive process cannot be recommended.

We now return to equation (2.19) and indicate how $f'(x)$ can be evaluated at a point using the b_is which are generated in finding $f(\lambda)$. Differentiating (2.19) formally we have

$$f'(x) = (x - \lambda)g'(x) + g(x)$$

where $g(x)$ is defined by (2.20). Thus, $f'(\lambda) = g(\lambda)$ and we use the nested form of $g(x)$ in order to find $f'(\lambda)$.

EXAMPLE 2.13

To evaluate the derivative of $f(x) \equiv x^5 - 6x^3 + x^2 + 7x - 4$ at the point $x = 2$ we evaluate the coefficients of the deflated equation as shown in Example 2.12 and form the following table

	(x^4)	(x^3)	(x^2)	(x^1)	(x^0)
	1	2	-2	-3	1
$x=2$	1	$2\times 1+2$	$2\times 4-2$	$2\times 6-3$	$2\times 9+1$
		$=4$	$=6$	$=9$	$=19$
	(c_3)	(c_2)	(c_1)	(c_0)	$g(2)$

where the c_is are the coefficients in the next deflated equation. We deduce that $f'(2) = g(2) = 19$ so that if $x = 2$ is a first estimate of a root of the equation $f(x) = 0$ and Newton–Raphson iteration is employed, the next iterate is given by

$$x_2 = 2 - (-2)/19 = 2.105$$

to four significant figures.

2.8.3 A program for polynomial evaluation

The programs of Case Study 2.3 employ the techniques of the previous subsection to evaluate a polynomial and its first derivative at the point $x = \alpha$. The coefficients of the polynomial and the value of α are read from the data.

The Algol 68 program contains the code:

```
read (m) ;
[0 : m] REAL a ;
```

This means that the size, or length, of *a* is determined by the first value in the data. Such an array is called a *dynamic array* because its bounds are determined when the program is executed.

The overheads associated with dynamic arrays are considerable and so, in the interests of economy, Pascal does not permit the declaration of dynamic arrays. The usual method that is used to circumvent this problem is to set aside a large enough array to handle most of the likely values of *m*. For example, the Pascal program of Case Study 2.3 contains:

```
CONST
   maxm = 100;
VAR
   a: ARRAY [0..maxm] OF real;
```

Thus, the array *a* has room for 101 coefficients and the program will fail if the value that is read for *m* is greater than 100. Often, the value of *m* will be much less than that of *maxm* in which case a large number of the elements of *a* will not be used. For this program, the amount of space which is wasted is small and so it is unimportant. However, if this approach is used when a program has quite a few arrays of a reasonably large size then the program will be asking for a large amount of space for its variables in order for it to run. Depending on the computer system being used, there may be some difficulties in running such a program. This is rather unfortunate if most of the space is not actually used.

The Algol 68 program contains the code

```
read ( ( a , alpha ) ) ;
```

This fills the array *a* with the next $m + 1$ real numbers in the data and then reads a real number into the variable *alpha*. Thus the values in the data must appear in the order $a_0, a_1, \ldots, a_m, \alpha$ and these values are stored in the variables $a[0], a[1], \ldots, a[m]$ and *alpha*. It is not so easy to do this in Pascal: the program must contain code which reads in the elements of *a* an element at a time

```
FOR i := 0 TO m DO
   read(a[i]);
read(alpha);
```

Note that in the Algol 68 program it is not necessary to declare the loop control parameter *i* whereas, in Pascal, control variables have to be declared (in the block which immediately contains the *FOR* statement).

Pascal permits the programmer to indicate the range of values that a variable is intended to take. This is done by using a *subrange type*. In the Pascal program of Case Study 2.3 the variables *m* and *i* are declared to be of the types $0..maxm$ and $-1..maxm$ respectively. In this way, the variables *m* and *i* are constrained to subranges of the type integer. There are two principal reasons for using subranges. First, the intended range of the values that are to be assigned to a variable is made clear to the reader of a program. This aids the reading and understanding of a program. Second, most Pascal systems allow programs to be run in such a way that an error message is printed if the user has unintentionally written code which assigns to a variable a value which is outside the intended range. Subrange declarations are not provided by Algol 68. However, the fact that *FOR i* declares the loop control parameter *i* means that, within the loops in Case Study 2.3, the value of *i* is constrained to lie within specified bounds.

CASE STUDY 2.3

Algol 68 Program

```
BEGIN
  COMMENT
  a program to evaluate a polynomial and its derivative at a point
  COMMENT
  INT m ;
  REAL alpha , b , c ;
  read ( m ) ;
  [ 0 : m ] REAL a ;
  COMMENT
  the polynomial coefficients are expected constant term first
  COMMENT
  read ( ( a , alpha ) ) ;
  print ( ( newline , "the polynomial coefficients are ( constant term first )"
           newline ) ) ;
  FOR i FROM 0 TO m DO
    print ( ( a [ i ] , newline ) )
  OD ;
  print ( ( "evaluation point is  " , alpha , newline ) ) ;
  b := a [ m ] ;
  c := 0.0 ;
  FOR i FROM m - 2 BY - 1 TO - 1 DO
    c := b + alpha * c ;
    b := a [ i + 1 ] + alpha * b
  OD ;
  print ( ( "value of the polynomial at this point is          " , b ,
            newline ,
            "value of the polynomial derivative at this point is  " , c ,
            newline ) )
END
```

Pascal Program

```
PROGRAM polynomialevaluation(input, output);
    { a program to evaluate a polynomial and its derivative at a point }
    CONST
       maxm=100;
```

```
    VAR
        m: 0..maxm;
        i: -1..maxm;
        alpha, b, c: real;
        a: ARRAY [ 0..maxm ] OF real;
BEGIN
    read(m);
    { the polynomial coefficients are expected constant term first }
    FOR i:= 0 TO m DO
        read(a[i]);
    read(alpha);
    writeln;
    writeln('the polynomial coefficients are ( constant term first )');
    FOR i:= 0 TO m DO
        writeln(a[i]);
    writeln('evaluation point is  ', alpha);
    b:= a[m];
    c:= 0.0;
    FOR i:= m - 2 DOWNTO -1 DO
    BEGIN
        c:= b + alpha*c;
        b:= a[i + 1] + alpha*b
    END ;
    writeln('value of the polynomial at this point is            ', b);
    writeln('value of the polynomial derivative at this point is  ', c)
END { polynomialevaluation } .
```

Sample Data

```
5
-4.0   7.0   1.0   -6.0   0.0   1.0
2.0
```

Sample Output

```
the polynomial coefficients are ( constant term first )
-4.00000e  0
 7.00000e  0
 1.00000e  0
-6.00000e  0
 0.00000e  0
 1.00000e  0
evaluation point is    2.00000e  0
value of the polynomial at this point is            -2.00000e  0
value of the polynomial derivative at this point is  1.90000e  1
```

The programs of Case Study 2.3 avoid storing the coefficients of the first deflated equation explicitly. Each time a new b_i is found it overwrites the value currently held in the real variable b, but not before the old value has itself been used to update c. The real variable c is used to store the coefficients of the second deflated equation in a similar manner.

2.8.4 Synthetic division of a polynomial by a quadratic

We can extend the ideas introduced in subsection 2.8.2 to the synthetic division of a given polynomial by a polynomial of degree two. Let $x^2 + px + q$ be the given quadratic. Then we wish to find the coefficients c_i and remainders R and S in the identity

$$f(x) \equiv \sum_{i=0}^{m} a_i x^i \equiv (x^2 + px + q) \sum_{i=0}^{m-2} c_i x^i + Rx + S.$$

On equating coefficients of $x^m, x^{m-1}, \ldots$ in turn we obtain the following results

$$c_{m-2} = a_m$$
$$c_{m-3} = a_{m-1} - pc_{m-2}$$
$$c_i = a_{i+2} - pc_{i+1} - qc_{i+2} \qquad i = m-4, m-5, \ldots, 0$$
$$R = a_1 - pc_0 - qc_1$$
$$S = a_0 - qc_0.$$

If we introduce fictitious values $c_m = c_{m-1} = 0$ and let $c_{-1} = R$, these results can be expressed compactly as

$$c_i = a_{i+2} - pc_{i+1} - qc_{i+2} \qquad i = m-2, m-3, \ldots, -1$$
$$S = a_0 - qc_0.$$

The coefficients c_i and the remainders R and S are, of course, functions of p and q. If we can find values of p and q such that $R(p,q) = S(p,q) = 0$, then $x^2 + px + q$ will be a quadratic factor of $f(x)$ and on solving $x^2 + px + q = 0$ we will have two of the (possibly complex) roots of $f(x) = 0$. The way in which appropriate values of p and q can be found is contained in Bairstow's method to which we make brief reference in subsection 2.9.4. The discussion has to be deferred until we have given an introduction to the problem of solving a pair of non-linear simultaneous equations.

2.8.5 Conditioning

In section 1.8 we introduced the concept of the conditioning of a problem. We observed that small changes in the initial data could produce relatively large changes in the solution and, in such a case, we say that the problem is 'ill-conditioned'. In the context of the polynomial equation

$$\sum_{i=0}^{m} a_i x^i = 0 \qquad (2.21)$$

the initial data are the coefficients $a_0, a_1, \ldots, a_m$. In general there will be a rounding error in the representation of each of these coefficients in a computer but there may be other uncertainties, for example if they were the result of some previous computations which are not exact. We suppose that a_i is in error by an amount $\delta a_i : i = 0, 1, \ldots, m$. If α is an exact root of (2.21) and $\alpha + \delta\alpha$ is the corresponding root of the perturbed equation

$$\sum_{i=0}^{m}(a_i+\delta a_i)x^i = 0$$

we would like to try to relate $\delta\alpha$ to the perturbations δa_i. We refer the reader to Ralston and Rabinowitz (1978) for complete details and merely observe here that even small perturbations can have a dramatic effect on the zeros of polynomials of high degree.

EXAMPLE 2.14
The polynomial of degree 20

$$f(x) \equiv (x+1)(x+2)\ldots(x+20)$$

has zeros at $-1, -2, \ldots, -20$. The perturbed polynomial $f(x) + 2^{-23}x^{19}$, which corresponds to a perturbation of about 10^{-7} in the coefficient of x^{19} has zeros which apart from those corresponding to $-1, -2, \ldots, -5$ are dramatically different from those of the original polynomial; indeed ten of the zeros become five conjugate complex pairs. This example, due originally to Wilkinson (1959), emphasises that there may be severe difficulties in estimating the roots of a polynomial equation of high degree.

2.9 NON-LINEAR SIMULTANEOUS EQUATIONS
2.9.1 Introduction
To date we have only considered estimating the root(s) of a single non-linear equation in one real variable. We now present some elementary ideas on how we might attempt to find a solution to the pair of simultaneous equations

$$\begin{aligned} f_1(x_1, x_2) &= 0 \\ f_2(x_1, x_2) &= 0 \end{aligned} \quad (2.22)$$

where both $f_1(x_1,x_2)$ and $f_2(x_1,x_2)$ are real-valued non-linear functions of the two variables x_1 and x_2. By a solution, we mean two numbers α_1 and α_2 which are such that $f_1(\alpha_1,\alpha_2) = f_2(\alpha_1,\alpha_2) = 0$.

EXAMPLE 2.15
The equations

$$\begin{aligned} f_1(x_1,x_2) &\equiv e^{x_1} + (2x_1 - x_2)^2 - e = 0 \\ f_2(x_1,x_2) &\equiv x_1^2 + x_2^2 - 5 = 0 \end{aligned}$$

have a solution $\alpha_1 = 1$ and $\alpha_2 = 2$.

As in the case of a single equation, iteration forms the basis of the methods discussed in this section. The methods generalise readily to the case of more than two simultaneous equations.

2.9.2 Functional iteration
Here, we set up a scheme which is a direct extension of that introduced in section 2.3. We suppose that the system (2.22) is rewritten in the form

$$x_1 = \phi_1(x_1, x_2)$$
$$x_2 = \phi_2(x_1, x_2) \qquad (2.23)$$

and that we have available an initial approximation $(x_1^{[0]}, x_2^{[0]})$ to a solution of (2.23) (and hence of (2.22)). Note that the iteration number is shown as a superscript in square brackets. The iterative process based on (2.23) is

$$x_1^{[n]} = \phi_1(x_1^{[n-1]}, x_2^{[n-1]})$$
$$x_2^{[n]} = \phi_2(x_1^{[n-1]}, x_2^{[n-1]}) \qquad n = 1, 2, \ldots \qquad (2.24)$$

and we generate a sequence of pairs of numbers $\{x_1^{[n]}, x_2^{[n]}\}$. Then, under suitable circumstances, $x_1^{[n]} \to \alpha_1$ and $x_2^{[n]} \to \alpha_2$.

EXAMPLE 2.16
The equations

$$f_1(x_1, x_2) \equiv x_1^2 + x_2^2 - x_1 = 0$$
$$f_2(x_1, x_2) \equiv x_1^2 - x_2^2 - x_2 = 0$$

may be rewritten as

$$x_1 = x_1^2 + x_2^2$$
$$x_2 = x_1^2 - x_2^2$$

so $\phi_1(x_1, x_2) = x_1^2 + x_2^2$ and $\phi_2(x_1, x_2) = x_1^2 - x_2^2$. This system clearly has the solution $x_1 = x_2 = 0$ and there is another solution near $x_1 = 0.78$ and $x_2 = 0.42$. From (2.24) functional iteration gives

$$x_1^{[n]} = x_1^{[n-1]^2} + x_2^{[n-1]^2}$$
$$x_2^{[n]} = x_1^{[n-1]^2} - x_2^{[n-1]^2}.$$

Starting with $x_1^{[0]} = x_2^{[0]} = 0.2$ and $x_1^{[0]} = 0.8$, $x_2^{[0]} = 0.4$ we seek the two solutions and list in Table 2.2 results obtained by working to six decimal place accuracy.

The iterates in the first case are clearly converging to the solution $x_1 = x_2 = 0$ but there is an obvious divergence in the second case.

Table 2.2

n	$x_1^{[n]}$	$x_2^{[n]}$	$x_1^{[n]}$	$x_2^{[n]}$
0	0.200 000	0.200 000	0.800 000	0.400 000
1	0.080 000	0.000 000	0.800 000	0.480 000
2	0.006 400	0.006 400	0.870 400	0.409 600
3	0.000 082	0.000 000	0.925 368	0.589 824
4	0.000 000	0.000 000	1.204 198	0.508 414

The question of the convergence of the scheme (2.24) is covered by a theorem (the *contraction mapping theorem*) which is an extension of Theorem 2.1. We shall not formally state this theorem but note in passing that the interval I of Theorem 2.1 is replaced by a circular region centred on a solution (α_1, α_2) and that the condition on the derivative of the iteration function becomes

$$\left\{\left(\frac{\partial \phi_1}{\partial x_1}\right)^2 + \left(\frac{\partial \phi_1}{\partial x_2}\right)^2 + \left(\frac{\partial \phi_2}{\partial x_1}\right)^2 + \left(\frac{\partial \phi_2}{\partial x_2}\right)^2\right\}^{\frac{1}{2}} \leq L < 1.$$

It is clear that the problem of trying to establish whether a particular iterative scheme will converge is difficult and we do not pursue the matter further.

2.9.3 Newton's method

We now consider the adaptation of Newton–Raphson iteration as described in section 2.5 to the solution of the system (2.22). Suppose that $(x_1^{[n-1]}, x_2^{[n-1]})$ is the most recently computed approximation to the solution (α_1, α_2) and let $\alpha_1 = x_1^{[n-1]} + h_1$ and $\alpha_2 = x_2^{[n-1]} + h_2$. Then

$$f_1(x_1^{[n-1]} + h_1, x_2^{[n-1]} + h_2) = 0$$
$$f_2(x_1^{[n-1]} + h_1, x_2^{[n-1]} + h_2) = 0.$$

We now use the two-variable Taylor series expansion (Theorem 1.2) to give

$$f_1(x_1^{[n-1]}, x_2^{[n-1]}) + h_1\left(\frac{\partial f_1}{\partial x_1}\right)_{n-1} + h_2\left(\frac{\partial f_1}{\partial x_2}\right)_{n-1} + \ldots = 0$$

$$f_2(x_1^{[n-1]}, x_2^{[n-1]}) + h_1\left(\frac{\partial f_2}{\partial x_1}\right)_{n-1} + h_2\left(\frac{\partial f_2}{\partial x_2}\right)_{n-1} + \ldots = 0$$

(2.25)

where the suffix $n-1$ on the partial derivatives means that they are evaluated at the point $(x_1^{[n-1]}, x_2^{[n-1]})$. Ignoring the higher order terms in (2.25) we can obtain approximations $h_1^{[n-1]}$ and $h_2^{[n-1]}$ to h_1 and h_2 respectively by solving the pair of linear simultaneous equations

$$h_1^{[n-1]}\left(\frac{\partial f_1}{\partial x_1}\right)_{n-1} + h_2^{[n-1]}\left(\frac{\partial f_1}{\partial x_2}\right)_{n-1} = -f_1(x_1^{[n-1]}, x_2^{[n-1]})$$

$$h_1^{[n-1]}\left(\frac{\partial f_2}{\partial x_1}\right)_{n-1} + h_2^{[n-1]}\left(\frac{\partial f_2}{\partial x_2}\right)_{n-1} = -f_2(x_1^{[n-1]}, x_2^{[n-1]}).$$

The new approximation to the solution is then taken to be

$$x_1^{[n]} = x_1^{[n-1]} + h_1^{[n-1]}$$

$$x_2^{[n]} = x_2^{[n-1]} + h_2^{[n-1]}.$$

For those familiar with vector and matrix notation, this process (*Newton's method*) may be formally stated as

$$\begin{aligned} J_{n-1}\mathbf{h}^{[n-1]} &= -\mathbf{f}^{[n-1]} \\ \mathbf{x}^{[n]} &= \mathbf{x}^{[n-1]} + \mathbf{h}^{[n-1]} \end{aligned} \quad n = 1, 2, \ldots \quad (2.26)$$

where $\mathbf{x}^T = (x_1, x_2)$, $\mathbf{h}^{[n-1]T} = (h_1^{[n-1]}, h_2^{[n-1]})$, $\mathbf{f}^T = (f_1(x_1^{[n-1]}, x_2^{[n-1]}), f_2(x_1^{[n-1]}, x_2^{[n-1]}))$ and J_{n-1} is the Jacobian matrix, J, of the functions $f_1(x_1, x_2)$ and $f_2(x_1, x_2)$ evaluated at the point $(x_1^{[n-1]}, x_2^{[n-1]})$ where

$$J = \begin{pmatrix} \partial f_1/\partial x_1 & \partial f_1/\partial x_2 \\ \partial f_2/\partial x_1 & \partial f_2/\partial x_2 \end{pmatrix}.$$

EXAMPLE 2.17
Consider the pair of equations

$$f_1(x_1, x_2) \equiv e^{x_1 - x_2} - 4x_2 = 0$$

$$f_2(x_1, x_2) \equiv \cos(x_1 + x_2) + 4x_1 = 0.$$

Let $x_1^{[0]} = x_2^{[0]} = 0$ be starting values for Newton's method. We need to find, analytically, the elements of the Jacobian matrix. Now

$$\frac{\partial f_1}{\partial x_1} = e^{x_1-x_2} \qquad \frac{\partial f_1}{\partial x_2} = -e^{x_1-x_2} - 4$$

$$\frac{\partial f_2}{\partial x_1} = -\sin(x_1+x_2) + 4 \qquad \frac{\partial f_2}{\partial x_2} = -\sin(x_1+x_2)$$

so that at $x_1^{[0]} = x_2^{[0]} = 0$

$$\left(\frac{\partial f_1}{\partial x_1}\right)_0 = 1, \left(\frac{\partial f_1}{\partial x_2}\right)_0 = -5, \left(\frac{\partial f_2}{\partial x_1}\right)_0 = 4 \text{ and } \left(\frac{\partial f_2}{\partial x_2}\right)_0 = 0.$$

The equations corresponding to the first iteration are therefore

$$h_1^{[0]} - 5h_2^{[0]} = -f_1(0,0) = -1$$
$$4h_1^{[0]} = -f_2(0,0) = -1$$

giving $h_1^{[0]} = -\frac{1}{4}$ and $h_2^{[0]} = \frac{3}{20}$. Hence, $x_1^{[1]} = -\frac{1}{4}$ and $x_2^{[1]} = \frac{3}{20}$.

The question of the convergence of Newton's method is very difficult to deal with and we do not pursue the matter further. We can assert, however, that it will converge provided the initial approximation is 'sufficiently close' to the solution which is sought; the assertion stems from the use of the generalised contraction mapping theorem (see Ortega and Rheinboldt, 1970).

Apart from the important question of convergence, there may be severe difficulties in implementing Newton's method in practice, although these may not be apparent from a consideration of the problem of finding a solution of just two non-linear simultaneous equations. More generally, we have to think in terms of m equations in m unknowns

$$f_i(x_1, x_2, \ldots, x_m) = 0 \qquad i = 1, 2, \ldots, m$$

which we can write as

$$\mathbf{f}(\mathbf{x}) = \mathbf{0}.$$

The derivation of Newton's method in this case is a straightforward extension of the approach for a pair of equations and the statement of the method is given by (2.26) where now J is the $m \times m$ matrix defined by

$$J_{ij} = \frac{\partial f_i}{\partial x_j}$$

and $\mathbf{x}$, $\mathbf{h}$ and $\mathbf{f}$ are m-dimensional vectors. In implementing the method it will first be necessary to determine the m^2 elements of J through formal differentiation and then to program them. If $m = 10$, for example, this is no light task and one which is liable to human error. At each iteration the elements of the Jacobian have to be re-evaluated and this could be expensive in computer time; further, the solution of a set of linear simultaneous equations is involved and difficulties can arise. To these problems we must add the uncertainty over convergence and the reader will not be surprised to learn that there has been much recent work aimed at removing or lessening the severity of some of these difficulties. A discussion of these modifications is beyond the scope of this book and the reader is referred to Ralston and Rabinowitz (1978). We can, however, indicate briefly some of the ideas behind a number of these modifications.

If the elements of J can be found analytically without too much difficulty and can be programmed, the problem of the convergence of Newton's method may be resolved by the use of so-called *damped iterates*. Here, the idea is that after an iteration we examine the value $F^{[n]}$, defined by

$$F^{[n]} = \sum_{i=1}^{m} f_i^{[n]2}$$

and compare this with $F^{[n-1]}$, the corresponding quantity at the previous iteration. If $F^{[n]} < F^{[n-1]}$ we accept the new point $\mathbf{x}^{[n]}$ and proceed to the next iteration. Otherwise, we try to find a new point

$$\mathbf{x}^{[n]} = \mathbf{x}^{[n-1]} + \rho \mathbf{h}^{[n-1]}$$

where $0 < \rho < 1$ which is such that $F^{[n]} < F^{[n-1]}$. It is usually possible to find a suitable value of ρ and a commonly used strategy is to use successively $\rho = \frac{1}{2}, \frac{1}{4}, \frac{1}{8}, \ldots$ until the condition is satisfied.

If the programmed elements of J are available without too much difficulty but the computation of them is expensive in computer time, the device in which the same numerical Jacobian is used for several successive iterations may be of value. Thus, suppose J_{n-1} is available. This could be used for p iterations:

$$J_{n-1} \mathbf{h}^{[n-1+k]} = -\mathbf{f}^{[n-1+k]}$$
$$\mathbf{x}^{[n+k]} = \mathbf{x}^{[n-1+k]} + \mathbf{h}^{[n-1+k]}. \qquad k = 0, 1, \ldots, p-1$$

After these p iterations the Jacobian is re-evaluated and the device used again. Of course, there is the problem of the choice of p and a number of strategies have been proposed.

There are modified Newton methods in which the elements of J are approximated numerically so that only the computation of the functions $f_i(x_1, x_2, \ldots, x_m): i = 1, 2, \ldots, m$ is involved at each iteration. One of the possibilities corresponds to a generalisation of Steffensen iteration which was introduced in section 2.6. We give some indication in Chapter 5 of how the derivative of a function may be estimated at a point in terms of values of the function near the point. The ideas can be extended to functions of more than one variable but we shall not pursue the details.

If we combine the system (2.26) formally we obtain

$$\mathbf{x}^{[n]} = \mathbf{x}^{[n-1]} - J_{n-1}^{-1} \mathbf{f}^{[n-1]}.$$

There is a class of methods known as *quasi-Newton* methods in which an approximation to J_{n-1}^{-1} is used and, letting H_{n-1} denote this approximation, the iteration takes the form

$$\mathbf{x}^{[n]} = \mathbf{x}^{[n-1]} - H_{n-1} \mathbf{f}^{[n-1]}.$$

Having found $\mathbf{x}^{[n]}$, H_{n-1} is *updated* to give H_n, the approximation to the inverse Jacobian at $\mathbf{x}^{[n]}$, through a *correction term* D_{n-1}:

$$H_n = H_{n-1} + D_{n-1} \tag{2.27}$$

so that we are ready for the next iteration. We shall not discuss how D_{n-1} is found but we can say that its formation in practice will involve only standard operations on arrays and is therefore straightforward. The form of equation (2.27) should be noted; forming $\mathbf{x}^{[n]}$ does not involve the solution of a set of linear simultaneous equations.

2.9.4 Bairstow iteration

Having introduced the idea of Newton's method for a pair of non-linear simultaneous equations we return to subsection 2.8.4 where the synthetic division of a polynomial by a quadratic was considered. We showed there how to find the coefficients c_i and the remainder $Rx + S$ in the identity

$$f(x) = \sum_{i=0}^{m} a_i x^i \equiv (x^2 + px + q) \sum_{i=0}^{m-2} c_i x^i + Rx + S.$$

We observed that the c_is, R and S are functions of p and q and that if we could find values of p and q such that $R(p,q) = S(p,q) = 0$, then $x^2 + px + q$ is a quadratic factor of $f(x)$ and two of the roots of $f(x) = 0$ can be found. The remaining roots are those of the equation

$$\sum_{i=0}^{m-2} c_i x^i = 0. \tag{2.28}$$

Bairstow iteration is essentially the application of Newton's method to the pair of equations

$$R(p,q) = 0$$
$$S(p,q) = 0.$$

The derivatives $\partial R/\partial p$, $\partial R/\partial q$, $\partial S/\partial p$ and $\partial S/\partial q$ are required and, although we do not pursue the details, it turns out that these are quite easily obtained algebraically and can be easily evaluated. Initially, estimates $p^{[0]}$ and $q^{[0]}$ are required to define an approximation, $x^2 + p^{[0]}x + q^{[0]}$, to a quadratic factor of $f(x)$. Newton's method can then be implemented and continued until some convergence criterion, such as $(R(p^{[n]}, q^{[n]}))^2 + (S(p^{[n]}, q^{[n]}))^2$ being sufficiently small, is satisfied. Of course, there is no guarantee of convergence but if it is achieved we have estimates of two of the roots of $f(x) = 0$ and these may form a conjugate complex pair. In principle we can now consider the deflated equation (2.28) and attempt to estimate a quadratic factor of this. However, for reasons we have met before, this can be a dangerous process; the coefficients c_i will not be exact and the quadratic factor obtained may be a poor approximation. Thus, a process for obtaining all the roots of a polynomial equation which depends on estimating a quadratic factor of successive deflated equations is not to be recommended. Although there are modifications of Bairstow's method which are more attractive from a computational point of view, the approach is one which is not now used widely.

2.10 THE IMPLEMENTATION OF ITERATIVE PROCESSES

During this chapter, we have discussed a number of iterative processes in the context of non-linear algebraic equations and have made comments about implementing them on a computer. It is perhaps useful to recapitulate these comments and to add one or two more and to attempt, as far as this is possible, to identify methods which are likely to prove useful.

For equations in one variable, it may be useful to distinguish between polynomial and non-polynomial equations. Although all the methods we have discussed are applicable in principle to polynomial equations, special methods are available. One such method is that due to Grant and Hitchens (1971); this depends on minimisation techniques which are too sophisticated to describe here. We recommend, however, that any worker faced with the problem of finding the roots of a polynomial equation should find out whether there is an implementation of the method on the computing facilities to which he has access. The method is available in the NAG library in Chapter c02. However, there are circumstances in which the algorithm breaks down and hence the NAG routines should be handled with a certain degree of caution. See the Chapter's introduction for further details. It may be that other special methods are available but if the user has to implement, say, Newton–Raphson iteration, the dangers of successive deflations should be kept in mind. It is perhaps useful to regard a root obtained from a deflated equation as an approximation which can then be improved by iterating with the original polynomial.

For the general non-linear algebraic equation, there is no 'best' method to which we can point. We would not usually think of using a first order process unless Aitken acceleration (section 2.4), introduced at an early stage, is also employed. Newton–Raphson iteration often represents a good compromise between the very rapid convergence which might be expected near the root from a higher order process and the additional work which is required at each iteration for such a process. However, Newton–Raphson iteration usually requires a good initial approximation and this could be provided by a method such as bisection (section 2.2) initially. It may be possible without too much difficulty to find an interval such that the conditions of Theorem 2.2 (subsection 2.5.2) are satisfied but this is unlikely if $f(x)$ is complicated and not much is known about the roots. The possible benefits of using the secant method instead of Newton–Raphson iteration were briefly discussed in subsection 2.7.4. Implementations of some of the interpolation methods introduced in section 2.7, together with the method of bisection, will be found in most numerical analysis software libraries.

If an iterative process has to be programmed, we have seen that a number of precautions should be taken. The use of a variable (*itermax* in the codes of subsections 2.3.4 and 2.5.4) to limit the number of iterations should convergence not be achieved is fundamental. In Newton–Raphson iteration it is also important to have warning of a small derivative. It is necessary to think carefully about an appropriate termination criterion, bearing in mind that functions can be evaluated only to a limited precision on a computer. Too stringent a condition will result in meaningless computations being carried out. This consideration is equally important when a library routine is used; the user will have to specify the value of one or more parameters which will determine the accuracy demanded.

In section 2.9, we gave a brief introduction to the problem of finding a solution of a system of non-linear simultaneous equations and we suggested that the reader should consult Ralston and Rabinowitz (1978) for further material. There are implementations of methods in the NAG library and the reader should consult Chapter c05 for a description of what is involved in using the routines.

EXERCISES

2.1 The equation $x + \ln(x) = 0$ has a root near $x = 0.55$. Among others, the following iterations are possible.

(i) $x_n = -\ln(x_{n-1})$;
(ii) $x_n = e^{-x_{n-1}}$; and
(iii) $x_n = (x_{n-1} + e^{-x_{n-1}})/2$.

Which of the processes will converge to the root, given a suitable starting value? Which do you expect to give the fastest rate of convergence? Will convergence be monotonic or oscillatory?

If $k \neq -1$ is some chosen number, the iteration

(iv) $x_n = (kx_{n-1} + e^{-x_{n-1}})/(k+1)$

is also valid. Suggest a value for k which will give a rapid rate of convergence.

For each iterative scheme, obtain numerical results which support your theoretical observations.

2.2 The equation $x^3 - 3x + 1 = 0$ has roots near 0.3, 1.5 and -2. It can be rewritten as $x = (1-k)x + k(x^3+1)/3$, where $k \neq 0$ is some chosen value. Suggest three values for k (one for each root) which will ensure that the iteration $x_n = (1-k)x_{n-1} + k(x_{n-1}^3 + 1)/3$ will converge if x_0 is chosen appropriately. Verify that your values result in convergent iterations.

2.3 Using the starting value $x_0 = 0.25$, apply Newton–Raphson iteration directly to the equation $e^{2x} - 3 = 0$, continuing the process until the absolute difference between successive iterates is less than 5×10^{-6}. Given that the root to which the process converges is 0.549306 correct to six decimal places, compute the error in each iteration and verify that convergence is quadratic (that is, $\varepsilon_n \sim \varepsilon_{n-1}^2$). Using the same starting value, employ Newton–Raphson iteration to find the corresponding root of the equivalent equation $e^{-x} - \sinh(x) = 0$ and, from a table of errors, verify that convergence is much faster.

2.4 By considering the equation $f(x) \equiv 1/(x^2) - a = 0$ show that the square root of $1/a$ may be found using the Newton–Raphson iteration $x_n = x_{n-1}(3 - ax_{n-1}^2)/2$. Derive an equivalent expression for finding the mth root ($m > 1$) of $1/a$ and use it to find the cube root of 2 to five decimal places.

2.5 Verify that $x^2 - 3x + 1$ is a quadratic factor of the quintic polynomial $f(x) \equiv x^5 - 5x^4 + 11x^3 - 15x^2 + 7x - 1$ and hence find two of the roots of the equation $f(x) = 0$.

2.6 Carry out one iteration of Newton's method for the equations
$$f_1(x_1, x_2) \equiv e^{-x_1} - 3(x_1 - x_2)^2 = 0$$
$$f_2(x_1, x_2) \equiv e^{x_1 - x_2} - 5x_2 = 0$$
using the starting values $x_1^{[0]} = x_2^{[0]} = 0$.

2.7 Write programs which implement

(i) the method of bisection;
(ii) the rule of false position;
(iii) the Illinois method;   and
(iv) the secant method.

Using the equation $f(x) \equiv (x-1)e^{-10x} + x^{10} = 0$ and the two initial values $x_0 = 0.0$ and $x_1 = 1.0$ find the root which lies in $[0.0, 1.0]$. By counting the number of iterations required by each method for convergence, determine which is the more efficient for this particular problem.

2.8 Write a program to find a multiple root of a polynomial equation using Newton–Raphson iteration. (You may care to make use of the code of Case Study 2.2 and of Case Study 2.3.) Assuming that all roots are simple, find the root of $f(x) \equiv x^4 - 0.92x^3 - 0.7884x^2 + 0.92x - 0.2116 = 0$ which is close to 0.5. Show that the assumption that this is a double root leads to a faster rate of convergence.

2.9 Write a program which implements Newton's method for a pair of non-linear simultaneous equations. You should assume that it is possible to obtain analytic expressions for the Jacobian. Employ a convergence criterion based on the size of $|x_1^{[n]} - x_1^{[n-1]}| + |x_2^{[n]} - x_2^{[n-1]}|$.

3

Linear Simultaneous Equations

3.1 INTRODUCTION

Most readers will have, at some time, solved a small set of linear simultaneous equations by hand and will be familiar with the approach in which the unknowns are successively eliminated until an equation is obtained in which only one of the unknowns is involved. This unknown can then be determined and the remaining unknowns found successively by a process of backward substitution.

EXAMPLE 3.1
The system

(i) $\quad x_1 - x_2 - 2x_3 = -5$
(ii) $\quad x_1 + 2x_2 + x_3 = 7$
(iii) $\quad x_1 + 3x_2 - x_3 = 2$

is a set of three equations in the three unknowns x_1, x_2 and x_3. If we subtract equation (i) from equations (ii) and (iii) we have

(iv) $\qquad 3x_2 + 3x_3 = 12$

(v) $\qquad 4x_2 + x_3 = 7$.

Now, (iv) $- 3 \times$ (v) yields

$$-9x_2 = -9$$

which means that $x_2 = 1$. Substituting this value into (v) we find that $x_3 = 3$ and then from (i) we have $x_1 = 2$.

In practice it is often necessary to solve large sets of equations in which tens, hundreds and possibly thousands of unknowns are involved. There are a number of reasons why it is important to analyse methods for the solution of large sets. One is that we must consider the computational effort involved; for example, in elimination methods the number of elementary arithmetic operations increases very rapidly with the number of unknowns and so an efficient algorithm must be used. Further, this very large number of arithmetic operations presents a potentially serious problem: each operation in general will not be exact and, unless we organise the calculation properly, the cumulative effect of these rounding errors could lead to serious inaccuracies in the results. In other words, we may have a serious problem of numerical instability (see section 1.8). In addition, the problem of ill-conditioning may be present (again, see section 1.8). Here the initial data of the problem are the coefficients of the unknowns and the numbers on the right-hand side of each equation. These will often be subject to uncertainty and small changes in them may well cause relatively large changes in the computed solution. Similar effects can be produced by rounding errors introduced during the computation. It is therefore necessary to ask whether the reliability of a computed solution can be assessed in some way.

We shall see that there are two different classes of method for the solution of linear simultaneous equations. Methods such as elimination followed by back-substitution are referred to as *direct* methods, the solution being obtained within the limits imposed by rounding errors in a predetermined number of stages. For a direct method it is, therefore, possible to state beforehand the number of elementary arithmetic operations required to obtain a solution as a function of n, the number of equations (or, equivalently, the number of unknowns). The other methods that we consider are *iterative* in nature. For such methods it is, of course, necessary to try to establish the conditions under which they will converge (cf. the discussions on convergence in Chapter 2). A sequence of successive approximations to the solution will be produced from a given set of initial values. Hence it will not be possible to predict the number of arithmetic operations that will be involved; this will depend on the accuracy being demanded and on how good the initial approximation is.

3.2 NOTATION

A set of n linear simultaneous equations in the n unknowns $x_1, x_2, \ldots, x_n$ can be written in the form

$$
\begin{aligned}
a_{11}x_1 + a_{12}x_2 + \ldots + a_{1n}x_n &= b_1 \\
a_{21}x_1 + a_{22}x_2 + \ldots + a_{2n}x_n &= b_2 \\
&\vdots \\
a_{n1}x_1 + a_{n2}x_2 + \ldots + a_{nn}x_n &= b_n
\end{aligned}
\tag{3.1}
$$

so that a_{ij} is the coefficient of x_j in the ith equation. The system (3.1) may be expressed more compactly as

$$\sum_{j=1}^{n} a_{ij}x_j = b_i \qquad i = 1, 2, \ldots, n \tag{3.2}$$

(so that if $i = 1$, (3.2) gives the first equation in (3.1), and so on) or, making use of matrix notation, as

$$A\mathbf{x} = \mathbf{b}$$

where A is an $n \times n$ square matrix containing the coefficients on the left-hand side of (3.1) and is defined by

$$A = \begin{bmatrix} a_{11} & a_{12} & \ldots & a_{1n} \\ a_{21} & a_{22} & \ldots & a_{2n} \\ \ldots & \ldots & & \ldots \\ a_{n1} & a_{n2} & \ldots & a_{nn} \end{bmatrix}.$$

The right hand side vector

$$\mathbf{b}^T = (b_1, b_2, \ldots, b_n)$$

consists of the right-hand sides of the system (3.1) and

$$\mathbf{x}^T = (x_1, x_2, \ldots, x_n)$$

is the vector of unknowns.

It is useful at this stage to identify a number of special matrices. It is hoped that the reader who is not yet familiar with matrix algebra will be able to relate the following definitions to the system (3.1) without much difficulty. The matrix A is said to be *symmetric* if $a_{ij} = a_{ji}$: $i, j = 1, 2, \ldots, n$. It is said to be *tridiagonal* if there are at least some non-zero elements on the principal diagonal, on the co-diagonal immediately above and on the co-diagonal immediately below, but all other elements are zero. Thus, for a tridiagonal matrix, $a_{ij} = 0$ for $|i - j| > 1$: $i, j = 1, 2, \ldots, n$. A is said to be *lower* (*upper*) *triangular* if $a_{ij} = 0$ for $j > i$ ($j < i$): $i, j = 1, 2, \ldots, n$. Thus, in a lower triangular matrix all the elements above the principal diagonal are zero; in an upper triangular matrix all the elements below the principal diagonal are zero. A matrix A for which the condition

$$|a_{ii}| > \sum_{\substack{j=1 \\ j \neq i}}^{n} |a_{ij}| \qquad i = 1, 2, \ldots, n$$

is satisfied is said to be *strictly diagonally dominant*. This definition means that for each row of the matrix, the magnitude of the element on the diagonal exceeds the sum of the magnitudes of the remaining elements in the row. If the condition

$$|a_{ii}| \geq \sum_{\substack{j=1 \\ j \neq i}}^{n} |a_{ij}| \qquad i = 1, 2, \ldots, n$$

is satisfied, with inequality holding for at least one value of i, then A is said to be *diagonally dominant*.

EXAMPLE 3.2

The foregoing definitions are exemplified by the following 4×4 matrices.

$$\begin{bmatrix} 1 & 2 & 3 & 4 \\ 2 & 5 & 6 & 7 \\ 3 & 6 & 8 & 9 \\ 4 & 7 & 9 & 10 \end{bmatrix} \qquad \begin{bmatrix} 1 & 2 & 0 & 0 \\ 3 & 4 & 5 & 0 \\ 0 & 6 & 7 & 8 \\ 0 & 0 & 9 & 10 \end{bmatrix} \qquad \begin{bmatrix} 1 & 2 & 0 & 0 \\ 2 & 3 & 4 & 0 \\ 0 & 4 & 5 & 6 \\ 0 & 0 & 6 & 7 \end{bmatrix}$$

(a) Symmetric   (b) Tridiagonal   (c) Symmetric and tridiagonal

$$\begin{bmatrix} 1 & 0 & 0 & 0 \\ 2 & 3 & 0 & 0 \\ 4 & 5 & 6 & 0 \\ 7 & 8 & 9 & 10 \end{bmatrix} \quad \begin{bmatrix} 1 & 2 & 3 & 4 \\ 0 & 5 & 6 & 7 \\ 0 & 0 & 8 & 9 \\ 0 & 0 & 0 & 10 \end{bmatrix} \quad \begin{bmatrix} 10 & -1 & -2 & 4 \\ 3 & 9 & 0 & -5 \\ 1 & -1 & -6 & -2 \\ 1 & 1 & -2 & 5 \end{bmatrix}$$

(d) Lower triangular (e) Upper triangular (f) Strictly diagonally dominant

$$\begin{bmatrix} 10 & -1 & -2 & 4 \\ 4 & 9 & 0 & -5 \\ 1 & -1 & -6 & -2 \\ 1 & 1 & -2 & 5 \end{bmatrix} \quad \begin{bmatrix} 10 & -2 & 0 & 0 \\ -2 & -6 & -1 & 0 \\ 0 & -1 & 3 & 1 \\ 0 & 0 & 1 & 4 \end{bmatrix}$$

(g) Diagonally dominant (h) Symmetric, tridiagonal and strictly diagonally dominant

A lower (upper) triangular matrix in which every element on the diagonal is equal to unity is said to be *unit lower (upper) triangular*. A lower (upper) triangular matrix in which every element on the diagonal is equal to zero is said to be *strictly lower (upper) triangular*. A matrix which has non-zero entries on the diagonal only is said to be *diagonal*. The diagonal matrix in which every non-zero entry is equal to unity is referred to as the *identity* (or *unit*) matrix and is denoted I. (Using the normal rules of matrix multiplication, we have that, for any square matrix A, $AI = IA = A$.)

Finally, we introduce two qualitative terms. We refer to a matrix as being *full* if it has at most a small number of zero elements. We refer to it as *sparse* if it has a large number of zero entries. These zero elements may occur in a regular pattern in the sense that there is a relation involving the row and column indices which defines the zero elements. Thus a 100×100 tridiagonal matrix (which contains at most 298 non-zero entries) could be described as a sparse matrix in which there is a regular pattern ($a_{ij} = 0$ for $|i-j| > 1$: $i, j = 1, 2, \ldots, n$).

3.3 GAUSS ELIMINATION AND BACK-SUBSTITUTION

3.3.1 Discussion of the basic method

Gauss elimination is a systematic way of solving the system (3.1) using the techniques employed in Example 3.1. We describe the method in the context of the set of four linear simultaneous equations

$$\sum_{j=1}^{4} a_{ij} x_j = b_i \quad i = 1, 2, 3, 4 \tag{3.3}$$

the generalisation being obvious. Writing the system (3.3) in its full form we have

$$a_{11}^{(1)}x_1 + a_{12}^{(1)}x_2 + a_{13}^{(1)}x_3 + a_{14}^{(1)}x_4 = b_1^{(1)}$$
$$a_{21}^{(1)}x_1 + a_{22}^{(1)}x_2 + a_{23}^{(1)}x_3 + a_{24}^{(1)}x_4 = b_2^{(1)}$$
$$a_{31}^{(1)}x_1 + a_{32}^{(1)}x_2 + a_{33}^{(1)}x_3 + a_{34}^{(1)}x_4 = b_3^{(1)} \qquad (3.4)$$
$$a_{41}^{(1)}x_1 + a_{42}^{(1)}x_2 + a_{43}^{(1)}x_3 + a_{44}^{(1)}x_4 = b_4^{(1)}$$

where we have added a superscript (1) to each coefficient a_{ij} and each right-hand side b_i. This superscript is not an iteration number but its value will change as the elimination process proceeds. Its purpose will soon become apparent.

We begin by eliminating x_1 from the second, third and fourth equations. Assuming $a_{11}^{(1)} \neq 0$ we subtract the multiple $a_{21}^{(1)}/a_{11}^{(1)}$ of the first equation from the second equation. Similarly, we take $a_{31}^{(1)}/a_{11}^{(1)}$ times the first equation from the third equation and $a_{41}^{(1)}/a_{11}^{(1)}$ times the first equation from the fourth equation. This transforms the system (3.4) to

$$a_{11}^{(1)}x_1 + a_{12}^{(1)}x_2 + a_{13}^{(1)}x_3 + a_{14}^{(1)}x_4 = b_1^{(1)}$$
$$a_{22}^{(2)}x_2 + a_{23}^{(2)}x_3 + a_{24}^{(2)}x_4 = b_2^{(2)}$$
$$a_{32}^{(2)}x_2 + a_{33}^{(2)}x_3 + a_{34}^{(2)}x_4 = b_3^{(2)} \qquad (3.5)$$
$$a_{42}^{(2)}x_2 + a_{43}^{(2)}x_3 + a_{44}^{(2)}x_4 = b_4^{(2)}$$

where

$$a_{ij}^{(2)} = a_{ij}^{(1)} - \frac{a_{i1}^{(1)}}{a_{11}^{(1)}} a_{1j}^{(1)} \qquad i,j = 2,3,4 \qquad (3.6)$$

and

$$b_i^{(2)} = b_i^{(1)} - \frac{a_{i1}^{(1)}}{a_{11}^{(1)}} b_1^{(1)} \qquad i = 2,3,4 \ . \qquad (3.7)$$

In the first stage of the elimination process the first equation in (3.4) remains unchanged. We now enter the second stage in which the first two equations of (3.5) are unaltered but x_2 is eliminated from the last two equations. This is achieved by subtracting $a_{i2}^{(2)}/a_{22}^{(2)}$ times the second equation from the ith equation for $i = 3, 4$ (provided that $a_{22}^{(2)} \neq 0$). This yields the system

$$a_{11}^{(1)}x_1 + a_{12}^{(1)}x_2 + a_{13}^{(1)}x_3 + a_{14}^{(1)}x_4 = b_1^{(1)}$$
$$a_{22}^{(2)}x_2 + a_{23}^{(2)}x_3 + a_{24}^{(2)}x_4 = b_2^{(2)}$$
$$a_{33}^{(3)}x_3 + a_{34}^{(3)}x_4 = b_3^{(3)}$$
$$a_{43}^{(3)}x_3 + a_{44}^{(3)}x_4 = b_4^{(3)} \quad (3.8)$$

where

$$a_{ij}^{(3)} = a_{ij}^{(2)} - \frac{a_{i2}^{(2)}}{a_{22}^{(2)}} a_{2j}^{(2)} \quad i,j = 3,4 \quad (3.9)$$

and

$$b_i^{(3)} = b_i^{(2)} - \frac{a_{i2}^{(2)}}{a_{22}^{(2)}} b_2^{(2)} \quad i = 3,4 \ . \quad (3.10)$$

Finally, we eliminate x_4 from the last equation in (3.8) by taking $a_{43}^{(3)}/a_{33}^{(3)}$ times the third equation away from it to give

$$a_{11}^{(1)}x_1 + a_{12}^{(1)}x_2 + a_{13}^{(1)}x_3 + a_{14}^{(1)}x_4 = b_1^{(1)}$$
$$a_{22}^{(2)}x_2 + a_{23}^{(2)}x_3 + a_{24}^{(2)}x_4 = b_2^{(2)}$$
$$a_{33}^{(3)}x_3 + a_{34}^{(3)}x_4 = b_3^{(3)}$$
$$a_{44}^{(4)}x_4 = b_4^{(4)} \quad (3.11)$$

where

$$a_{44}^{(4)} = a_{44}^{(3)} - \frac{a_{43}^{(3)}}{a_{33}^{(3)}} a_{34}^{(3)} \quad (3.12)$$

and

$$b_4^{(4)} = b_4^{(3)} - \frac{a_{43}^{(3)}}{a_{33}^{(3)}} b_3^{(3)} \ . \quad (3.13)$$

To summarise, the effect of the basic Gauss elimination process is to transform the original system (3.4) into the equivalent system (3.11) in which the matrix of coefficients is upper triangular. The solution is now obtained by a process of *back-substitution*: x_4 is obtained from the final equation in (3.11), x_3 from the last but one, and so on.

The equation which is used for elimination purposes at a particular stage is known as the *pivotal* equation. (Thus, in (3.4), the first equation, which is used to eliminate x_1 from the other equations, is the pivotal equation). In the pivotal equation, the coefficient of the term which is to be eliminated elsewhere is known as the *pivot* (so that in (3.4) $a_{11}^{(1)}$ is the pivot). The equations (3.6), (3.7), (3.9), (3.10), (3.12) and (3.13) are referred to as the *updating formulae*. The generalisation of these formulae to the case of n linear simultaneous equations in n unknowns is quite straightforward. We have

$$a_{ij}^{(k)} = a_{ij}^{(k-1)} - \frac{a_{i,k-1}^{(k-1)}}{a_{k-1,k-1}^{(k-1)}} a_{k-1,j}^{(k-1)} \quad i,j = k, k+1, \ldots, n \quad (3.14)$$

$$\left. \begin{array}{l} \\ \\ \end{array} \right\} k = 2, 3, \ldots, n$$

$$b_i^{(k)} = b_i^{(k-1)} - \frac{a_{i,k-1}^{(k-1)}}{a_{k-1,k-1}^{(k-1)}} b_{k-1}^{(k-1)} \quad i = k, k+1, \ldots, n \quad (3.15)$$

whilst

$$x_i = \left(b_i^{(i)} - \sum_{j=i+1}^{n} a_{ij}^{(i)} x_j \right) / a_{ii}^{(i)} \quad i = n, n-1, \ldots, 1 \quad (3.16)$$

is the general form for back-substitution.

EXAMPLE 3.3
Gauss elimination applied to the system

(i) $x_1 - x_2 - 2x_3 = -5$
(ii) $x_1 + 2x_2 + x_3 = 7$
(iii) $x_1 + 3x_2 - x_3 = 2$

of Example 3.1 yields

Stage 1:

(i)' $x_1 - x_2 - 2x_3 = -5$ $(= (i))$
(ii)' $3x_2 + 3x_3 = 12$ $(= (ii) - \frac{1}{1} \times (i))$
(iii)' $4x_2 + x_3 = 7$ $(= (iii) - \frac{1}{1} \times (i))$

Stage 2:

$x_1 - x_2 - 2x_3 = -5$ $(= (i)')$
$3x_2 + 3x_3 = 12$ $(= (ii)')$
$-3x_3 = -9$ $(= (iii)' - \frac{4}{3}(ii)')$

Back-substitution:

$$x_3 = -9/-3 = 3$$
$$x_2 = (12 - 3x_3)/3 = 1$$
$$x_1 = (-5 + x_2 + 2x_3)/1 = 2 \ .$$

As mentioned in the introduction to this chapter, for a direct method it is possible to predict the total number of elementary arithmetic operations involved. It can be shown that the *operations count* for Gauss elimination is dominated by $n^3/3$ multiplications and additions/subtractions. Hence the effort in solving a linear system of equations by this method increases rapidly with n.

Although the numerical evaluation of a determinant is an operation which is rarely required, we make the observation that the value of the determinant of the matrix in the final system (3.11) is the same as that for the original coefficient matrix A. Moreover, it follows immediately that this value is just $a_{11}^{(1)} a_{22}^{(2)} a_{33}^{(3)} a_{44}^{(4)}$. The extension of this result to the general case is obvious.

At each stage of the elimination process we have made the assumption that the pivot, $a_{ii}^{(i)}$, is non-zero. If $a_{11}^{(1)} = 0$ the process just described will break down. However, if this situation arises, at least one of $a_{21}^{(1)}$, $a_{31}^{(1)}$ and $a_{41}^{(1)}$ must be non-zero and one of these may be used as the pivot instead of $a_{11}^{(1)}$. This new pivot then defines a new pivotal equation. At the end of the first stage, $a_{22}^{(2)}$ may be zero but, if this is so, at least one of a $a_{32}^{(2)}$ and $a_{42}^{(2)}$ will be non-zero and so it will be possible to select a new non-zero pivot. Finally, if $a_{33}^{(3)} = 0$, $a_{43}^{(3)}$, which must be non-zero, can be taken as the pivot. Even if none of the pivots is identically zero, it is possible that they may be very small in magnitude and consequently the multipliers $a_{i,k-1}^{(k-1)}/a_{k-1,k-1}^{(k-1)}$ may be large. This is undesirable from the point of view of numerical stability and we return to this point in subsection 3.3.4.

3.3.2 A program for Gauss elimination

Subsections 3.3.3 and 3.3.4 will describe refinements to the Gauss elimination process and a program which includes these refinements will be presented in subsection 3.3.5. However, in this subsection we discuss a program which implements the basic Gauss elimination process.

The code for this program is somewhat longer than those we have met so far and it makes use of a number of important devices. It is often very difficult to write a program of this length from scratch and a simpler approach is to break down the problem into a number of subproblems. The following subproblems can be identified:

read the input data
print the input data
do Gauss elimination by using (3.14) and (3.15)
do back-substitution by using (3.16)
print the solution

It is useful to print the input data as well as the solution in order to check that we have solved the correct problem. However, the input data will have to be printed before the elimination process begins as the values stored in the coefficient matrix and the right-hand side vector get overwritten in the elimination process.

It would also be useful to find out how good the solution is. This can be done by computing the *residual vector*

$$\mathbf{r} = \mathbf{b} - A\mathbf{x}$$

but we omit this further development for the moment and return to it in subsection 3.4.1. In the meantime we must be content to check the results by hand.

The code for each of the above subproblems can be easily located in both the Algol 68 and the Pascal programs of Case Study 3.1.

Once again the Algol 68 program demonstrates the ease with which complete arrays can be read. The code

read ((a , b))

fills the array a with the next n^2 real numbers in the data and then fills b with the next n real numbers. (The array a is filled row by row, starting with the first row.) In Pascal, *FOR* loops have to be used to read in the elements of a and b an element at a time.

CASE STUDY 3.1

Algol 68 Program

```
BEGIN
  COMMENT
  a program for gauss elimination
  COMMENT
  COMMENT
  declaration of arrays and input
  COMMENT
  INT n ; read ( n ) ;
  [ 1 : n , 1 : n ] REAL a ;
  [ 1 : n ] REAL x , b ;
  read ( ( a , b ) ) ;
  print ( ( newline , "the coefficient matrix is" , newline ) ) ;
  FOR i TO n
  DO
     print ( ( a [ i , ] , newline ) )
  OD ;
```

```
             print ( ( newline , "the right hand side vector is" , newline , b ,
                     newline ) ) ;
COMMENT
elimination
COMMENT
REAL mult ;
FOR k FROM 2 TO n
DO
   FOR i FROM k TO n
   DO
     mult := a [ i , k - 1 ] / a [ k - 1 , k - 1 ] ;
     b[ i ] -:= mult * b [ k - 1 ] ;
     FOR j FROM k TO n
     DO
        a [ i , j ] -:= mult * a [ k - 1 , j ]
     OD
   OD
OD ;
COMMENT
back - substitution
COMMENT
REAL sum ;
FOR i FROM n BY - 1 TO 1
DO
   sum := b [ i ] ;
   FOR j FROM i+1 TO n
   DO
      sum -:= a [ i , j ] * x [ j ]
   OD ;
   x [ i ] := sum / a [ i , i ]
OD ;
   print ( ( newline , "the solution vector is" , newline , x , newline ) )
END
```

Pascal Program

```
PROGRAM basicgausselimination(input, output);
    {  a program for gauss elimination   }
    CONST
       maxn=50;
    VAR
       i, j, k, n:1..maxn;
       a:ARRAY [ 1..maxn, 1..maxn ] OF real;
       x, b:ARRAY [ 1..maxn ] OF real;
       mult, sum:real;
BEGIN
   {  read input data  }
   read(n);
   FOR i:= 1 TO n DO
      FOR j:= 1 TO n DO
         read(a[i, j]);
   FOR i:= 1 TO n DO
      read(b[i]);
   {  print input data  }
   writeln;
   writeln('the coefficient matrix is');
   FOR i:= 1 TO n DO
   BEGIN
      FOR j:= 1 TO n DO
         write(a[i, j]);
      writeln
   END ;
   writeln;
   writeln('the right hand side vector is');
   FOR i:= 1 TO n DO
      write(b[i]);
   writeln;
```

```
{ elimination }
FOR k:= 2 TO n DO
   FOR i:= k TO n DO
   BEGIN
      mult:= a[i, k - 1]/a[k - 1, k - 1];
      b[i]:= b[i] - mult*b[k - 1];
      FOR j:= k TO n DO
         a[i, j]:= a[i, j] - mult*a[k - 1, j]
   END ;
{ back-substitution }
FOR i:= n DOWNTO 1 DO
BEGIN
   sum:= b[i];
   FOR j:= i + 1 TO n DO
      sum:= sum - a[i, j]*x[j];
   x[i]:= sum/a[i, i]
END ;
{ print solution }
writeln;
writeln('the solution vector is');
FOR i:= 1 TO n DO
   write(x[i]);
writeln
END { basicgausselimination } .
```

Sample Data

```
3
  1.0  -1.0  -2.0
  1.0   2.0   1.0
  1.0   3.0  -1.0
 -5.0   7.0   2.0
```

Sample Output

```
the coefficient matrix is
  1.00000e  0 -1.00000e  0 -2.00000e  0
  1.00000e  0  2.00000e  0  1.00000e  0
  1.00000e  0  3.00000e  0 -1.00000e  0

the right hand side vector is
 -5.00000e  0  7.00000e  0  2.00000e  0

the solution vector is
  2.00000e  0  1.00000e  0  3.00000e  0
```

When printing a two-dimensional array, it is natural to require that the rows of the array be printed on separate lines. To do this we must repeatedly execute statements which print the *i*th row of the array followed by a newline. In Algol 68 this can be achieved by:

```
FOR i TO n
DO
    print ( ( a [ i , ] , newline ) )
OD
```

whereas, in Pascal, we need to print the elements in each row separately:

```
FOR i:= 1 TO n DO
BEGIN
  FOR j:= 1 TO n DO
    write(a[i,j]);
  writeln
END
```

When coding mathematical equations, one ought to be aware that the code produced by a straightforward translation into a programming language may be inefficient. Equations (3.14) and (3.15) refer to the value $a[i, k-1]/a[k-1, k-1]$. A direct coding of these equations would result in code like:

```
b[i] := b[i] - a[i, k - 1]/a[k - 1, k - 1] * b[k - 1];
FOR j FROM k TO n
DO
  a[i,j] := a[i,j] - a[i, k - 1]/a[k - 1, k - 1] * a[k - 1, j]
OD
```

Inside the j loop the term $a[i, k-1]/a[k-1, k-1]$ is independent of j but is being evaluated each time the j loop is cycled. The programs of Case Study 3.1 avoid this inefficiency by using the variable *mult* which is assigned the value of this multiplier before the j loop is executed. Moreover, the same quantity is employed in the evaluation of $b[i]$.

As in Case Study 2.3, arrays of a fixed size have again been used in the Pascal program of Case Study 3.1. However, the problem is rather more severe. Since a is a two-dimensional array, the number of unused elements can be very large indeed. Even with a modest limit of 50, if the value of n is 3, only 9 out of 2500 elements (less than 0.4%) will actually be used.

3.3.3 Several right-hand sides

In practical situations it sometimes happens that we wish to solve several sets of linear simultaneous equations in which the same matrix of coefficients is involved but the right-hand sides are different. Formally we can state the problem as wishing to find the m solutions $x_1, x_2, \ldots, x_m$ of the m sets of equations

$$A x_p = b_p \qquad p = 1, 2, \ldots, m . \qquad (3.17)$$

The basic Gauss elimination process can be adapted to this situation and there is no need to solve each set independently. All we need to do is to carry out the elimination process on the m right-hand sides simultaneously and, if $n = 4$, say, instead of (3.11) we have

$$a_{11}^{(1)}x_1 + a_{12}^{(1)}x_2 + a_{13}^{(1)}x_3 + a_{14}^{(1)}x_4 = b_{11}^{(1)}, b_{21}^{(1)}, \ldots, b_{m1}^{(1)}$$
$$a_{22}^{(2)}x_2 + a_{23}^{(2)}x_3 + a_{24}^{(2)}x_4 = b_{12}^{(2)}, b_{22}^{(2)}, \ldots, b_{m2}^{(2)}$$
$$a_{33}^{(3)}x_3 + a_{34}^{(3)}x_4 = b_{13}^{(3)}, b_{23}^{(3)}, \ldots, b_{m3}^{(3)}$$
$$a_{44}^{(4)}x_4 = b_{14}^{(4)}, b_{24}^{(4)}, \ldots, b_{m4}^{(4)}$$

where the first suffix, p, in $b_{pi}^{(k)}$ is the suffix p of (3.17). We then have

$$x_{p4} = b_{p4}/a_{44}^{(4)} \quad p = 1, 2, \ldots, m$$

and hence we can now determine the fourth components of each of the m solution vectors. The back-substitution stage proceeds in the usual way, m solutions being generated at each stage.

EXAMPLE 3.4
Suppose that we wish to solve the system

$$x_1 - x_2 - 2x_3 = 1$$
$$x_1 + 2x_2 + x_3 = 7$$
$$x_1 + 3x_2 - x_3 = 6.$$

in which the matrix of coefficients is the same as that for Example 3.3. If we apply the same operations to the vector $\mathbf{b}^T = (1, 7, 6)$ as we did to the right-hand side of Example 3.3, it gets transformed first to $(1, 6, 5)$ and then to $(1, 6, -3)$. Back-substitution yields the solution $x_3 = 1$, $x_2 = 1$ and $x_1 = 4$.

We shall refer briefly in section 3.7 to the problem of finding the inverse of a given $n \times n$ matrix. One method is to use the present device with n right-hand sides where $\mathbf{b}_p = \mathbf{e}_p$, the pth unit vector (or the pth column of the identity matrix): $p = 1, 2, \ldots, n$. By doing this, we are solving

$$A(\mathbf{x}_1, \mathbf{x}_2, \ldots, \mathbf{x}_n) = (\mathbf{e}_1, \mathbf{e}_2, \ldots, \mathbf{e}_n)$$

and so the solution vectors constitute the columns of the inverse matrix.

3.3.4 Pivoting strategies and scaling
At the end of subsection 3.3.1 we made reference to the undesirability of having small pivots $a_{ii}^{(i)}$ at any stage of the elimination process. The reason for this is that the multipliers may then be large which implies that any errors in the initial data (the coefficients and right-hand sides) and accumulated

rounding error will get magnified. What we would like to do is ensure, as far as possible, that these errors do not get magnified and we can hope to make some progress in this respect if we can arrange that the multipliers do not exceed unity in magnitude. Referring to the system (3.4), in the elimination of x_1 this can be achieved by selecting as the pivotal equation the i_1th equation, where i_1 is such that

$$|a_{i_1,1}^{(1)}| \geq |a_{i,1}^{(1)}| \quad i = 1, 2, \ldots, n \ .$$

That is, the i_1th equation has the largest coefficient of x_1 in absolute value. Similar considerations apply to later stages of the elimination process and (3.11) gives the set of pivotal equations selected by the process. This strategy is known as *partial pivoting* and computer library routines which implement Gauss elimination usually make use of the device. It might be thought necessary physically to interchange equations but, as we shall see in the next subsection, the use of an index avoids this.

EXAMPLE 3.5
After the first stage of Gauss elimination applied to the three linear simultaneous equations of Example 3.3 we have

(i)' $\quad x_1 - x_2 - 2x_3 = -5$
(ii)' $\quad 3x_2 + 3x_3 = 12$
(iii)'$\quad 4x_2 + x_3 = 7 \ .$

(The coefficients of x_1 in the original system are all 1.) Using partial pivoting, equation (iii)' is the next pivotal equation and the final stage of the elimination process gives

$$x_1 - x_2 - 2x_3 = -5 \quad (= \text{(i)}')$$
$$ 4x_2 + x_3 = 7 \quad (= \text{(iii)}')$$
$$ \tfrac{9}{4}x_3 = \tfrac{27}{4} \quad (= \text{(ii)}' - \tfrac{3}{4}\text{(iii)}') \ .$$

The reader should now verify that back-substitution again yields $x_3 = 3$, $x_2 = 1$ and $x_1 = 2$.

The partial pivoting strategy can be extended to give a more sophisticated strategy known as *full pivoting*. Here, all of the equations which have not yet been used as pivotal equations are scanned to find the largest coefficient (ignoring the sign). If this occurs in the pth row and qth column, say, the pth equation is used as the pivotal equation and the unknown x_q is eliminated. Thus, the unknowns are not necessarily eliminated in the order $x_1, x_2, \ldots, x_{n-1}$ and some additional bookkeeping must be made. This strategy can be expensive in computer time since, at each stage, a search through a two-dimensional array must be made. Partial pivoting is usually considered to be sufficient.

Although we shall not pursue the details here, it is necessary to point out that pivoting can usually be made more effective by a device known as *scaling* (or *equilibration*). Here we divide each row by the element of largest magnitude within the row. The reader is referred to Ralston and Rabinowitz (1978) for further details.

3.3.5 An improved version of the program for Gauss elimination
We now consider a Gauss elimination program that includes code for partial pivoting and for handling several right-hand sides.

First we consider the problem of partial pivoting. We could include code which physically interchanges rows when necessary but this is rather expensive in terms of computer time. Instead, we keep a note of the row numbers and when we want to swap rows we simply swap the row numbers themselves. We have to declare and initialise the array *rownum*. In Algol 68, the code for initialisation is

```
FOR i TO n
DO
    rownum [ i ] := i
OD
```

Each reference to the first index of the matrix must now be in terms of the appropriate element of the array *rownum* (so that, for example, $a[i,j]$ must now be expressed as $a[rownum[i], j]$). Similar considerations apply to the right-hand side vector. Inside the k loop of the elimination process we include the code

```
kminus1 := k − 1 ;
abspivot := ABS a [ rownum [ kminus1 ] , kminus1 ] ;
pivotrow := kminus1 ;
FOR i FROM k TO n
DO
    absa := ABS a [ rownum [ i ] , kminus1 ] ;
    IF absa > abspivot
    THEN
        abspivot := absa ;
        pivotrow := i
    FI
OD ;
```

```
IF pivotrow # kminus1
THEN
  dummy := rownum [ kminus1 ] ;
  rownum [ kminus1 ] := rownum [ pivotrow ] ;
  rownum [ pivotrow ] := dummy
FI
```

which first of all finds where the largest coefficient of the unknown to be next eliminated is and then 'interchanges' the rows via the array *rownum*. (The symbol '#' is used to denote the 'not equal to' operator.)

Before proceeding the reader should make use of the two sections of code given above to transform the code of Case Study 3.1 into a program which implements partial pivoting. A suitable example should be constructed to check that the pivoting process works. Note that the variable *kminus1* has been introduced to avoid unnecessary recomputation of the quantity $k - 1$. This and other new variables will need to be declared.

Finally, we deal with the problem of several right-hand sides. This is fairly straightforward since all that we need to do is to add an extra dimension to the vectors x and b. The current references to these vectors will then need to be enclosed within suitable loops.

Case Study 3.2 gives the final version of the program. If we modify the program of Case Study 3.1 so that it implements both partial pivoting and several right-hand sides, a reasonably long and unmanageable program would be produced. Instead of doing this, we have recognised that the tasks of elimination and back-substitution form identifiable subproblems and have chosen to put the code for these subproblems within the procedures *gausselim* and *backsub* respectively. So the details of how these procedures work are hidden from the main part of the program. This is good because in the main part of the program we are interested, not in how the procedures work, but in what they do.

Note that Pascal uses a different word-symbol in order to distinguish procedures from functions, whereas, in Algol 68, procedures are effectively functions that return a *VOID* (that is, no) result.

The programs illustrate the way in which arrays are passed to procedures. In Pascal, we must use an identifier to specify the type of a value or variable parameter. So the types *matrixtype*, *rhstype* and *rownumtype* have been introduced in order to be able to pass arrays like a, b and *rownum* as parameters. We saw earlier that we only use a portion of these arrays. It is for this reason that we are unable to specify the parameter as a *conformant array parameter* (as defined by Level 1 of the Pascal standard, see BSI, 1982). Instead, the number of elements that are to be used is also passed to the procedures (by means of the parameters n and m). In Algol 68, we can use dynamic arrays and the size of an array that is passed to a procedure can be determined from within the procedure by using the upper bound operator

UPB. (Further, we can pass across portions of arrays using trimming and slicing.) Thus, unlike Pascal, there is no need for extra parameters that indicate the size of the arrays.

The parameter *a* of the *backsub* procedure has been specified to be of mode REF [,] $REAL$ in the Algol 68 program and to be a variable parameter in the Pascal program. In principle, it is possible for the code of the procedure to alter the contents of this array and we have to search the code in order to see that this does not in fact happen. To guard against this, we could specify the parameter to be of mode [,] $REAL$ in Algol 68 or to be a value parameter in Pascal. We have chosen not to do this because this second way of specifying the array normally involves a copy of the whole array being made. This is considered to be wasteful of computer time. Throughout this book we will specify any array parameter using a REF mode in Algol 68 and a variable parameter in Pascal. (But we shall find it convenient not to adhere to this convention when we declare Algol 68 operators.)

CASE STUDY 3.2

Algol 68 Program

```
BEGIN
  COMMENT
  a program for gauss elimination with partial pivoting
  this code is capable of handling the case of several right hand sides
  COMMENT
  PROC gausselim = ( REF [ , ] REAL a , b, REF [ ] INT rownum ) VOID :
  BEGIN
    COMMENT
    this procedure performs the elimination process
    COMMENT
    INT n = UPB a , m = 2 UPB b ;
    FOR i TO n
    DO
       rownum [ i ] := i
    OD ;
    REAL mult , abspivot , absa ;
    INT kminus1 , pivotrow , dummy ;
    FOR k FROM 2 TO n
    DO
      kminus1 := k - 1 ;
      COMMENT
      find the pivot
      COMMENT
      abspivot := ABS a [ rownum [ kminus1 ] , kminus1 ] ;
      pivotrow := kminus1 ;
      FOR i FROM k TO n
      DO
         absa := ABS a [ rownum [ i ] , kminus1 ] ;
         IF absa > abspivot
         THEN
            abspivot := absa ;
            pivotrow := i
         FI
      OD ;
      COMMENT
      interchange if necessary
      COMMENT
```

102 LINEAR SIMULTANEOUS EQUATIONS [Ch. 3

```
      IF pivotrow # kminus1
      THEN
        dummy := rownum [ kminus1 ] ;
        rownum [ kminus1 ] := rownum [ pivotrow ] ;
        rownum [ pivotrow ] := dummy
      FI ;
      FOR i FROM k TO n
      DO
        mult := a [ rownum [ i ] , kminus1 ]
                / a [ rownum [ kminus1 ] , kminus1 ] ;
        FOR p TO m
        DO
          b[ rownum [ i ] , p ] -:= mult * b [ rownum [ kminus1 ] , p ]
        OD ;
        FOR j FROM k TO n
        DO
          a [ rownum [ i ] , j ] -:= mult * a [ rownum [ kminus1 ] , j ]
        OD
      OD
    OD
END ;
PROC backsub = ( REF [ , ] REAL a, x, b, REF [ ] INT rownum ) VOID :
BEGIN
  COMMENT
  this procedure performs back - substitution
  COMMENT
  INT n = UPB a , m = 2 UPB b ;
  REAL sum ;
  FOR i FROM n BY - 1 TO 1
  DO
    FOR p TO m
    DO
      sum := b [ rownum [ i ] , p ] ;
      FOR j FROM i+1 TO n
      DO
        sum -:= a [ rownum [ i ] , j ] * x [ j , p ]
      OD ;
      x [ i , p ] := sum / a [ rownum [ i ] , i ]
    OD
  OD
END ;
COMMENT
declaration of arrays and input
COMMENT
INT n ; read ( n ) ;
[ 1 : n , 1 : n ] REAL a ;
read ( a ) ;
INT m ; read ( m ) ;
COMMENT
x [ i , p ] is regarded as the ith element of the pth solution vector
b [ i , p ] is regarded as the ith element of the pth right hand side vector
COMMENT
[ 1 : n , 1 : m ] REAL x , b ;
FOR p TO m
DO
  read ( b [ , p ] )
OD ;
print ( ( newline , "the coefficient matrix is" , newline ) ) ;
FOR i TO n
DO
  print ( ( a [ i , ] , newline ) )
OD ;
print ( ( newline , "the right hand side vectors are" , newline ) ) ;
FOR p TO m
DO
  print ( ( b [ , p ] , newline ) )
OD ;
[ 1 : n ] INT rownum ;
gausselim ( a , b , rownum ) ;
backsub ( a, x, b, rownum ) ;
```

```
        print ( ( newline , "the solution vectors are" , newline ) ) ;
        FOR p TO m
        DO
           print ( ( x [ , p ] , newline ) )
        OD
END
```

Pascal Program

```
PROGRAM exampleofgausselimination(input, output);

   { a program for gauss elimination with partial pivoting
       this code is capable of handling the case of several right hand sides  }

   CONST
      maxm=10;
      maxn=50;

   TYPE
      rhsindextype=1..maxm;
      matrixindextype=1..maxn;
      rhstype=ARRAY[ matrixindextype, rhsindextype ] OF real;
      matrixtype=ARRAY[ matrixindextype, matrixindextype ] OF real;
      rownumtype=ARRAY[ matrixindextype ] OF matrixindextype;

   VAR
      i, j, n:matrixindextype;
      m, p:rhsindextype;
      a:matrixtype;
      {  x[i, p] is regarded as the ith element of the pth solution vector
         and b[i, p] is regarded as the ith element of the pth right hand
         side vector  }
      x, b:rhstype;
      rownum:rownumtype;

   PROCEDURE gausselim(n:matrixindextype; VAR a:matrixtype;
                       m:rhsindextype; VAR b:rhstype; VAR rownum:rownumtype);
      VAR
         i, j, k, kminus1, pivotrow, dummy:matrixindextype;
         p:rhsindextype;
         mult, abspivot, absa:real;
   BEGIN
      FOR i:= 1 TO n DO
         rownum[i]:= i;
      FOR k:= 2 TO n DO
      BEGIN
         kminus1:= k - 1;
         { find the pivot  }
         abspivot:= abs( a[ rownum[kminus1], kminus1 ] );
         pivotrow:= kminus1;
         FOR i:= k TO n DO
         BEGIN
            absa:= abs( a[ rownum[i], kminus1 ] );
            IF absa>abspivot
            THEN
               BEGIN
                  abspivot:= absa;
                  pivotrow:= i
               END
         END ;
         { interchange if necessary  }
         IF pivotrow<>kminus1
         THEN
            BEGIN
               dummy:= rownum[kminus1];
               rownum[kminus1]:= rownum[pivotrow];
               rownum[pivotrow]:= dummy
            END ;
```

```
            FOR i:= k TO n DO
            BEGIN
               mult:= a[ rownum[i], kminus1 ]/a[ rownum[kminus1], kminus1 ];
               FOR p:= 1 TO m DO
                  b[ rownum[i] , p ]:=
                           b[ rownum[i] , p ] - mult*b[ rownum[kminus1] , p ];
               FOR j:= k TO n DO
                  a[ rownum[i], j ]:=
                           a[ rownum[i], j ] - mult*a[ rownum[kminus1], j ]
            END
         END
   END { gausselim } ;

   PROCEDURE backsub(n:matrixindextype; VAR a:matrixtype;
                     m:rhsindextype; VAR x, b:rhstype; VAR rownum:rownumtype);
      VAR
         i, j:matrixindextype;
         p:rhsindextype;
         sum:real;
   BEGIN
      FOR i:= n DOWNTO 1 DO
         FOR p:= 1 TO m DO
            BEGIN
               sum:= b[ rownum[i], p ];
               FOR j:= i + 1 TO n DO
                  sum:= sum - a[ rownum[i], j ]*x[j, p];
               x[i, p]:= sum/a[ rownum[i], i ]
            END
   END { backsub } ;

BEGIN
   { read input data }
   read(n);
   FOR i:= 1 TO n DO
      FOR j:= 1 TO n DO
         read(a[i, j]);
   read(m);
   FOR p:= 1 TO m DO
      FOR i:= 1 TO n DO
         read(b[i, p]);

   { print input data }
   writeln;
   writeln('the coefficient matrix is');
   FOR i:= 1 TO n DO
   BEGIN
      FOR j:= 1 TO n DO
         write(a[i, j]);
      writeln
   END ;

   writeln;
   writeln('the right hand side vectors are');
   FOR p:= 1 TO m DO
   BEGIN
      FOR i:= 1 TO n DO
         write(b[i, p]);
      writeln
   END ;

   gausselim(n, a, m, b, rownum);
   backsub(n, a, m, x, b, rownum);

   { print solution vectors }
   writeln;
```

```
writeln('the solution vectors are');
FOR p:= 1 TO m DO
BEGIN
    FOR i:= 1 TO n DO
        write(x[i, p]);
    writeln
END

END { exampleofgausselimination } .
```

Sample Data

```
4
 0.0  12.0   8.0   7.0
 2.0   4.0   6.0   3.0
 6.0  12.0   7.0   7.0
-6.0 -12.0  -1.0  -4.0
2
27.0  15.0  33.0 -23.0
53.0  35.0  58.0 -22.0
```

Sample Output

```
the coefficient matrix is
  0.00000e  0  1.20000e  1  8.00000e  0  7.00000e  0
  2.00000e  0  4.00000e  0  6.00000e  0  3.00000e  0
  6.00000e  0  1.20000e  1  7.00000e  0  7.00000e  0
 -6.00000e  0 -1.20000e  1 -1.00000e  0 -4.00000e  0

the right hand side vectors are
  2.70000e  1  1.50000e  1  3.30000e  1 -2.30000e  1
  5.30000e  1  3.50000e  1  5.80000e  1 -2.20000e  1

the solution vectors are
  1.12698e  0  6.86508e -1  7.61905e -1  1.80952e  0
  1.38095e  0 -9.40475e -1  3.28571e  0  5.42857e  0
```

3.4 CONDITIONING AND ERROR ANALYSIS

3.4.1 Conditioning

We have already made several references to the notion of the conditioning of a problem. Recall that a problem is said to be ill-conditioned if small changes in the initial data of the problem produce relatively large changes in the answers. An example of this phenomenon was given in section 1.8 and a more substantial example is discussed in subsection 4.3.4. Suppose that the coefficients a_{ij} are perturbed by the quantities ε_{ij}: $i,j = 1, 2, \ldots, n$ and the right-hand sides by δ_i: $i = 1, 2, \ldots, n$. We would like to be able to predict the effect of these small changes on the solution. Although it is possible to do this, the results obtained are not usually of practical use. Knowledge of the exact inverse matrix A^{-1} is required and, in any case, the bound on the perturbation in the solution is often over-pessimistic.

In practice, it will be difficult to recognise a poor solution. An obvious device is to form the residual vector **r** from the computed solution $\mathbf{x}_c$, say, as

$$\mathbf{r} = \mathbf{b} - A\mathbf{x}_c .$$

(That is, **r** represents the amount by which the computed solution fails to satisfy the original system of equations.) We would intuitively expect that if the r_is are all small in magnitude then $\mathbf{x}_c$ is an accurate solution. Unfortunately this is not necessarily the case and the residual vector cannot be used as a reliable indicator of accuracy. If a poor solution is suspected, it may be desirable for the user deliberately to perturb the initial data, rerun the problem and see if there is a large change in the solution. However, the device known as iterative refinement may be much more useful and we now discuss it briefly.

3.4.2 Iterative refinement of an approximate solution

Suppose **X** is a known approximate solution of the system (3.1) and let

$$\mathbf{v} = \mathbf{x} - \mathbf{X}$$

be the difference between the true and approximate solutions. Now

$$\mathbf{r} = \mathbf{b} - A\mathbf{X} \qquad (3.18)$$
$$= A\mathbf{x} - A\mathbf{X}$$
$$= A\mathbf{v}$$

so that, theoretically, the correction **v** to **X** can be found as the solution of the set of linear simultaneous equations

$$A\mathbf{v} = \mathbf{r} \qquad (3.19)$$

where the elements of **r** are obtained from (3.18). Having found **v** the true solution to (3.1) is found as

$$\mathbf{x} = \mathbf{X} + \mathbf{v} .$$

There are a number of comments which have to be made about this simple idea. The first is that if **X** has been obtained by the use of Gauss elimination and back-substitution, we merely have to apply the formula (3.15) to the residual vector since the coefficient matrix in (3.19) is the same as in (3.1). This means that in the elimination process we must store the multipliers $a_{i,k-1}^{(k-1)}/a_{k-1,k-1}^{(k-1)}$ (see subsection 3.6.2). Equation (3.16) may then be used to compute the solution. However, because of rounding errors, this process will yield only an approximation to **v**, **V** say, and consequently the quantity $\mathbf{X} + \mathbf{V}$ is still only an approximation (hopefully better than $\mathbf{X}$!) to $\mathbf{x}$. This now suggests that we should form the residual vector based on the new approximation $\mathbf{X} + \mathbf{V}$ and repeat the process. These ideas can now be formulated as an iterative scheme with $\mathbf{X}^{[0]}$ as the initial approximate solution

> *while* some suitable convergence criterion is not satisfied
> *do*
>   increment m;
>   form $\mathbf{r}^{[m]} = \mathbf{b} - A\mathbf{X}^{[m]}$;
>   solve $A\mathbf{V}^{[m]} = \mathbf{r}^{[m]}$ for $\mathbf{V}^{[m]}$;
>   form $\mathbf{X}^{[m+1]} = \mathbf{X}^{[m]} + \mathbf{V}^{[m]}$
> *od*

Criteria for terminating the iterative process are based on the use of *vector norms* (see subsection 3.5.3).

A most important consideration arises in the evaluation of $\mathbf{r}^{[m]}$. The elements of this residual vector must be computed accurately and double length facilities must be used. The reason for this is that the components of $\mathbf{r}^{[m]}$ will usually be small and the accumulated single length rounding error could be of the same order of magnitude.

3.4.3 Error analysis

Here we are concerned with trying to assess the effect of the very large number of rounding errors which are introduced by the elementary arithmetic operations involved in using a direct method such as Gauss elimination followed by back-substitution. The most sophisticated analysis of this sort is known as *backward error analysis* (see Wilkinson, 1963). It depends on being able to show that the computed solution $\mathbf{x}_c$ of the system (3.1) satisfies

$$(A + E)\mathbf{x}_c = \mathbf{b}$$

where the matrix E is a perturbation of the original matrix A. For Gauss elimination with partial pivoting it is possible to derive an upper bound on the 'size' of E in terms of the number of equations, the word length of the machine and the largest, in absolute value, computed $a_{ij}^{(k)}$. What we mean by the size of a matrix involves the concept of a *matrix norm* and is beyond the scope of this book. However, having derived this quantity it is then possible to derive an upper bound on the 'size' of the vector $\mathbf{x} - \mathbf{x}_c$. This last result really requires knowledge of the inverse matrix A^{-1} and may not therefore be directly useful. The reader is referred to Ralston and Rabinowitz (1978) for further details.

3.5 ITERATION

3.5.1 Derivation of the Jacobi and Gauss–Seidel schemes

Before considering direct methods for the solution of the system (3.1) further we examine ways of setting up iterative processes which will, hopefully, converge to the required solution. These methods can be particularly useful if the coefficient matrix is sparse.

Suppose that we wish to solve the set of three linear simultaneous equations

$$\sum_{j=1}^{3} a_{ij}x_j = b_i \quad i = 1,2,3 \tag{3.20}$$

for the three unknowns x_1, x_2 and x_3. Let $x_1^{[0]}$, $x_2^{[0]}$ and $x_3^{[0]}$ be initial approximations to x_1, x_2 and x_3 respectively. Then, assuming that a_{11}, a_{22} and a_{33} are all non-zero, the system (3.20) may be rewritten as

$$x_1 = (b_1 - a_{12}x_2 - a_{13}x_3)/a_{11}$$
$$x_2 = (b_2 - a_{21}x_1 - a_{23}x_3)/a_{22} \tag{3.21}$$
$$x_3 = (b_3 - a_{31}x_1 - a_{32}x_2)/a_{33}$$

and this suggests the iterative scheme (*Jacobi iteration*)

$$x_1^{[k]} = (b_1 - a_{12}x_2^{[k-1]} - a_{13}x_3^{[k-1]})/a_{11}$$
$$x_2^{[k]} = (b_2 - a_{21}x_1^{[k-1]} - a_{23}x_3^{[k-1]})/a_{22} \quad k = 1,2,\ldots \tag{3.22}$$
$$x_3^{[k]} = (b_3 - a_{31}x_1^{[k-1]} - a_{32}x_2^{[k-1]})/a_{33}$$

where the superscript indicates an iteration number. Thus, a sequence of approximations $\{(x_1^{[k]}, x_2^{[k]}, x_3^{[k]}): k = 0,1,\ldots\}$ is generated and under suitable circumstances it will converge to the required solution.

EXAMPLE 3.6
The system

$$4x_1 + x_2 - x_3 = 12$$
$$-x_1 + 3x_2 + x_3 = 6$$
$$2x_1 + 2x_2 + 5x_3 = 5$$

has the solution $x_1 = 2$, $x_2 = 3$, $x_3 = -1$. If the initial approximation is $x_1^{[0]} = x_2^{[0]} = x_3^{[0]} = 0$, the first application of Jacobi iteration yields $x_1^{[1]} = 12/4 = 3$, $x_2^{[1]} = 6/3 = 2$, $x_3^{[1]} = 5/5 = 1$. The next iteration gives

$$x_1^{[2]} = (12 - 1 \times 2 - (-1) \times 1)/4 = 11/4 = 2.75000$$
$$x_2^{[2]} = (6 - (-1) \times 3 - 1 \times 1)/3 = 8/3 = 2.66667$$
$$x_3^{[2]} = (5 - 2 \times 3 - 2 \times 2)/5 = -5/5 = -1.00000 \ .$$

Further iterations are listed in Table 3.1 correct to six significant figures.

Table 3.1

	Jacobi			Gauss–Seidel		
k	$x_1^{[k]}$	$x_2^{[k]}$	$x_3^{[k]}$	$x_1^{[k]}$	$x_2^{[k]}$	$x_3^{[k]}$
0	0.00000	0.00000	0.00000	0.00000	0.00000	0.00000
1	3.00000	2.00000	1.00000	3.00000	3.00000	−1.40000
2	2.75000	2.66667	−1.00000	1.90000	3.10000	−1.00000
3	2.08333	3.25000	−1.16667	1.97500	2.99167	−0.986667
4	1.89583	3.08333	−1.13333	2.00542	2.99736	−1.00111
5	1.94583	3.00972	−0.991667	2.00038	3.00050	−1.00035
6	1.99965	2.97917	−0.982222	1.99979	3.00005	−0.999933
7	2.00965	2.99396	−0.991528	2.00000	2.99998	−0.999994
8	2.00363	3.00039	−1.00144	2.00001	3.00000	−1.00000
9	1.99954	3.00169	−1.00161	2.00000	3.00000	−1.00000
10	1.99918	3.00038	−1.00049			
11	1.99978	2.99989	−0.999823			
12	2.00007	2.99987	−0.999868			
13	2.00007	2.99998	−0.999976			
14	2.00001	3.00001	−1.00002			
15	1.99999	3.00001	−1.00001			

An immediate extension of the scheme (3.22) is to make use of the new iterates as soon as they become available. This gives the process (*Gauss–Seidel iteration*)

$$x_1^{[k]} = (b_1 - a_{12}x_2^{[k-1]} - a_{13}x_3^{[k-1]})/a_{11}$$
$$x_2^{[k]} = (b_2 - a_{21}x_1^{[k]} - a_{23}x_3^{[k-1]})/a_{22} \qquad k = 1, 2, \ldots$$
$$x_3^{[k]} = (b_3 - a_{31}x_1^{[k]} - a_{32}x_2^{[k]})/a_{33}$$

so that, for example, in the second equation of the system (3.22) the most recently computed estimate, $x_1^{[k]}$, of x_1 is used to find $x_2^{[k]}$.

EXAMPLE 3.7
Consider the system of equations of Example 3.6 and again let $x_1^{[0]} = x_2^{[0]} = x_3^{[0]} = 0$. Then, as before, $x_1^{[1]} = 3$ but now

$$x_2^{[1]} = (6 - (-1) \times 3 - 1 \times 0)/3 = 9/3 = 3$$

and

$$x_3^{[1]} = (5 - 2 \times 3 - 2 \times 3)/5 = -7/5 = -1.4 \ .$$

Table 3.1 lists further iterations and we note that Gauss–Seidel iteration has obtained the solution (2.00000, 3.00000, −1.00000) after just 9 iterations whereas, at the same stage, Jacobi iteration still has some way to go. Further, after two iterations both the Jacobi and Gauss–Seidel schemes give $x_3^{[2]} = -1$ and then move away from this value before converging to it ultimately. We conclude that when monitoring iterates it is important that we measure the overall convergence of the values $\{(x_1^{[k]}, x_2^{[k]}, x_3^{[k]}): k = 0, 1, \ldots\}$ and we investigate ways of doing this in subsection 3.5.3.

Jacobi and Gauss–Seidel iteration can be readily extended to the case of n linear simultaneous equations in n unknowns. Jacobi iteration takes the form

$$x_i^{[k]} = \left(b_i - \sum_{\substack{j=1 \\ j \neq i}}^{n} a_{ij} x_j^{[k-1]}\right)/a_{ii} \qquad \begin{array}{l} i = 1, 2, \ldots, n \ ; \\ k = 1, 2, \ldots \end{array} \qquad (3.23)$$

whilst Gauss–Seidel iteration is

$$x_i^{[k]} = \left(b_i - \sum_{j=1}^{i-1} a_{ij} x_j^{[k]} - \sum_{j=i+1}^{n} a_{ij} x_j^{[k-1]}\right)/a_{ii} \qquad \begin{array}{l} i = 1, 2, \ldots, n \ ; \\ k = 1, 2, \ldots \end{array} \qquad (3.24)$$

In both cases we have assumed $a_{ii} \neq 0: i = 1, 2, \ldots, n$.

As in the iterative schemes discussed in Chapter 2, it is necessary to examine the conditions under which a suitable scheme will converge and, if convergent, to formulate a suitable termination criterion.

3.5.2 A convergence theorem
THEOREM 3.1
In the set of n linear simultaneous equations

$$\sum_{j=1}^{n} a_{ij}x_j = b_i \qquad i = 1, 2, \ldots, n$$

suppose the matrix of coefficients is strictly diagonally dominant, that is,

$$|a_{ii}| > \sum_{\substack{j=1 \\ j \neq i}}^{n} |a_{ij}| \qquad i = 1, 2, \ldots, n \ .$$

Then Jacobi and Gauss–Seidel iteration converge from arbitrary initial estimates of the unknowns, that is, for any $x_1^{[0]}, x_2^{[0]}, \ldots, x_n^{[0]}$.

The proof of this theorem is not particularly difficult but, without some formal concepts from linear algebra, it is very lengthy (see Johnson and Riess (1982) for further details). The theorem shows the importance of the class of strictly diagonally dominant matrices. However, we note that strict diagonal dominance is a sufficient but not necessary condition for convergence and consequently these iterative schemes may converge for other sets of equations.

EXAMPLE 3.8
The coefficient matrix of Example 3.6 is strictly diagonally dominant and hence we would expect Jacobi and Gauss–Seidel iteration to converge for this particular problem. The results contained in Table 3.1 bear this out. In Example 3.1, the coefficient matrix is not strictly diagonally dominant and we can say there is no guarantee that Jacobi and Gauss–Seidel iteration will converge. We cannot guarantee that they will both diverge but, starting with $x_1^{[0]} = x_2^{[0]} = x_3^{[0]} = 0$, Jacobi iteration gives $x_1^{[10]} = 302.125$, $x_2^{[10]} = -106.188$, $x_3^{[10]} = 217.375$ whilst Gauss–Seidel iteration gives $x_1^{[10]} = 85011.3$, $x_2^{[10]} = -58568.9$, $x_3^{[10]} = -90697.4$ and both processes are clearly diverging from the true solution. In the system

$$4x_1 - 2x_2 + x_3 = 12$$
$$2x_1 + 3x_2 - x_3 = 7$$
$$2x_1 - 2x_2 + 2x_3 = 8$$

the coefficient matrix is again not strictly diagonally dominant. However, starting with $x_1^{[0]} = x_2^{[0]} = x_3^{[0]} = 0$, Jacobi iteration gives $x_1^{[18]} = 3.00000$, $x_2^{[18]} = 1.00001$, $x_3^{[18]} = 2.00002$ and Gauss–Seidel iteration gives $x_1^{[15]} = 3.00000$, $x_2^{[15]} = 1.00000$, $x_3^{[15]} = 2.00000$. (All figures here have been quoted correct to six significant figures.) Since the true solution is $x_1 = 3, x_2 = 1, x_3 = 2$ we can say that both Jacobi and Gauss–Seidel iteration converge for this problem given the starting value $\mathbf{x}^{[0]T} = (0,0,0)$.

3.5.3 Termination criteria

We first consider ways of determining the size of a vector $\mathbf{y}$. Three possible measures are

(i) the sum of the absolute values of the components of $\mathbf{y}$

$$\sum_{i=1}^{n} |y_i|, \tag{3.25}$$

(ii) the square root of the sum of the squares of the components of $\mathbf{y}$

$$\left(\sum_{i=1}^{n} y_i^2\right)^{\frac{1}{2}}, \tag{3.26}$$

(iii) the maximum absolute component of $\mathbf{y}$

$$\underset{1 \leq i \leq n}{\text{maximum}} |y_i|. \tag{3.27}$$

The quantities (3.25), (3.26) and (3.27) are, in turn, referred to as the 1-*norm*, 2-*norm* and ∞-*norm* of $\mathbf{y}$. Whilst these are not the only ways of quantifying the size of a vector, they are the ones most commonly used in numerical analysis. (See Broyden (1975) for further details of vector norms.)

Now, in Example 3.7, we noted that a convergence criterion for an iterative process must be based on the convergence of each of the iterates $x_i^{[k]}$: $k = 0, 1, \ldots, i = 1, 2, \ldots$. If we use (3.27) to test for convergence an iterative method such as Jacobi or Gauss–Seidel would be continued until

$$\max_{1 \leq i \leq n} |x_i^{[k]} - x_i^{[k-1]}| < \varepsilon \tag{3.28}$$

where ε is some chosen tolerance. It may be preferable to use a relative, as opposed to absolute, accuracy criterion.

If the chosen iteration is slowly convergent there are likely to be difficulties. In subsection 2.3.3 we observed that a convergence criterion based on ensuring that the absolute value of the difference between successive iterates is less than some tolerance does not guarantee that the root has been found to this accuracy. Here, if the iterative process is continued until (3.28) is satisfied, there is no guarantee that

$$\max_{1 \leq i \leq n} |x_i^{[k]} - x_i| < \varepsilon$$

holds. In such a case it may be preferable to examine the residual vector $\mathbf{r}^{[k]}$ whose components are defined by

$$r_i^{[k]} = b_i - \sum_{j=1}^{n} a_{ij} x_j^{[k]} \qquad i = 1, 2, \ldots, n$$

and formulate a criterion based on some measure of this vector. However, in the case of a badly conditioned set of equations we observed in section 3.4 that small residuals do not necessarily imply an accurate solution.

3.5.4 A program for Gauss–Seidel iteration

In subsection 3.3.2 we saw how to develop a complex program by breaking down the underlying algorithm into a number of subproblems. We use the same approach here to produce code which implements Gauss–Seidel iteration (equation (3.24)). The programs are given in Case Study 3.3.

We observe that, once again, a two-dimensional array can be used to store the coefficient matrix and one-dimensional arrays used for the right-hand side and solution vectors. Therefore, those sections of the code of Case Study 3.1 which declare, read and print these quantities form the basis of the new program. Code to read in the starting value $\mathbf{x}^{[0]}$ will also be required.

The body of the procedure *gaussseidel* contains a loop which implements the Gauss–Seidel iteration. This loop is iterated until $|x[i] - xold[i]| < \varepsilon$ for all values of i. In effect, we are using condition (3.28) to decide whether or not to stop the iteration. To make the procedure as general as possible, it has an additional two parameters, namely *tol*, the accuracy tolerance ε, and *itermax*, the maximum number of iterations allowed. The purpose of the quantity *itermax* was discussed in detail in subsection 2.3.4.

One of the first tasks that has to be performed in the iteration loop is the copying of the current values of the vector x into the vector *xold*. In the Pascal program this is done by the loop

```
FOR i:= 1 TO n DO
    xold[i]:= x[i]
```

Although Pascal permits array assignments we cannot in fact use xold:= x

here because only the first *n* of the *maxn* elements of *x* have values; the remaining elements of *x* are undefined. Because the Algol 68 program uses dynamic arrays, the arrays *x* and *xold* have exactly *n* elements and so the Algol 68 version of the program implements the copying of the array by an array assignment.

Before leaving these programs we note that a very simple modification to the code will result in Jacobi (equation (3.23)), rather than Gauss–Seidel, iteration being performed. All that we need to do is to replace *x[j]* by *xold[j]* in both the Algol 68 clause

sum +:= a [i , j] * x [j]

and the Pascal statement

sum:= sum + a[i,j]*x[j] .

CASE STUDY 3.3

```
Algol 68 Program
BEGIN
  COMMENT
  a program for gauss - seidel iteration
  COMMENT
  PROC gaussseidel = ( REF [ , ] REAL a , REF [ ] REAL b , x ,
                       REAL tol , INT itermax ) VOID :
  BEGIN
    COMMENT
    this procedure implements the iterative process itself
    COMMENT
    INT n = UPB b ;
    [ 1 : n ] REAL xold ;
    BOOL converged := FALSE ;
    INT iter := 0 ;
    REAL sum ;
    WHILE NOT converged AND iter < itermax
    DO
      iter +:= 1 ;
      xold := x ;
      FOR i TO n
      DO
        sum := 0.0 ;
        FOR j TO n
        DO
          IF i # j
          THEN
            sum +:= a [ i , j ] * x [ j ]
          FI
        OD ;
        x [ i ] := ( b [ i ] - sum ) / a [ i , i ]
      OD ;
```

```
          converged := TRUE ;
          FOR i TO n WHILE converged
          DO
             converged := ABS ( x [ i ] - xold [ i ] ) < tol
          OD
       OD ;
       print ( ( newline , IF converged THEN "" ELSE "no " FI ,
                 "convergence to specified tolerance after  " , iter ,
                 " iterations" , newline ) )
   END ;
   INT n ; read ( n ) ;
   [ 1 : n , 1 : n ] REAL a ;
   [ 1 : n ] REAL x , b ;
   REAL tol ;
   INT itermax ;
   read ( ( a , b , x , tol , itermax ) ) ;
   print ( ( newline , "the coefficient matrix is" , newline ) ) ;
   FOR i TO n
   DO
      print ( ( a [ i , ] , newline ) )
   OD ;
   print ( ( newline , "the right hand side vector is" , newline , b ,
             newline , newline , "the initial approximation is" , newline , x ,
             newline , newline , "accuracy tolerance                   " , tol ,
             newline , "maximum number of iterations allowed  " , itermax ,
             newline ) ) ;
   gaussseidel ( a , b , x , tol , itermax ) ;
   print ( ( newline , "the solution vector is" , newline , x ,
             newline ) )
END
```

Pascal Program

```
PROGRAM performgaussseideliteration(input, output);

   CONST
      maxn=50;

   TYPE
      matrixtype=ARRAY [ 1..maxn, 1..maxn ] OF real;
      rhstype=ARRAY [ 1..maxn ] OF real;
      itermaxtype=1..maxint;

   VAR
      a:matrixtype;
      x, b:rhstype;
      tol:real;
      i, j, n:1..maxn;
      itermax:itermaxtype;

   PROCEDURE gaussseidel(n:integer; VAR a:matrixtype; VAR b, x:rhstype;
                         tol:real; itermax:itermaxtype);
      { this procedure implements the iterative process itself }
      VAR
         xold:rhstype;
         converged:boolean;
         iter:0..maxint;
         sum:real;
         i:0..maxn;  j:1..maxn;
   BEGIN
      iter:= 0;
      REPEAT
         iter:= iter + 1;
         FOR i:= 1 TO n DO
            xold[i]:= x[i];
```

```
                    FOR i:= 1 TO n DO
                    BEGIN
                       sum:= 0.0;
                       FOR j:= 1 TO n DO
                          IF i<>j
                          THEN
                             sum:= sum + a[i,j]*x[j];
                       x[i]:= (b[i] - sum)/a[i,i]
                    END ;
                    i:= 0;
                    REPEAT
                       i:= i + 1;
                       converged:= abs(x[i] - xold[i])<tol
                    UNTIL (i=n) OR NOT converged
                 UNTIL converged OR (iter=itermax);
              writeln;
              IF NOT converged
              THEN
                 write('no ');
              writeln('convergence to specified tolerance after  ', iter,
                      ' iterations')
           END { gaussseidel } ;

BEGIN
   { read input data }
   read(n);
   FOR i:= 1 TO n DO
      FOR j:= 1 TO n DO
         read(a[i,j]);
   FOR i:= 1 TO n DO
      read(b[i]);
   FOR i:= 1 TO n DO
      read(x[i]);
   read(tol, itermax);

   { print input data }
   writeln;
   writeln('the coefficient matrix is');
   FOR i:= 1 TO n DO
   BEGIN
      FOR j:= 1 TO n DO
         write(a[i,j]);
      writeln
   END ;
   writeln;
   writeln('the right hand side vector is');
   FOR i:= 1 TO n DO
      write(b[i]);
   writeln;
   writeln;
   writeln('the initial approximation is');
   FOR i:= 1 TO n DO
      write(x[i]);
   writeln;
   writeln;
   writeln('accuracy tolerance                     ', tol);
   writeln('maximum number of iterations allowed   ', itermax);

   gaussseidel(n, a, b, x, tol, itermax);

   writeln;
   writeln('the solution vector is');
   FOR i:= 1 TO n DO
      write(x[i]);
   writeln

END { performgaussseideliteration } .
```

Sample Data

```
3
   4.0   -2.0    1.0
   2.0    3.0   -1.0
   2.0   -2.0    2.0
  12.0    7.0    8.0
   0.0    0.0    0.0
   0.00005
  25
```

Sample Output

```
the coefficient matrix is
  4.00000e  0 -2.00000e  0  1.00000e  0
  2.00000e  0  3.00000e  0 -1.00000e  0
  2.00000e  0 -2.00000e  0  2.00000e  0

the right hand side vector is
  1.20000e  1  7.00000e  0  8.00000e  0

the initial approximation is
  0.00000e  0  0.00000e  0  0.00000e  0

accuracy tolerance                          5.00000e -5
maximum number of iterations allowed        25

convergence to specified tolerance after    13 iterations

the solution vector is
  3.00000e  0  9.99986e -1  1.99998e  0
```

3.5.5 Computational efficiency

The number of elementary arithmetic operations involved in each iteration, whether Jacobi or Gauss–Seidel, is proportional to n^2. The overall cost of each scheme is then determined by the number of iterations required to achieve convergence. Using the same initial approximation it is possible to compare the efficiency of these two schemes and, in this sense, it is usually the case that Gauss–Seidel iteration is more efficient than Jacobi (see Table 3.1). Other iterative schemes have been proposed which aim to keep the number of iterations required as low as possible.

Consider the generalised form of the system (3.21)

$$x_i = \left(b_i - \sum_{\substack{j=1 \\ j \neq i}}^{n} a_{ij} x_j\right) / a_{ii} \qquad i = 1, 2, \ldots, n \ .$$

Suppose that we choose some value p and multiply through by this number. Adding x_i to both sides and rearranging we have

$$x_i = (1-p)x_i + p\left(b_i - \sum_{\substack{j=1 \\ j \neq i}}^{n} a_{ij} x_j\right) / a_{ii} \qquad i = 1, 2, \ldots, n$$

and this suggests the iterative scheme

$$x_i^{[k]} = (1-p)x_i^{[k-1]} + p\left(b_i - \sum_{j=1}^{i-1} a_{ij}x_j^{[k]} - \sum_{j=i+1}^{n} a_{ij}x_j^{[k-1]}\right)/a_{ii} \quad (3.29)$$

$$i = 1, 2, \ldots, n \ ;$$
$$k = 1, 2, \ldots$$

(compare this with equation (3.24)) known as *successive over-relaxation (SOR)*. The choice of the parameter p is a matter of advanced analysis and we can do no more than state here that, for certain classes of matrix, it is possible to find the optimum value of p in order to achieve rapid convergence, and that for other classes it may be possible to find a value for p which is near the optimum. The reader with a moderate knowledge of linear algebra is referred to Smith (1978) where a very readable introductory account of this problem and iteration in general is given.

We conclude this subsection by observing that it is possible to formulate and use the Aitken acceleration scheme described in section 2.4 in the present context. Three successive iterates, $\mathbf{x}^{[k]}$, $\mathbf{x}^{[k+1]}$ and $\mathbf{x}^{[k+2]}$ are required and the components $x_i^{[k]}$, $x_i^{[k+1]}$ and $x_i^{[k+2]}$ can be combined to give a new initial approximation which is used to generate two further iterates and so on. The stage at which the device should be introduced is a matter of some importance and the reader is again referred to Smith (1978) where other 'dynamic' acceleration devices are also discussed.

3.5.6 A further discussion of iterative methods

Before leaving iterative methods we give a very brief indication of the way in which a fuller analysis of some iterative schemes might be undertaken. Matrix algebra is used extensively in this subsection which may be omitted at a first reading.

Consider the system of equations (3.1) and assume that $a_{ii} \neq 0$: $i = 1, 2, \ldots, n$. The coefficient matrix A may be split as

$$A = L + D + U$$

where D is a diagonal matrix with elements $a_{11}, a_{22}, \ldots, a_{nn}$ on the diagonal and L and U are strictly lower and upper triangular matrices, respectively, defined by

$$L = \begin{bmatrix} 0 & 0 & \ldots & 0 \\ a_{21} & 0 & \ldots & 0 \\ \ldots & & & \\ a_{n1} & a_{n2} & \ldots & 0 \end{bmatrix}, \quad U = \begin{bmatrix} 0 & a_{12} & \ldots & a_{1n} \\ 0 & 0 & \ldots & a_{2n} \\ \ldots & & & \\ 0 & 0 & \ldots & 0 \end{bmatrix}.$$

This means that the original system may be expressed as

$$(L + D + U)\mathbf{x} = \mathbf{b}$$

and this can be rewritten in a number of ways to provide different iterative schemes. In particular, we have

$$D\mathbf{x} = -(L + U)\mathbf{x} + \mathbf{b}$$

which suggests the scheme

$$\mathbf{x}^{[k]} = -D^{-1}(L + U)\mathbf{x}^{[k-1]} + D^{-1}\mathbf{b}$$

and this is Jacobi iteration. It follows immediately that Gauss–Seidel iteration takes the form

$$\mathbf{x}^{[k]} = -D^{-1}L\mathbf{x}^{[k]} - D^{-1}U\mathbf{x}^{[k-1]} + D^{-1}\mathbf{b} \ . \tag{3.30}$$

However, on premultiplying (3.30) by D and rearranging, we have

$$(D + L)\mathbf{x}^{[k]} = -U\mathbf{x}^{[k-1]} + \mathbf{b}$$

and so Gauss–Seidel iteration may also be written as

$$\mathbf{x}^{[k]} = -(D + L)^{-1}U\mathbf{x}^{[k-1]} + (D + L)^{-1}\mathbf{b} \ .$$

Now, consider the system (3.29). Multiplying through by a_{ii}, these equations may be written, in matrix form, as

$$D\mathbf{x}^{[k]} = (1 - p)D\mathbf{x}^{[k-1]} + p\mathbf{b} - pL\mathbf{x}^{[k]} - pU\mathbf{x}^{[k-1]} \ .$$

Rearranging, we have

$$(D + pL)\mathbf{x}^{[k]} = ((1 - p)D - pU)\mathbf{x}^{[k-1]} + p\mathbf{b}$$

or

$$\mathbf{x}^{[k]} = (D + pL)^{-1}((1 - p)D - pU)\mathbf{x}^{[k-1]} + p(D + pL)^{-1}\mathbf{b} \ .$$

Hence, it is possible to express each of the iterative schemes that we have considered in the form

$$\mathbf{x}^{[k]} = M\mathbf{x}^{[k-1]} + \mathbf{g} \qquad (3.31)$$

where

(i) for Jacobi iteration $M = -D^{-1}(L+U)$ and $\mathbf{g} = D^{-1}\mathbf{b}$;

(ii) for Gauss–Seidel iteration $M = -(D+L)^{-1}U$ and $\mathbf{g} = (D+L)^{-1}\mathbf{b}$; and

(iii) for SOR $M = (D+pL)^{-1}((1-p)D - pU)$ and $\mathbf{g} = p(D+pL)^{-1}\mathbf{b}$.

Equation (3.31) allows us to study the convergence of all of our iterative methods within a general framework. Let ε_k denote the error in $\mathbf{x}^{[k]}$, so that

$$\varepsilon_k = \mathbf{x} - \mathbf{x}^{[k]}$$

where $\mathbf{x}$ is the solution to the original system (3.1). Then from (3.31) we must have

$$\mathbf{x} = M\mathbf{x} + \mathbf{g} \qquad (3.32)$$

so that subtracting (3.31) from (3.32) we find that

$$\varepsilon_k = M\varepsilon_{k-1}.$$

Continuing the process

$$\varepsilon_k = M^2 \varepsilon_{k-2} = \ldots = M^k \varepsilon_0$$

where ε_0 is the error in the initial approximation $\mathbf{x}^{[0]}$. Hence, the iterative process will converge if M^k tends to the zero matrix (that is, the matrix whose entries are all zero) as k increases. We can go no further than this with the mathematical tools currently at our disposal.

3.6 FACTORISATION METHODS

3.6.1 Introduction

In this section we return to the development of direct methods for the solution of a system of linear simultaneous equations. We shall need to use matrix notation and some elementary results of matrix algebra.

We begin by considering the solution of three equations in three unknowns for which the coefficient matrix, A, is given by

$$A = \begin{bmatrix} a_{11} & a_{12} & a_{13} \\ a_{21} & a_{22} & a_{23} \\ a_{31} & a_{32} & a_{33} \end{bmatrix}.$$

Let L, D and U be the three matrices

$$L = \begin{bmatrix} 1 & 0 & 0 \\ l_{21} & 1 & 0 \\ l_{31} & l_{32} & 1 \end{bmatrix} \quad D = \begin{bmatrix} d_1 & 0 & 0 \\ 0 & d_2 & 0 \\ 0 & 0 & d_3 \end{bmatrix} \quad U = \begin{bmatrix} 1 & u_{12} & u_{13} \\ 0 & 1 & u_{23} \\ 0 & 0 & 1 \end{bmatrix}$$

that is, L and U are unit lower and upper triangular respectively and D is diagonal. We now investigate whether it is possible to find values for the components of L, D and U such that

$$LDU = A .\qquad(3.33)$$

If we perform the matrix multiplications on the left-hand side of (3.33) and then equate corresponding elements row-by-row, we obtain the following results.

Row 1: $d_1 = a_{11}$; $d_1 u_{12} = a_{12}$; $d_1 u_{13} = a_{13}$.
Row 2: $l_{21} d_1 = a_{21}$; $l_{21} d_1 u_{12} + d_2 = a_{22}$; $l_{21} d_1 u_{13} + d_2 u_{23} = a_{23}$.
Row 3: $l_{31} d_1 = a_{31}$; $l_{31} d_1 u_{12} + l_{32} d_2 = a_{32}$; $l_{31} d_1 u_{13} + l_{32} d_2 u_{23} + d_3 = a_{33}$.

The results for row 1 enable d_1, u_{12} and u_{13} to be found in turn; those for row 2 enable l_{21}, d_2 and u_{23} to be found in turn and those for row 3 enable l_{31}, l_{32} and d_3 to be found in turn.

EXAMPLE 3.9
Consider the system of equations of Example 3.1. Then the coefficient matrix for this problem is

$$A = \begin{bmatrix} 1 & -1 & -2 \\ 1 & 2 & 1 \\ 1 & 3 & -1 \end{bmatrix}$$

so that

$$d_1 = a_{11} = 1$$

$u_{12} = a_{12}/d_1 = -1/1 = -1$

$u_{13} = a_{13}/d_1 = -2/1 = -2$

$l_{21} = a_{21}/d_1 = 1/1 = 1$

$d_2 = a_{22} - l_{21}d_1 u_{12} = 2 - 1 \times 1 \times (-1) = 3$

$u_{23} = (a_{23} - l_{21}d_1 u_{13})/d_2 = (1 - 1 \times 1 \times (-2))/3 = 1$

$l_{31} = a_{31}/d_1 = 1/1 = 1$

$l_{32} = (a_{32} - l_{31}d_1 u_{12})/d_2 = (3 - 1 \times 1 \times (-1))/3 = 4/3$

$d_3 = a_{33} - l_{31}d_1 u_{13} - l_{32}d_2 u_{23} = -1 - 1 \times 1 \times (-2) - 4/3 \times 3 \times 1 = -3$.

Hence

$$A = \begin{bmatrix} 1 & 0 & 0 \\ 1 & 1 & 0 \\ 1 & 4/3 & 1 \end{bmatrix} \begin{bmatrix} 1 & 0 & 0 \\ 0 & 3 & 0 \\ 0 & 0 & -3 \end{bmatrix} \begin{bmatrix} 1 & -1 & -2 \\ 0 & 1 & 1 \\ 0 & 0 & 1 \end{bmatrix} .$$

This result should now be checked by first forming $Z = DU$ and then $A = LZ$.

If $a_{11} = 0$ then $d_1 = 0$ which means u_{12}, u_{13}, l_{21} and l_{31} cannot be found and the process breaks down. If $a_{11} \neq 0$ but $d_2 = 0$, it will not be possible to compute u_{23} and l_{32}. Now, $d_2 = a_{22} - l_{21}d_1 u_{12} = a_{22} - (a_{21}/d_1)d_1(a_{12}/d_1) = a_{22} - a_{21}a_{12}/a_{11}$. Hence the condition $d_2 \neq 0$ is the same as $a_{11}a_{22} - a_{21}a_{12} \neq 0$. But this means that the determinant of the minor

$$\begin{bmatrix} a_{11} & a_{12} \\ a_{21} & a_{22} \end{bmatrix}$$

of A must be non-zero. Extending these results further we have the following theorem.

THEOREM 3.2
In the $n \times n$ matrix A suppose that the determinant of each sub-matrix A_1, A_2, ..., A_n is non-zero where

$$A_1 = (a_{11}) , \quad A_2 = \begin{bmatrix} a_{11} & a_{12} \\ a_{21} & a_{22} \end{bmatrix} , \quad A_3 = \begin{bmatrix} a_{11} & a_{12} & a_{13} \\ a_{21} & a_{22} & a_{23} \\ a_{31} & a_{32} & a_{33} \end{bmatrix}$$

and so on. Then there exists a unique factorisation $A = LDU$, where L and U are respectively unit lower and upper triangular and D is diagonal with non-zero diagonal elements.

We shall not give the proof of this theorem (which may be found in Phillips and Taylor, 1973) but rather proceed to two of its consequences. The first is that the product DU is an upper triangular matrix so a unique factorisation of A into a unit lower triangular matrix and an upper triangular matrix exists. Similarly, the product LD is a lower triangular matrix so a unique factorisation into a lower triangular matrix and a unit upper triangular matrix exists.

EXAMPLE 3.10
The coefficient matrix of Example 3.9 (and hence of Example 3.1) can be expressed as the product

$$A = \begin{bmatrix} 1 & 0 & 0 \\ 1 & 1 & 0 \\ 1 & 4/3 & 1 \end{bmatrix} \begin{bmatrix} 1 & -1 & -2 \\ 0 & 3 & 3 \\ 0 & 0 & -3 \end{bmatrix}$$

or as the product

$$A = \begin{bmatrix} 1 & 0 & 0 \\ 1 & 3 & 0 \\ 1 & 4 & -3 \end{bmatrix} \begin{bmatrix} 1 & -1 & -2 \\ 0 & 1 & 1 \\ 0 & 0 & 1 \end{bmatrix}.$$

If A is a symmetric matrix, $A^T = A$ where by A^T (the *transpose* of A) we mean the matrix whose elements are given by

$$(A^T)_{ji} = a_{ij} \quad i,j = 1, 2, \ldots, n \ .$$

Now if A is symmetric and satisfies the conditions of Theorem 3.2 it can be shown (see Phillips and Taylor, 1973) that A possesses a unique factorisation $A = LDL^T$, where L is unit lower triangular and D is diagonal with non-zero diagonal elements. Suppose E is the diagonal matrix with elements $e_i = \sqrt{d_i}$: $i = 1, 2, \ldots, n$. Then $D = E^2$ and

$$A = LEEL^T = LEE^TL^T = (LE)(LE)^T \ .$$

Now, LE is a lower triangular matrix so a unique factorisation of a symmetric matrix into a lower triangular matrix and its transpose exists. We note that if any of the d_i are negative, the corresponding e_i will be a complex number with only the imaginary part non-zero. Consequently some of the elements of the matrix LE will also be imaginary numbers. In contrast if all the d_i are positive, then each element of LE will be real. (The condition $d_i > 0: i = 1, 2, \ldots, n$ holds if A is *positive-definite*; that is, if $\mathbf{x}^T A \mathbf{x} > 0$ for all non-zero vectors $\mathbf{x}$. For a symmetric matrix, this is equivalent to the condition $\det(A_k) > 0$ where A_k is defined in Theorem 3.2. See Phillips and Taylor (1973) for further details.)

EXAMPLE 3.11
Consider the symmetric matrix

$$A = \begin{bmatrix} 1 & 2 & 1 \\ 2 & 6 & 4 \\ 1 & 4 & 5 \end{bmatrix}.$$

First form the LDU factorisation of A (with $U = L^T$).

$d_1 = a_{11} = 1$
$u_{12} = a_{12}/d_1 = 2/1 = 2 = l_{21}$
$u_{13} = a_{13}/d_1 = 1/1 = 1 = l_{31}$
$d_2 = a_{22} - l_{21} d_1 u_{12} = 6 - 2 \times 1 \times 2 = 2$
$u_{23} = (a_{23} - l_{21} d_1 u_{13})/d_2 = (4 - 2 \times 1 \times 1)/2 = 1 = l_{32}$
$d_3 = a_{33} - l_{31} d_1 u_{13} - l_{32} d_2 u_{23} = 5 - 1 \times 1 \times 1 - 1 \times 2 \times 1 = 2$.

Hence

$$A = \begin{bmatrix} 1 & 0 & 0 \\ 2 & 1 & 0 \\ 1 & 1 & 1 \end{bmatrix} \begin{bmatrix} 1 & 0 & 0 \\ 0 & 2 & 0 \\ 0 & 0 & 2 \end{bmatrix} \begin{bmatrix} 1 & 2 & 1 \\ 0 & 1 & 1 \\ 0 & 0 & 1 \end{bmatrix}.$$

Now split the matrix D so that

$$A = \begin{bmatrix} 1 & 0 & 0 \\ 2 & 1 & 0 \\ 1 & 1 & 1 \end{bmatrix} \begin{bmatrix} 1 & 0 & 0 \\ 0 & \sqrt{2} & 0 \\ 0 & 0 & \sqrt{2} \end{bmatrix} \begin{bmatrix} 1 & 0 & 0 \\ 0 & \sqrt{2} & 0 \\ 0 & 0 & \sqrt{2} \end{bmatrix} \begin{bmatrix} 1 & 2 & 1 \\ 0 & 1 & 1 \\ 0 & 0 & 1 \end{bmatrix}$$

and then form LE and $E^T L^T$ to give

$$A = \begin{bmatrix} 1 & 0 & 0 \\ 2 & \sqrt{2} & 0 \\ 1 & \sqrt{2} & \sqrt{2} \end{bmatrix} \begin{bmatrix} 1 & 2 & 1 \\ 0 & \sqrt{2} & \sqrt{2} \\ 0 & 0 & \sqrt{2} \end{bmatrix}.$$

So far we have looked at the factorisation (or *decomposition*) of a matrix from an abstract viewpoint only. We now show that the above results are of practical use in solving a system of linear simultaneous equations.

3.6.2 The methods of Doolittle, Crout and Choleski

In the previous subsection we observed that if the matrix A satisfies the conditions of Theorem 3.2 a unique resolution $A = LU$ exists where L is unit lower triangular and U is upper triangular. If A has three rows and three columns we have

$$\begin{bmatrix} 1 & 0 & 0 \\ l_{21} & 1 & 0 \\ l_{31} & l_{32} & 1 \end{bmatrix} \begin{bmatrix} u_{11} & u_{12} & u_{13} \\ 0 & u_{22} & u_{23} \\ 0 & 0 & u_{33} \end{bmatrix} = \begin{bmatrix} a_{11} & a_{12} & a_{13} \\ a_{21} & a_{22} & a_{23} \\ a_{31} & a_{32} & a_{33} \end{bmatrix} . \quad (3.34)$$

The entries in L and U can either be found by forming the LDU decomposition of A (where, here, U is unit upper triangular) and computing DU or directly by multiplying out the left-hand side of (3.34) and equating corresponding elements in the equation to find u_{11}, u_{12}, u_{13}, l_{21}, u_{22}, u_{23}, l_{31}, l_{32} and u_{33} in that order.

Given that the LU decomposition of a matrix A exists, the system of equations (3.1) may be written as

$$LU\mathbf{x} = \mathbf{b}$$

or

$$L\mathbf{y} = \mathbf{b} \quad (3.35)$$

where $\mathbf{y}$ satisfies the equation

$$U\mathbf{x} = \mathbf{y} . \quad (3.36)$$

Hence, the solution vector $\mathbf{x}$ may be found by factorising A into the product LU and then

(i) forming the intermediate vector $\mathbf{y}$ from (3.35) using forward substitution,
(ii) forming $\mathbf{x}$ from (3.36) using back-substitution.

126 LINEAR SIMULTANEOUS EQUATIONS   [Ch. 3

In detail, for a system of three equations in three unknowns we have, for stage (i)

$$\begin{bmatrix} 1 & 0 & 0 \\ l_{21} & 1 & 0 \\ l_{31} & l_{32} & 1 \end{bmatrix} \begin{bmatrix} y_1 \\ y_2 \\ y_3 \end{bmatrix} = \begin{bmatrix} b_1 \\ b_2 \\ b_3 \end{bmatrix}$$

so that

$$y_1 = b_1$$
$$l_{21}y_1 + y_2 = b_2$$
$$l_{31}y_1 + l_{32}y_2 + y_3 = b_3$$

and y_1, y_2 and y_3 can be found in turn. For stage (ii), we have

$$\begin{bmatrix} u_{11} & u_{12} & u_{13} \\ 0 & u_{22} & u_{23} \\ 0 & 0 & u_{33} \end{bmatrix} \begin{bmatrix} x_1 \\ x_2 \\ x_3 \end{bmatrix} = \begin{bmatrix} y_1 \\ y_2 \\ y_3 \end{bmatrix}$$

giving

$$u_{33}x_3 = y_3$$
$$u_{22}x_2 + u_{23}x_3 = y_2$$
$$u_{11}x_1 + u_{12}x_2 + u_{13}x_3 = y_1$$

and now x_3, x_2 and x_1 can be found in turn.

The process based on the LU factorisation of A and the subsequent solution of equations (3.35) and (3.36) is known as *Doolittle's* method. Formally, for a system of n equations in n unknowns we have

$$u_{1j} = a_{1j} \qquad\qquad j = 1, 2, \ldots, n$$
$$l_{i1} = a_{i1}/u_{11} \qquad\qquad i = 2, 3, \ldots, n$$

$$\left. \begin{aligned} u_{ij} &= a_{ij} - \sum_{k=1}^{i-1} l_{ik} u_{kj} \qquad j = i, i+1, \ldots, n \\[1em] l_{ji} &= \left(a_{ji} - \sum_{k=1}^{i-1} l_{jk} u_{ki} \right)/u_{ii} \quad j = i+1, i+2, \ldots, n \end{aligned} \right\} \; i = 2, 3, \ldots, n$$

and then

$$y_i = b_i - \sum_{j=1}^{i-1} l_{ij} y_j \quad i = 1, 2, \ldots, n \quad (3.37)$$

$$x_i = \left(y_i - \sum_{j=i+1}^{n} u_{ij} x_j\right)/u_{ii} \quad i = n, n-1, \ldots, 1. \quad (3.38)$$

EXAMPLE 3.12
In Example 3.10 we saw that the coefficient matrix of Example 3.1 may be expressed as

$$\begin{bmatrix} 1 & -1 & -2 \\ 1 & 2 & 1 \\ 1 & 3 & -1 \end{bmatrix} = \begin{bmatrix} 1 & 0 & 0 \\ 1 & 1 & 0 \\ 1 & 4/3 & 1 \end{bmatrix} \begin{bmatrix} 1 & -1 & -2 \\ 0 & 3 & 3 \\ 0 & 0 & -3 \end{bmatrix}.$$

Hence, if $\mathbf{b}^T = (-5, 7, 2)$

$y_1 = -5$

$y_2 = 7 - 1 \times (-5) = 12$

$y_3 = 2 - 1 \times (-5) - 4/3 \times 12 = -9$

and

$x_3 = -9/-3 = 3$

$x_2 = (12 - 3 \times 3)/3 = 1$

$x_1 = (-5 - (-2) \times 3 - (-1) \times 1)/1 = 2$.

There are three observations which must be made. The first is that there is a close connection between the Doolittle method as we have described it and the basic Gauss elimination process described in subsection 3.3.1. The upper triangular matrix U is the matrix of coefficients defined by (3.14) and the off-diagonal components of the matrix L are the multipliers appearing in the same equation. In fact, we have

$$l_{i,k-1} = \frac{a^{(k-1)}_{i,k-1}}{a^{(k-1)}_{k-1,k-1}} \qquad i = k, k+1, \ldots, n; \; k = 2, 3, \ldots, n \; .$$

The intermediate vector is the vector defined by equation (3.15). It is not difficult to verify this association in the case of three equations in three unknowns but the general proof is lengthy and we shall not pursue the matter further. The second observation is that it will be necessary to consider some reformulation of the method so that pivoting strategies can be included. The reader is referred to Ralston and Rabinowitz (1978) for details of the way in which permutation matrices are employed to take account of row and column interchanges. Finally, we note that once we have the LU factorisation of a matrix A any problem containing this coefficient matrix may be solved by making use of equations (3.37) and (3.38) only and we refer back to the process of the iterative refinement of an approximate solution described in subsection 3.4.2.

The method of *Crout* can be developed in a similar manner to that of Doolittle. It depends on the resolution $A = LU$, where L is lower triangular and U is unit upper triangular. *Choleski's* method makes use of the resolution $A = LL^T$ for a symmetric matrix, where L is lower triangular. Complete details of both methods may be found in Ralston and Rabinowitz (1978).

Factorisation methods can be of special value when the matrix A has a special form and we give an example of this by considering the case of a tridiagonal matrix.

3.6.3 Factorisation of a tridiagonal matrix
Let A be the tridiagonal matrix

$$A = \begin{bmatrix} a_1 & c_1 & & & & 0 \\ d_2 & a_2 & c_2 & & & \\ & \ldots & \ldots & \ldots & & \\ & & & d_{n-1} & a_{n-1} & c_{n-1} \\ 0 & & & & d_n & a_n \end{bmatrix}$$

with the a_is on the diagonal, the c_is on the co-diagonal immediately above and the d_is on the co-diagonal immediately below. All other entries are zero. We seek a resolution of A of the form $A = LU$ where

$$L = \begin{bmatrix} \alpha_1 & & & & 0 \\ d_2 & \alpha_2 & & & \\ \dots & \dots & & & \\ & & d_{n-1} & \alpha_{n-1} & \\ 0 & & & d_n & \alpha_n \end{bmatrix}, \quad U = \begin{bmatrix} 1 & \gamma_1 & & & 0 \\ & 1 & \gamma_2 & & \\ & & \dots & \dots & \\ & & & 1 & \gamma_{n-1} \\ 0 & & & & 1 \end{bmatrix}.$$

In this definition of the lower triangular matrix L, the zeros indicate that all entries except those on the diagonal and the co-diagonal immediately below are identically zero. Similarly, the upper triangular matrix U has non-zero entries only on the diagonal and the co-diagonal immediately above.

On multiplying out and equating elements we find that

$$\left. \begin{array}{l} \alpha_1 = a_1 \\ \alpha_{i-1}\gamma_{i-1} = c_{i-1} \\ d_i\gamma_{i-1} + \alpha_i = a_i \end{array} \right\} \quad i = 2, 3, \dots, n \tag{3.39}$$

From this system we can find $\alpha_1, \gamma_1, \alpha_2, \gamma_2, \dots$ and so on in turn provided that no $\alpha_i = 0$: $i = 1, 2, \dots, n$ (and this condition is satisfied if A is positive definite). Now, let A be the coefficient matrix in the system (3.1). The solution vector $\mathbf{x}$ may be obtained by first solving

$$L\mathbf{y} = \mathbf{b}$$

for the intermediate vector $\mathbf{y}$ and then solving

$$U\mathbf{x} = \mathbf{y} \ .$$

The forward substitution process for $\mathbf{y}$ is simply

$$\begin{aligned} y_1 &= b_1/\alpha_1 \\ y_i &= (b_i - d_i y_{i-1})/\alpha_i \quad i = 2, 3, \dots, n \end{aligned} \tag{3.40}$$

whilst the back-substitution for $\mathbf{x}$ is

$$\begin{aligned} x_n &= y_n \\ x_i &= y_i - \gamma_i x_{i+1} \quad i = n-1, n-2, \dots, 1 \ . \end{aligned} \tag{3.41}$$

We note that the number of multiplications involved in each of the stages (3.39), (3.40) and (3.41) is proportional to n. This compares very favourably with the basic Gauss elimination process where the operations count is proportional to n^3. In addition, if the tridiagonal matrix is diagonally dominant the method can be shown to be numerically stable, so that no pivoting strategy is required.

EXAMPLE 3.13
Consider the system of equations

$$\begin{bmatrix} 2 & 2 & 0 \\ -4 & -3 & 2 \\ 0 & 1 & 3 \end{bmatrix} \begin{bmatrix} x_1 \\ x_2 \\ x_3 \end{bmatrix} = \begin{bmatrix} 2 \\ 3 \\ 11 \end{bmatrix}$$

in which the coefficient matrix A is tridiagonal. Then, we have

$\alpha_1 = a_1 = 2$

$\gamma_1 = c_1/\alpha_1 = 2/2 = 1$

$\alpha_2 = a_2 - d_2\gamma_1 = -3 - (-4) \times 1 = 1$

$\gamma_2 = c_2/\alpha_2 = 2/1 = 2$

$\alpha_3 = a_3 - d_3\gamma_2 = 3 - 1 \times 2 = 1$

so that

$$A = \begin{bmatrix} 2 & 0 & 0 \\ -4 & 1 & 0 \\ 0 & 1 & 1 \end{bmatrix} \begin{bmatrix} 1 & 1 & 0 \\ 0 & 1 & 2 \\ 0 & 0 & 1 \end{bmatrix}.$$

Using forward substitution we have

$y_1 = 2/2 = 1$

$y_2 = (3 - (-4) \times 1)/1 = 7$

$y_3 = (11 - 1 \times 7)/1 = 4$

and back-substitution yields

$x_3 = 4$

$x_2 = 7 - 2 \times 4 = -1$

$x_1 = 1 - 1 \times (-1) = 2$.

Sec. 3.6] FACTORISATION METHODS

If the matrix A is symmetric as well as tridiagonal, that is

$$A = \begin{bmatrix} a_1 & c_1 & & & & 0 \\ c_1 & a_2 & c_2 & & & \\ \ldots & \ldots & \ldots & & & \\ & & c_{n-2} & a_{n-1} & c_{n-1} \\ 0 & & & c_{n-1} & a_n \end{bmatrix}$$

and the conditions of Theorem 3.2 are satisfied, the factorisation $A = LL^T$ exists where

$$L = \begin{bmatrix} p_1 & & & & 0 \\ q_1 & p_2 & & & \\ \ldots & \ldots & & & \\ & & q_{n-2} & p_{n-1} & \\ 0 & & & q_{n-1} & p_n \end{bmatrix}$$

and

$$p_1 = \sqrt{a_1}$$
$$\left.\begin{array}{l} q_{i-1} = c_{i-1}/p_{i-1} \\ p_i = \sqrt{(a_i - q_{i-1}^2)} \end{array}\right\} \quad i = 2, 3, \ldots n \ .$$

If A is positive definite the values p_i: $i = 1, 2, \ldots, n$ are guaranteed to be real, and not complex, numbers.

A tridiagonal matrix is an example of a *band matrix*. For such matrices, $a_{ij} = 0$: $|i - j| > k$ for some k. The value $2k + 1$ is known as the *band width* of the matrix. Clearly the ideas introduced here may be readily extended to the case $k > 1$.

3.6.4 A program for tridiagonal coefficient matrices

The final program in this chapter implements the solution technique outlined in the previous subsection for a system of linear simultaneous equations in which the coefficient matrix is tridiagonal.

The main part of the program consists of a sequence of calls to predefined and user-defined procedures. These calls describe the algorithm on which the process is based. In turn, they are as follows.

(i) A call of the predefined procedure *read* to read in the size of the system into the variable n.
(ii) A call of *readtridiag* (user-defined) to read in the tridiagonal matrix. The non-zero elements only are read in and these are stored as

described in the comments within the declaration of this procedure. Storing the matrix as three one-dimensional arrays represents a significant saving in terms of storage requirements over the basic Gauss elimination program.

(iii) *read* is used to read in the elements of the right-hand side vector **b**.

(iv) A call of the user-defined procedure *printtridiag* is used to print the tridiagonal matrix in full.

(v) The elements of the right-hand side vector are printed.

(vi) A call of *factorise* (user-defined) to form the LU decomposition of the coefficient matrix. Here the α_is as defined in subsection 3.6.3 overwrite the a_is as they are computed, making further savings in storage requirements. Similarly, the γ_is overwrite the c_is.

(vii) A call of *forwardsub* (user-defined) to form the intermediate vector **y**. The y_is overwrite the b_is as they are computed.

(viii) A call of *backsub* (user-defined) to find the solution vector **x**. The x_is overwrite the y_is as they are computed.

(ix) Predefined printing procedures are called to print a heading and the solution vector.

This program indicates that a careful analysis of the underlying algorithm can lead to a program with significantly smaller storage requirements than a direct implementation would incur. However, this saving is achieved at the expense of an arguable loss of clarity in the final program. Such drawbacks can be overcome, to some extent, if the program is well commented. Further documentation is also useful and although this phase of program development is tedious it should always be given due attention. The documentation given in the NAG manuals is a good example of the standards attainable. In the programs of Case Study 3.4 we have indicated the purpose of each procedure and the effect that each has on its parameters.

CASE STUDY 3.4

Algol 68 Program

```
BEGIN
  COMMENT
  a program for the solution of a system of linear equations in which the
  coefficient matrix is tridiagonal
  COMMENT
  PROC readtridiag = ( REF [ ] REAL a , c , d ) VOID :
  BEGIN
    COMMENT
    this procedure reads a tridiagonal matrix
    on exit
      a         contains the diagonal
      c         contains the co - diagonal immediately above
      d         contains the co - diagonal immediately below
    COMMENT
    INT n = UPB a ;
```

Sec. 3.6]   FACTORISATION METHODS   133

```
      FOR i TO n
      DO
        IF i # 1
        THEN
          read ( d [ i ] )
        FI ;
        read ( a [ i ] ) ;
        IF i # n
        THEN
          read ( c [ i ] )
        FI
      OD
    END ;
    PROC printtridiag = ( REF [ ] REAL a , c , d ) VOID :
    BEGIN
      COMMENT
      this procedure prints a tridiagonal matrix
      on entry
      a       contains the diagonal
      c       contains the co - diagonal immediately above
      d       contains the co - diagonal immediately below
      COMMENT
      INT n = UPB a ;
      FOR i TO n
      DO
        FOR j TO i - 2
        DO
          print ( 0.0 )
        OD ;
        IF i # 1
        THEN
          print ( d [ i ] )
        FI ;
        print ( a [ i ] ) ;
        IF i # n
        THEN
          print ( c [ i ] )
        FI ;
        FOR j FROM i + 2 TO n
        DO
          print ( 0.0 )
        OD ;
        print ( newline )
      OD
    END ;
    PROC factorise = ( REF [ ] REAL a , c , d ) VOID :
    BEGIN
      COMMENT
      this procedure produces the lu factorisation of a triangular matrix
      on entry
      a       contains the diagonal of the tridiagonal matrix
      c       contains the co - diagonal immediately above
      d       contains the co - diagonal immediately below
      on exit
      a       contains the diagonal of the lower triangular factor
      c       contains the co - diagonal immediately above the diagonal of the
              upper triangular factor
      d       remains unchanged
      COMMENT
      FOR i FROM 2 TO UPB a
      DO
        c [ i - 1 ] /:= a [ i - 1 ] ;
        a [ i ] -:= d [ i ] * c [ i - 1 ]
      OD
    END ;
    PROC forwardsub = ( REF [ ] REAL a , d , b ) VOID :
    BEGIN
      COMMENT
      this procedure performs forward substitution
      on entry
      a       contains the diagonal of the lower triangular matrix
      d       contains the co - diagonal immediately below
```

```
        b       contains the right hand sides
      on exit
        a       remains unchanged
        d       remains unchanged
        b       contains the intermediate vector
      COMMENT
      b [ 1 ] := b [ 1 ] / a [ 1 ] ;
      FOR i FROM 2 TO UPB a
      DO
         b [ i ] := ( b [ i ] - d [ i ] * b [ i - 1 ] ) / a [ i ]
      OD
   END ;
   PROC backsub = ( REF [ ] REAL b , c ) VOID :
   BEGIN
      COMMENT
      this procedure performs back - substitution
      on entry
        b       contains the intermediate vector
        c       contains the co - diagonal immediately above the diagonal of the
                upper triangular factor
      on exit
        b       contains the solution vector
        c       remains unchanged
      COMMENT
      FOR i FROM UPB b - 1 BY - 1 TO 1
      DO
         b [ i ] -:= c [ i ] * b [ i + 1 ]
      OD
   END ;
   INT n ; read ( n ) ;
   [ 1 : n ] REAL a , b ;
   [ 1 : n - 1 ] REAL c ;
   [ 2 : n ] REAL d ;
   readtridiag ( a , c , d ) ;
   read ( b ) ;
   print ( ( newline , "the coefficient matrix is" , newline ) ) ;
   printtridiag ( a , c , d ) ;
   print ( ( newline , "the right hand side vector is" , newline , b ,
            newline ) ) ;
   factorise ( a , c , d ) ;
   forwardsub ( a , d , b ) ;
   backsub ( b , c ) ;
   print ( ( newline , "the solution vector is" , newline , b , newline ) )
END
```

Pascal Program

```
PROGRAM tridiagonalmatrices(input, output);

   {  a program for the solution of a system of linear equations
      in which the coefficient matrix is tridiagonal  }

   CONST
      maxn=50;

   TYPE
      indextype=1..maxn;
      vector=ARRAY [ indextype ] OF real;

   VAR
      a, b, c, d:vector;
      i, n:indextype;
```

```
PROCEDURE readtridiag(n:indextype; VAR a, c, d:vector);
    { this procedure reads a tridiagonal matrix
        on exit
        a       contains the diagonal
        c       contains the co - diagonal immediately above
        d       contains the co - diagonal immediately below
    }
    VAR
        i:indextype;
BEGIN
    FOR i:= 1 TO n DO
    BEGIN
        IF i<>1
        THEN
            read(d[i]);
        read(a[i]);
        IF i<>n
        THEN
            read(c[i])
    END
END { readtridiag } ;

PROCEDURE printtridiag(n:indextype; VAR a, c, d:vector);
    { this procedure prints a tridiagonal matrix
        on entry
        a       contains the diagonal
        c       contains the co - diagonal immediately above
        d       contains the co - diagonal immediately below
    }
    VAR
        i, j:indextype;
BEGIN
    FOR i:= 1 TO n DO
    BEGIN
        FOR j:= 1 TO i - 2 DO
            write(0.0);
        IF i<>1
        THEN
            write(d[i]);
        write(a[i]);
        IF i<>n
        THEN
            write(c[i]);
        FOR j:= i + 2 TO n DO
            write(0.0);
        writeln
    END
END { printtridiag } ;

PROCEDURE factorise(n:indextype; VAR a, c, d:vector);
    { this procedure produces the lu factorisation of a triangular matrix
        on entry
        a       contains the diagonal of the tridiagonal matrix
        c       contains the co - diagonal immediately above
        d       contains the co - diagonal immediately below
        on exit
        a       contains the diagonal of the lower triangular factor
        c       contains the co - diagonal immediately above the diagonal
                of the upper triangular factor
        d       remains unchanged
    }
    VAR
        i:indextype;
BEGIN
    FOR i:= 2 TO n DO
    BEGIN
        c[i - 1]:= c[i - 1]/a[i - 1];
        a[i]:= a[i] - d[i]*c[i - 1]
    END
END { factorise } ;
```

```
        PROCEDURE forwardsub(n:indextype; VAR a, d, b:vector);
            { this procedure performs forward substitution
                on entry
                    a       contains the diagonal of the lower triangular matrix
                    d       contains the co - diagonal immediately below
                    b       contains the right hand sides
                on exit
                    a       remains unchanged
                    d       remains unchanged
                    b       contains the intermediate vector
            }
            VAR
                i:indextype;
        BEGIN
            b[1]:= b[1]/a[1];
            FOR i:= 2 TO n DO
                b[i]:= (b[i] - d[i]*b[i - 1])/a[i]
        END { forwardsub } ;

        PROCEDURE backsub(n:indextype; VAR b, c:vector);
            { this procedure performs back - substitution
                on entry
                    b       contains the intermediate vector
                    c       contains the co - diagonal immediately above the diagonal
                            of the upper triangular factor
                on exit
                    b       contains the solution vector
                    c       remains unchanged
            }
            VAR
                i:indextype;
        BEGIN
            FOR i:= n - 1 DOWNTO 1 DO
                b[i]:= b[i] - c[i]*b[i + 1]
        END { backsub } ;

BEGIN
    read(n);
    readtridiag(n, a, c, d);
    FOR i:= 1 TO n DO
        read(b[i]);
    writeln;
    writeln('the coefficient matrix is');
    printtridiag(n, a, c, d);
    writeln;
    writeln('the right hand side vector is');
    FOR i:= 1 TO n DO
        write(b[i]);
    writeln;
    factorise(n, a, c, d);
    forwardsub(n, a, d, b);
    backsub(n, b, c);
    writeln;
    writeln('the solution vector is');
    FOR i:= 1 TO n DO
        write(b[i]);
    writeln
END { tridiagonalmatrices } .
```

Sample Data

```
3
  2.0   2.0
 -4.0  -3.0   2.0
        1.0   3.0
  2.0   3.0  11.0
```

```
Sample Output
the coefficient matrix is
  2.00000e  0  2.00000e  0  0.00000e  0
 -4.00000e  0 -3.00000e  0  2.00000e  0
  0.00000e  0  1.00000e  0  3.00000e  0

the right hand side vector is
  2.00000e  0  3.00000e  0  1.10000e  1

the solution vector is
  2.00000e  0 -1.00000e  0  4.00000e  0
```

3.7 MATRIX INVERSION

Given a matrix A, we occasionally wish to find the inverse A^{-1}. (If A^{-1} exists A is said to be *invertible*.) This is a rare occurrence but in certain statistical methods, for example, it is necessary to know A^{-1} explicitly. We emphasise here, however, that the solution of a set of linear simultaneous equations, which can be written formally as

$$\mathbf{x} = A^{-1}\mathbf{b}$$

does not require the explicit knowledge of A^{-1}. The direct and iterative methods we have discussed are considerably more efficient. If it is necessary to form A^{-1} there are a number of ways that we can do this, one of which was described at the end of subsection 3.3.3.

It is possible to adapt the factorisation methods of the previous section to matrix inversion. If A is factored as $A = LU$ we have

$$A^{-1} = U^{-1}L^{-1} .$$

Now the inverse of a lower (upper) triangular matrix has lower (upper) triangular form and efficient algorithms for the inversion of such matrices can be derived (see Ralston and Rabinowitz, 1978). In particular if A is symmetric, the factorisation $A = LL^T$ can be of use since

$$A^{-1} = (LL^T)^{-1} = (L^T)^{-1}L^{-1} = (L^{-1})^T L^{-1}$$

where we have used the result $(L^T)^{-1} = (L^{-1})^T$. Now all that we need to do is to find L^{-1}.

The inverse of a matrix may also be found using an appropriate iterative process. For example, given an initial approximation X_0 to A^{-1} the iteration

$$X_n = X_{n-1}(2I - AX_{n-1}) \qquad n = 1, 2, \ldots$$

generates a sequence of matrices $\{X_n: n = 0, 1, \ldots\}$ which, in suitable circumstances, converges to A^{-1} (see Stummel and Hainer, 1980).

If the matrix A has a special structure it may be possible to devise useful results for A^{-1}. An important example of this situation is the case where $A = B + \mathbf{u}\mathbf{v}^T$ where $\mathbf{u}$ and $\mathbf{v}$ are each n-dimensional vectors and B^{-1} is known. The result for A^{-1} is known as the *Sherman-Morrison formula* (see Broyden, 1975) and has an important place in the theoretical development of methods for the solution of non-linear simultaneous equations (the quasi-Newton method briefly introduced in subsection 2.9.3) and for non-linear optimisation.

3.8 CONCLUDING REMARKS

We conclude this chapter by considering, in detail, the routines available in the current version (Mark 11) of the NAG library. There are a large number of routines in the relevant chapter (f04), most of which are capable of handling several right-hand sides at once. Some deal with the case in which there are more equations than unknowns and this extension to the basic problem is covered in Chapter 5 of this book.

We consider here only those routines which deal with problems for which the coefficient matrix is square (that is, the same number of rows as columns). We distinguish between routines which calculate an *approximate* solution (using, for example, Crout decomposition followed by forward and backward substitution) and those which calculate an *accurate* solution. In the latter case, iterative refinement is used to improve an initial approximate solution.

Although the underlying numerical methods are essentially the same, the workings, and in particular the user interfaces, of the NAG Fortran/Algol 60 and Algol 68 f04 routines are rather different. Further the same terminology is used to define rather different routine classifications in the respective documentation.

In the Fortran/Algol 60 library the routines are classified as either *black box* or *general purpose*. If one of the former, the routine will actually form the appropriate decomposition before solving, whereas in the latter case the user is expected to supply the appropriate decomposition computed by one of the routines from Chapter f03. Routines exist for problems possessing 'general' real and complex coefficient matrices and use LU decomposition. However, routines are also available for dealing with real symmetric positive definite and real symmetric positive definite banded coefficient matrices; here Choleski (LL^T) factorisation is employed.

In the Algol 68 library the terms black box and general purpose are used to denote a hierarchy. Black box routines allow the user virtually no control over how a solution is computed but general purpose routines permit more interaction. For example, in a general purpose routine the user may be able to specify the accuracy condition for iterative refinement and recover the decomposition on output. Also available are *special purpose* routines which are themselves called from the corresponding black box and general purpose routines. These allow the user considerably more control and, in particular, permit the user to specify that decomposition only is required, or that a decomposition is being supplied, or even that a special routine supplied by the user is to be employed when factorising the coefficient matrix. They make extensive use of the Algol 68 *MODE* declaration facility (considered further in Chapter 6) and also the ability to pass procedures as parameters. Note that whereas the NAG Fortran routines require *workspace* arrays to be passed across, Algol 68's dynamic array declaration facility may be exploited to obviate the need for such parameters. Further, the bounds of these arrays may, in many cases, be determined from those arrays which do need to be passed across using the *LWB* and *UPB* operators. In consequence the parameter lists of some routines is quite short.

Besides the matrix types covered by the Fortran/Algol 60 routines, Algol 68 codes exist for the real symmetric banded case, where LDL^T decomposition is employed.

EXERCISES

3.1 Solve the following system of equations using Gauss elimination (without pivoting) and back-substitution.

$$5x_1 + 3x_2 + 10x_3 = 5$$
$$2x_1 + 0.2x_2 + 3x_3 = -1$$
$$x_1 - 0.4x_2 + 3x_3 = 6 \ .$$

Repeat using full pivoting.

3.2 An approximate solution of the system of equations

$$2x_1 - x_2 + x_3 = 5$$
$$x_1 + x_2 + 3x_3 = 6$$
$$x_1 - 2x_2 - x_3 = 1$$

is $x_1 = 0.9$, $x_2 = -1$ and $x_3 = 2.1$. Obtain the elements of the residual vector and hence set up and solve (using Gauss elimination without pivoting) the equations which define the correction to this approximate solution. Determine a more accurate solution to the original system.

3.3 The system of equations

$$x_1 + 0.1x_2 = 1$$
$$0.1x_1 + x_2 + 0.1x_3 = 2$$
$$ 0.1x_2 + x_3 = 3$$

is strictly diagonally dominant. Solve this system using (i) Jacobi iteration, and (ii) Gauss–Seidel iteration. In each case use the criterion (3.28) to test for convergence with a value for ε of 0.0005. Use the starting value $\mathbf{x}^{[0]T} = (1.0, 1.0, 1.0)$.

3.4 A solution to the following system of equations is required.

$$-x_1 + 5x_2 + x_4 = 2$$
$$ - x_2 - 2x_3 + 4x_4 = 1$$
$$8x_1 - x_2 + x_3 = 1$$
$$x_1 + 4x_3 - 2x_4 = 0 \ .$$

Reorder these equations to obtain a system for which the coefficient matrix is strictly diagonally dominant. Starting with $\mathbf{x}^{[0]T} = (0,0,0,0)$ perform five iterations of Jacobi and Gauss–Seidel iteration on the transformed system. Using the measure (3.25) comment on the relative performance of the two schemes.

3.5 Decompose the matrix

$$A = \begin{bmatrix} 1 & 2 & 3 \\ 2 & 8 & 7 \\ 2 & 16 & 12 \end{bmatrix}$$

into the product LU, where L is unit lower triangular and U is upper triangular. Hence find the solution to the system of equations

$$x_1 + 2x_2 + 3x_3 = 2$$
$$2x_1 + 8x_2 + 7x_3 = 7$$
$$2x_1 + 16x_2 + 12x_3 = 10 \ .$$

3.6 Decompose the tridiagonal matrix

$$A = \begin{bmatrix} 2 & 2 & 0 \\ 2 & 4 & -2 \\ 0 & -2 & 3 \end{bmatrix}$$

into the product LU, where L has non-zero entries only on the diagonal and the co-diagonal immediately below and U has ones on the diagonal and zeros everywhere else except for the co-diagonal immediately above. Hence find A^{-1}.

3.7 Modify the program of Case Study 3.2 by writing a procedure *invert* which forms the inverse of a matrix. Amend the program so that it forms the inverse of a matrix read from data. Check your code with the matrix

$$A = \begin{bmatrix} 2 & 2 & -2 \\ 2 & -4 & 6 \\ -6 & 4 & 7 \end{bmatrix}.$$

You should find that

$$A^{-1} \approx \begin{bmatrix} 0.30 & 0.13 & -0.02 \\ 0.29 & -0.01 & 0.09 \\ 0.09 & 0.12 & 0.07 \end{bmatrix}.$$

3.8 Add code to that of Case Study 3.3 to test for strict diagonal dominance of the coefficient matrix. Further modify your program by employing the measure (3.26) to test for convergence of the iterates. Use your program to verify that the coefficient matrix in the system

$$-8x_1 - 3x_2 + x_3 = 15$$
$$2x_1 + 9x_2 - 6x_3 = -38$$
$$7x_1 + 8x_2 + 20x_3 = 37$$

is strictly diagonally dominant and find the solution to this system using $\mathbf{x}^{[0]T} = (0,0,0)$. Use the program to find the solution generated by Jacobi iteration given the same starting values.

3.9 Write a program which implements successive over-relaxation (equation 3.29). With values of p equal to 1.0, 0.5 and 0.94 use your code to find a solution to the system

$$12.3x_1 + 1.2x_2 - 3.5x_3 = 13.2$$
$$-2.2x_1 + 10.4x_2 - 1.4x_3 = 4.3$$
$$3.9x_1 - 1.2x_2 + 8.5x_3 = -9.7$$

using the measure (3.25) to test for convergence of the iterates and $x^{[0]T} = (0,0,0)$. Compare the rates of convergence of your results.

3.10 Write a program which forms the LL^T factorisation of a tridiagonal positive definite symmetric matrix. Use your program to compute a solution to the system of equations

$$2x_1 + 2x_2 = 6$$
$$2x_1 + 4x_2 - 2x_3 = 12$$
$$ -2x_2 + 3x_3 = -8 \; .$$

4

Approximation of Continuous Functions

4.1 INTRODUCTION

This chapter looks at the approximation of a continuous function over finite and infinite intervals. There are often very good reasons for wishing to do this, not least of which is that the function may be difficult, and hence expensive, to evaluate. If it were possible to replace such a function accurately with, say, a low degree polynomial then the situation would be somewhat improved.

There are two distinct ways of approaching approximation. First, the function could be evaluated at a set of points and the problem is then transformed to fitting an approximating function to a discrete set of data values. Second, an overall approximation could be sought and it is this class of problems that concerns us here. The wider implications of discrete approximation are considered, in detail, in Chapter 5.

Quite unashamedly most of the worked examples in this chapter are concerned with the approximation of the exponential function e^x, usually on the interval [0,1]. This choice is governed by the fact that the exponential function possesses a number of desirable properties which make it an ideal candidate for illustrating the main features of the methods discussed. In particular the mathematics involved (for example, differentiation) can be kept reasonably simple. The reader is encouraged to compare different solutions and to try out the methods on other functions.

4.2 TAYLOR SERIES APPROXIMATION

Perhaps the simplest way of obtaining an approximation to a function $f(x)$ is provided by its Taylor series expansion. Suppose that we are interested in the behaviour of $f(x)$ over the finite interval $[a,b]$ and let $x_0 \in [a,b]$ be some chosen point. Then Taylor's theorem (Theorem 1.1) gives

$$f(x) = f(x_0) + (x - x_0)f'(x_0) + \frac{(x - x_0)^2}{2} f''(x_0)$$

$$+ \ldots + \frac{(x - x_0)^n}{n!} f^{(n)}(x_0) + \frac{(x - x_0)^{n+1}}{(n+1)!} f^{(n+1)}(\xi) \quad (4.1)$$

where ξ is some interior point of the open interval $]x_0,x[$ (if $x > x_0$) or $]x,x_0[$ (if $x < x_0$). If this series is convergent then, provided that n is large enough, the polynomial

$$p_n(x) = f(x_0) + (x - x_0) f'(x_0) + \frac{(x - x_0)^2}{2} f''(x_0)$$

$$+ \ldots + \frac{(x - x_0)^n}{n!} f^{(n)}(x_0) \quad (4.2)$$

which is of degree at most n ($f^{(n)}(x_0)$ may be zero) can be used as an approximation to $f(x)$.

EXAMPLE 4.1
Consider the function $f(x) = e^x$ and let $[0,1]$ be the interval of interest. Since each derivative of e^x is e^x itself, the Taylor series for this function expanded about the point $x = 0$ gives

$$e^x = e^0 + xe^0 + \frac{x^2}{2} e^0 + \frac{x^3}{6} e^0 + \frac{x^4}{24} e^\xi \quad \xi \in]0,x[.$$

Hence

$$p_3(x) = 1 + x + \frac{x^2}{2} + \frac{x^3}{6}$$

gives a polynomial approximation of degree 3 to e^x.

Having obtained a polynomial approximation we would like to know how good it is. From equations (4.1) and (4.2) we have that the error $E_n(x)$ in the approximation of $f(x)$ by $p_n(x)$ is given by

$$E_n(x) = f(x) - p_n(x) = \frac{(x - x_0)^{n+1}}{(n+1)!} f^{(n+1)}(\xi). \quad (4.3)$$

On its own, the error function defined by (4.3) does not provide us with much information since we do not know the value of ξ. However, in principle, provided that successive derivatives of $f(x)$ do not grow at too fast a rate then, by increasing n, a point will be reached where the denominator in (4.3) will begin to dominate. (But there can be severe computational problems; see Example 4.3). This does mean that we need to be able to compute the appropriate derivative values which is likely to be a tedious operation (although it is possible that the derivatives will satisfy a recurrence relation). In addition algebraic errors can easily creep in when forming derivatives.

In function approximation we often look not for the actual error in approximation, which is usually not computable, but for an upper bound on this quantity. We shall see in Example 4.2 that such bounds are often over-pessimistic and should be treated with some caution. Nevertheless they do allow us to use the underlying numerical methods with some degree of confidence. Here, suppose that we can bound the $n+1$'st derivative of $f(x)$ on the interval $[a,b]$, that is, we can determine a value M_{n+1} such that

$$|f^{(n+1)}(x)| < M_{n+1} \quad x \in [a,b] \ .$$

Then $|x - x_0| \leq b - a$ so that

$$|E_n(x)| < \frac{(b-a)^{n+1}}{(n+1)!} M_{n+1}$$

provides a bound on the error in the approximation of $f(x)$ by $p_n(x)$.

EXAMPLE 4.2
For the approximation of e^x by the polynomial $1 + x + (x^2/2) + (x^3/6)$ (Example 4.1) we have

$$E_3(x) = \frac{x^4}{24} e^\xi$$

which is everywhere positive in $[0,1]$. We can therefore draw the qualitative conclusion that the polynomial approximation will be less than e^x for all $x \in [0,1]$. Further,

$$\max_{0 \leq x \leq 1} |e^x| = e \simeq 2.71828$$

and so

$$|E_3(x)| \leq \frac{1}{24} e \simeq 0.11326 \ . \tag{4.4}$$

In Table 4.1 we list some actual errors which support the above conclusions. Clearly, for small values of x the error bound (4.4) is a gross overestimate of the actual error incurred but it does at least guarantee that the error will be no greater than this quantity.

It is possible to use Taylor series approximations in a computer program to evaluate standard functions. We add successive terms until a point is reached at which all subsequent terms are small enough to be ignored. However, care must be taken as the next example shows.

Table 4.1

x	$f(x) = e^x$	$p_3(x) = 1 + x + x^2/2 + x^3/6$	$E_3(x) = f(x) - p_3(x)$
0	1.000 00	1.000 00	0.000 00
$\frac{1}{4}$	1.284 03	1.283 85	0.000 17
$\frac{1}{2}$	1.648 72	1.645 83	0.002 89
$\frac{3}{4}$	2.117 00	2.101 56	0.015 44
1	2.718 28	2.666 67	0.051 62

EXAMPLE 4.3
Suppose that we wish to determine an approximation to e^{-x} valid on $[0,10]$. Expanding about the point 0 we have

$$e^{-x} = 1 - x + \frac{x^2}{2} - \frac{x^3}{6} + \frac{x^4}{24} - \ldots \tag{4.5}$$

which is a convergent power series for all x. Now

$$e^{-10} = 4.53999 \times 10^{-5}$$

to six significant figures. Let us see what happens if we try to evaluate the Taylor series for $x = 10$, accumulating terms until the first one whose absolute value is less than 5×10^{-8}, say, is encountered. If we use T_i to denote the ith term then

$$T_i = \frac{x^i}{i!}(-1)^i \qquad i = 0, 1, \ldots \tag{4.6}$$

and we note that the recurrence relation

$$T_i = -\frac{x}{i} T_{i-1} \qquad i = 1, 2, \ldots \qquad (4.7)$$

holds with $T_0 = 1$. Using (4.7) to compute the individual terms instead of (4.6) will help to keep the operations count down. As soon as $i > 10$ the terms start to decrease in value. However, in order to reach such a stage we need to compute terms such as

$$T_3 = -\frac{10^3}{6} = -\frac{1000}{6}$$

and, to six significant figures, $T_3 = -166.667$. The rounding error here, 0.3×10^{-3}, is greater than the value we are trying to compute and hence if we work to this accuracy the result obtained will be completely meaningless. (If it were possible to work to greater precision we would find that the very large terms which occur early on in the series cancel each other out.) These difficulties may be overcome by either

(a) writing e^{-10} as $(e^{-1})^{10}$, evaluating (4.5) for $x = 1$ and raising the result to the tenth power, or
(b) writing e^{-10} as $1/e^{10}$, using the Taylor series expansion for e^x with $x = 10$ and then forming the reciprocal of the result.

Before leaving this section we observe that with Taylor series approximations we are effectively solving the problem of finding the polynomial of degree at most n whose first n derivatives agree with those of $f(x)$ at the point $x = x_0$. For example, from (4.2) we have

$$\frac{d^2}{dx^2} p_n(x) = f''(x_0) + (x - x_0) f'''(x_0) + \ldots + \frac{(x - x_0)^{n-2}}{(n-2)!} f^{(n)}(x_0)$$

and hence

$$\frac{d^2}{dx^2} p_n(x)\big|_{x=x_0} = f''(x_0) .$$

We shall make use of this fact in section 4.8 when considering certain types of rational approximation.

4.3 LEAST SQUARES APPROXIMATION

4.3.1 Introduction to best approximation

One of the most important concepts that we are concerned with in this chapter is the idea of a *best* approximation, that is, an approximating function which is, in some sense, the best of all of those in its class. For example, there are an infinite number of straight lines which we might choose to take as an approximation to a given function and we need some criterion for deciding which to accept. Let $f(x)$ be a continuous function on the finite interval $[a,b]$. Then we wish to determine values of α_1 (the intercept along the y-axis) and α_2 (the slope) which ensure that

$$p_1(x) = \alpha_1 + \alpha_2 x$$

is a good approximation to $f(x)$ everywhere in $[a,b]$. We need, therefore, to formulate a condition for deciding which are the optimal values, α_1^* and α_2^*, of the parameters α_1 and α_2; that is, we need to define a *measure* such that, with respect to it,

$$p_1^*(x) = \alpha_1^* + \alpha_2^* x \tag{4.8}$$

is the best possible straight line approximation to $f(x)$.

Consider the *residual function*

$$r(x) = f(x) - p_1(x).$$

Then, as a measure of the error in the approximation of $f(x)$ by $p_1(x)$ we could use

$$\underset{a \leq x \leq b}{\text{maximum}} |r(x)| \tag{4.9}$$

that is, the maximum absolute discrepancy between $f(x)$ and $p_1(x)$ in $[a,b]$. The best approximation (4.8) is therefore defined to be that which gives the minimum value of (4.9); that is, we aim to minimise (4.9) with respect to the intercept and slope. Clearly the optimal values α_1^* and α_2^* will be such that

$$\underset{a \leq x \leq b}{\text{maximum}} |f(x) - \alpha_1^* - \alpha_2^* x| \leq \underset{a \leq x \leq b}{\text{maximum}} |f(x) - \alpha_1 - \alpha_2 x|$$

for all values of α_1 and α_2. (There is a direct relationship between the measure (4.9) and the ∞ − norm of a vector (equation (3.27)). We do not pursue the details here but remark that (4.9) is known as the ∞ − *norm* of the residual function. This link will become clearer in Chapter 5.)

The measure (4.9) is not the only one that can be used and an alternative is

$$\int_a^b (r(x))^2 \, dx \ . \tag{4.10}$$

The optimal values of α_1 and α_2 which minimise (4.10) define the *least squares* straight line approximation to $f(x)$. Throughout this and the next two sections it is (4.10) which provides the measure on which our numerical methods are based. The use of (4.9) and other measures is discussed in section 4.7.

4.3.2 Least squares

In the previous subsection we considered the approximation of a continuous function $f(x)$ on a finite interval $[a,b]$ by a straight line. In the general case we seek values of the coefficients $\alpha_1, \alpha_2, \ldots, \alpha_n$ which ensure that

$$p_{n-1}(x) = \alpha_1 + \alpha_2 x + \ldots + \alpha_n x^{n-1}$$

$$= \sum_{j=1}^{n} \alpha_j x^{j-1}$$

is the best polynomial approximation of degree $n-1$. Using the least squares criterion this means that we aim to find values $\alpha_1^*, \alpha_2^*, \ldots, \alpha_n^*$ which give

$$\underset{\alpha_1, \alpha_2, \ldots, \alpha_n}{\text{minimum}} \int_a^b (f(x) - p_{n-1}(x))^2 dx.$$

Let $\boldsymbol{\alpha}^T = (\alpha_1, \alpha_2, \ldots, \alpha_n)$ and define

$$I(\boldsymbol{\alpha}) = \int_a^b (f(x) - p_{n-1}(x))^2 dx \ . \tag{4.11}$$

Then we wish to minimise $I(\boldsymbol{\alpha})$ with respect to the parameters $\alpha_1, \alpha_2, \ldots, \alpha_n$. We recall that for a function of one real variable, at a stationary point (that is, a maximum or minimum) the tangent (that is, the derivative) is equal to zero. For a function of more than one real variable the situation is similar except that now we have to talk in terms of partial, instead of full, derivatives. The necessary conditions are

$$\frac{\partial}{\partial \alpha_i} I(\boldsymbol{\alpha}) = 0 \quad i = 1, 2, \ldots, n. \tag{4.12}$$

The solution $\boldsymbol{\alpha}^*$ of (4.12) gives a turning point of $I(\boldsymbol{\alpha})$ and it is fairly easy to show (we need to look at second derivatives also) that we actually get a minimum. For simplicity, from here on, we drop the asterisk attached to the best approximation.

In order to see what the equations defining the least squares polynomial approximation look like, we return to the approximation of $f(x)$ by a straight line. Here $n = 2$ and we need to find α_1 and α_2 such that

$$\frac{\partial}{\partial \alpha_1} \int_a^b (f(x) - (\alpha_1 + \alpha_2 x))^2 \, dx = 0$$

$$\frac{\partial}{\partial \alpha_2} \int_a^b (f(x) - (\alpha_1 + \alpha_2 x))^2 \, dx = 0$$

that is,

$$-2 \int_a^b (f(x) - (\alpha_1 + \alpha_2 x)) \, dx = 0$$

$$-2 \int_a^b (f(x) - (\alpha_1 + \alpha_2 x)) x \, dx = 0.$$

Rearranging these equations and dividing by 2 we obtain

$$\alpha_1 \int_a^b 1 \, dx + \alpha_2 \int_a^b x \, dx = \int_a^b f(x) \, dx$$

$$\alpha_1 \int_a^b x \, dx + \alpha_2 \int_a^b x^2 \, dx = \int_a^b x f(x) \, dx$$

which is a system of two linear simultaneous equations in the two unknowns α_1 and α_2.

In the general case we need to solve the system of equations

$$0 = \frac{\partial}{\partial \alpha_i} I(\boldsymbol{\alpha})$$

$$= -2 \int_a^b \left[f(x) - \sum_{j=1}^n \alpha_j x^{j-1} \right] x^{i-1} \, dx \qquad i = 1, 2, \ldots, n$$

and so, rearranging, we get

$$\sum_{j=1}^{n} \alpha_j \int_a^b x^{i-1} x^{j-1} \, dx = \int_a^b x^{i-1} f(x) \, dx \qquad i = 1, 2, \ldots, n \quad (4.13)$$

or, in matrix form,

$$A\alpha = b \qquad (4.14)$$

where

$$b_i = \int_a^b x^{i-1} f(x) \, dx \qquad i = 1, 2, \ldots, n \qquad (4.15)$$

$$A_{ij} = \int_a^b x^{i-1} x^{j-1} \, dx \qquad i, j = 1, 2, \ldots, n. \qquad (4.16)$$

Usually it will be possible to evaluate the integrals (4.15) using integration by parts. The entries (4.16) in the coefficient matrix can be easily found since

$$A_{ij} = \left[\frac{x^{i+j-1}}{i+j-1} \right]_a^b = \frac{b^{i+j-1} - a^{i+j-1}}{i+j-1} \qquad i, j = 1, 2, \ldots, n. \qquad (4.17)$$

The equations (4.13) are known as the *normal equations*. Having set up the matrix equation the solution to the least squares approximation problem may be obtained using, for example, Gauss elimination (although, as we shall see shortly, the situation is not quite as straightforward as it appears at first sight). We observe that the coefficient matrix A is symmetric and this property should be borne in mind if a factorisation method is used (see section 3.6).

EXAMPLE 4.4
In this example we consider the approximation of the function e^x on the interval $[0,1]$ by a quadratic using the least squares criterion. To find the optimal values of the coefficients in

$$p_2(x) = \alpha_1 + \alpha_2 x + \alpha_3 x^2$$

we have to set up a system of three equations in three unknowns in which the coefficient matrix is, from (4.17), given by

$$A_{ij} = \frac{1}{i+j-1} \qquad i, j = 1, 2, 3 \qquad (4.18)$$

(since $a = 0$ and $b = 1$). To obtain the elements of the right-hand side vector we note that

$$b_1 = \int_0^1 e^x \, dx = e - 1 \tag{4.19}$$

and

$$b_2 = \int_0^1 xe^x \, dx = [xe^x]_0^1 - \int_0^1 e^x \, dx = e - b_1 = 1 \tag{4.20}$$

$$b_3 = \int_0^1 x^2 e^x \, dx = [x^2 e^x]_0^1 - 2\int_0^1 xe^x \, dx = e - 2b_2 = e - 2. \tag{4.21}$$

Hence, the system of normal equations for the unknowns α_1, α_2 and α_3 takes the form

$$\begin{bmatrix} 1 & 1/2 & 1/3 \\ 1/2 & 1/3 & 1/4 \\ 1/3 & 1/4 & 1/5 \end{bmatrix} \begin{bmatrix} \alpha_1 \\ \alpha_2 \\ \alpha_3 \end{bmatrix} = \begin{bmatrix} e - 1 \\ 1 \\ e - 2 \end{bmatrix}.$$

Eliminating α_1 from the second and third equations gives

$$\begin{bmatrix} 1 & 1/2 & 1/3 \\ 0 & 1/12 & 1/12 \\ 0 & 1/12 & 4/45 \end{bmatrix} \begin{bmatrix} \alpha_1 \\ \alpha_2 \\ \alpha_3 \end{bmatrix} = \begin{bmatrix} e - 1 \\ \frac{1}{2}(-e + 3) \\ \frac{1}{3}(2e - 5) \end{bmatrix}.$$

Finally, eliminating α_2 from the third equation, we have

$$\begin{bmatrix} 1 & 1/2 & 1/3 \\ 0 & 1/12 & 1/12 \\ 0 & 0 & 1/180 \end{bmatrix} \begin{bmatrix} \alpha_1 \\ \alpha_2 \\ \alpha_3 \end{bmatrix} = \begin{bmatrix} e - 1 \\ \frac{1}{2}(-e + 3) \\ \frac{1}{6}(7e - 19) \end{bmatrix}.$$

Now, using back-substitution, we obtain

$$\alpha_3 = \tfrac{180}{6}(7e-19) = 30(7e-19) \simeq 0.83918$$
$$\alpha_2 = 12(\tfrac{1}{2}(-e+3) - \tfrac{1}{12}\alpha_3) = -12(18e-49) \simeq 0.85113$$
$$\alpha_1 = e - 1 - \tfrac{1}{2}\alpha_2 - \tfrac{1}{3}\alpha_3 = 3(13e-35) \simeq 1.01299$$

so that, to five decimal place accuracy, the least squares quadratic approximation to e^x on $[0,1]$ is

$$p_2(x) = 1.01299 + 0.85113x + 0.83918x^2.$$

4.3.3 A program for least squares approximation

A program for least squares polynomial approximation is quite easy to construct; all that we need to do is to set up and solve the matrix equation (4.14). In the program of Case Study 4.1 we give code for the least squares approximation of e^x on the interval $[0,1]$. From (4.18) we have that the coefficient matrix is given by

$$A = \begin{bmatrix} 1 & 1/2 & 1/3 & \cdots & 1/n \\ 1/2 & 1/3 & 1/4 & \cdots & 1/(n+1) \\ 1/3 & 1/4 & 1/5 & \cdots & 1/(n+2) \\ \cdot & \cdot & \cdot & & \cdot \\ \cdot & \cdot & \cdot & & \cdot \\ \cdot & \cdot & \cdot & & \cdot \\ 1/n & 1/(n+1) & 1/(n+2) & \cdots & 1/(2n-1) \end{bmatrix} \quad (4.22)$$

where $n - 1$ is the order of the polynomial approximation (and n is the value given in the data for the program). The components of the right-hand side vector are also easy to compute. A generalisation of equations (4.20) and (4.21) gives

$$b_i = e - (i-1)b_{i-1} \qquad i = 2,3,\ldots \quad (4.23)$$

where b_1 is equal to $e - 1$ (see equation (4.19)). For reasons which will become apparent in the next subsection, we have chosen to solve the same problem 5 times with b_1 being correct to a different number of decimal places each time. Thus, instead of just having one right-hand side vector, we have a matrix of 5 right-hand sides. The first row of the matrix contains the values of b_1 correct to 1, 2,..., 5 decimal places. This can be expressed in Algol 68 by:

$$b\,[\,1\,,\,]\,:=\,(\,1.7\,,\,1.72\,,\,1.718\,,\,1.7183\,,\,1.71828\,)\,;$$

The rounding error in these values of b_1 is then propagated to the other elements of the right-hand side vectors by the code that implements equation (4.23).

To solve the matrix equation we have made use of the program of Case Study 3.2 which implements Gauss elimination for problems with several right-hand sides. The bodies of the procedures *gausselim* and *backsub* have been omitted from the listing of Case Study 4.1 so the reader should refer to Case Study 3.2 for the details.

CASE STUDY 4.1

Algol 68 Program

```
BEGIN
  COMMENT
  a program for the least squares approximation of exp(x) on [0,1] by a
  polynomial
  COMMENT
  OP * = ( [ , ] REAL a , [ ] REAL x ) [ ] REAL :
  BEGIN
    COMMENT
    an operator to form a matrix-vector product
    COMMENT
    INT n = UPB x ; [ 1 : n ] REAL result ; REAL sum ;
    FOR i TO n
    DO
      sum := 0.0 ;
      FOR j TO n
      DO
        sum +:= a [ i , j ] * x [ j ]
      OD ;
      result [ i ] := sum
    OD ;
    result
  END ;
  OP - = ( [ ] REAL a , b ) [ ] REAL :
  BEGIN
    COMMENT
    an operator to form the difference of two vectors
    COMMENT
    INT l = LWB a , u = UPB a ;
    [ l : u ] REAL result ;
    FOR i FROM l TO u
    DO
      result [ i ] := a [ i ] - b [ i ]
    OD ;
    result
  END ;
  PROC gausselim = ( REF [ , ] REAL a , b , REF [ ] INT rownum ) VOID :
  BEGIN
  COMMENT
  this procedure performs the elimination process
  we omit the details
  COMMENT
  END ;
  PROC backsub = ( REF [ , ] REAL a , x , b , REF [ ] INT rownum ) VOID :
  BEGIN
  COMMENT
  this procedure performs back - substitution
  we omit the details
```

```
              COMMENT
              END ;
              INT n ; read ( n ) ;
              INT m = 5 ;
              [ 1 : n , 1 : n ] REAL a , savea ;
              [ 1 : n , 1 : m ] REAL alpha , b , saveb ;
              [ 1 : n ] INT rownum ;
              REAL e = exp ( 1.0 ) ;
              COMMENT
              set up the components of the right hand side vectors
              COMMENT
              b [ 1 , ] := ( 1.7 , 1.72 , 1.718 , 1.7183 , 1.71828 ) ;
              FOR i FROM 2 TO n
              DO
                FOR p TO m
                DO
                  b [ i , p ] := e - ( i - 1.0 ) * b [ i - 1 , p ]
                OD
              OD ;
              COMMENT
              set up the components of the coefficient matrix
              COMMENT
              FOR i TO n
              DO
                a [ i , i ] := 1.0 / ( 2 * i - 1.0 ) ;
                FOR j FROM i + 1 TO n
                DO
                  a [ i , j ] := a [ j , i ] := 1.0 / ( i + j - 1.0 )
                OD
              OD ;
              COMMENT
              make copies of the coefficient matrix and the right hand side vectors so that
              the residual vector may be computed at a later stage
              COMMENT
              savea := a ; saveb := b ;
              COMMENT
              solve the system of equations
              COMMENT
              gausselim ( a , b , rownum ) ;
              backsub ( a , alpha , b , rownum ) ;
              FOR p TO m
              DO
                print ( ( newline , "using" , p , " decimal place accuracy" , newline ,
                         "the right hand side vector is" , newline , saveb [ , p ] ,
                         newline , newline , "and the solution vector is" , newline ,
                         alpha [ , p ] , newline , newline , "the residual is" , newline ,
                         saveb [ , p ] - savea * alpha [ , p ] , newline , newline ,
                         newline ) )
              OD
         END
```

<u>Pascal Program</u>

```
PROGRAM exampleofleastsquares(input, output);

   {  a program for the least squares approximation of
      exp(x) on [0,1] by a polynomial  }

   CONST
      m=5;
      maxn=50;

   TYPE
      rhsindextype=1..m;
      matrixindextype=1..maxn;
      rhstype=ARRAY[ matrixindextype, rhsindextype ] OF real;
```

```
           matrixtype=ARRAY[ matrixindextype, matrixindextype ] OF real;
           rownumtype=ARRAY[ matrixindextype ] OF matrixindextype;

VAR
    i, j, n:matrixindextype;
    p:rhsindextype;
    a, savea:matrixtype;
    alpha, b, saveb:rhstype;
    rownum:rownumtype;
    e, residual:real;

PROCEDURE gausselim(n:matrixindextype; VAR a:matrixtype;
                    m:rhsindextype; VAR b:rhstype; VAR rownum:rownumtype);
    { this procedure performs the elimination process
      we omit the details  }
END { gausselim } ;

PROCEDURE backsub(n:matrixindextype; VAR a:matrixtype;
                  m:rhsindextype; VAR x, b:rhstype; VAR rownum:rownumtype);
    { this procedure performs back - substitution
      we omit the details  }
END { backsub } ;

BEGIN
    read(n);
    { set up the components of the right hand side vectors  }
    b[1, 1]:= 1.7;
    b[1, 2]:= 1.72;
    b[1, 3]:= 1.718;
    b[1, 4]:= 1.7183;
    b[1, 5]:= 1.71828;
    e:= exp(1.0);
    FOR i:= 2 TO n DO
        FOR p:= 1 TO m DO
            b[i, p]:= e - (i - 1.0)*b[i - 1, p];
    { set up the components of the coefficient matrix  }
    FOR i:= 1 TO n DO
    BEGIN
        a[i, i]:= 1.0/(2*i - 1.0);
        FOR j:= i + 1 TO n DO
        BEGIN
            a[i, j]:= 1.0/(i + j - 1.0);
            a[j, i]:= a[i, j]
        END
    END ;
    { make copies of the coefficient matrix and the right hand side vectors
      so that the residual vector may be computed at a later stage  }
    FOR i:= 1 TO n DO
        FOR j:= 1 TO n DO
            savea[i, j]:= a[i, j];
    saveb:= b;
    { solve the system of equations  }
    gausselim(n, a, m, b, rownum);
    backsub(n, a, m, alpha, b, rownum);
    { print the results  }
    FOR p:= 1 TO m DO
    BEGIN
        writeln;
        writeln('using', p, ' decimal place accuracy');
        writeln('the right hand side vector is');
        FOR i:= 1 TO n DO
            write(saveb[i, p]);
        writeln;
        writeln;
        writeln('and the solution vector is');
        FOR i:= 1 TO n DO
            write(alpha[i, p]);
        writeln;
        writeln;
```

```
        writeln('the residual is');
        FOR i:= 1 TO n DO
        BEGIN
           residual:= saveb[i, p];
           FOR j:= 1 TO n DO
              residual:= residual - savea[i, j]*alpha[j, p];
           write(residual)
        END ;
        writeln;
        writeln;
        writeln
     END
END { exampleofleastsquares } .
```

Sample Data

   3

Sample Output

```
using    1  decimal place accuracy
the right hand side vector is
   1.70000e  0  1.01828e  0  6.81718e -1

and the solution vector is
  -9.06606e -1  1.16009e  1 -9.58148e  0

the residual is
   1.19209e -6  7.15256e -7  5.96046e -7


using    2  decimal place accuracy
the right hand side vector is
   1.72000e  0  9.98282e -1  7.21718e -1

and the solution vector is
   1.19341e  0 -1.59217e -1  1.81860e  0

the residual is
   4.76837e -7  2.38419e -7  2.38419e -7


using    3  decimal place accuracy
the right hand side vector is
   1.71800e  0  1.00028e  0  7.17718e -1

and the solution vector is
   9.83398e -1  1.01685e  0  6.78536e -1

the residual is
   2.38419e -7  2.38419e -7  2.38419e -7


using    4  decimal place accuracy
the right hand side vector is
   1.71830e  0  9.99981e -1  7.18318e -1
```

```
and the solution vector is
   1.01492e  0  8.40348e -1  8.49630e -1

the residual is
   4.76837e -7  2.38419e -7  2.38419e -7


using     5 decimal place accuracy
the right hand side vector is
   1.71828e  0  1.00000e  0  7.18278e -1

and the solution vector is
   1.01281e  0  8.52125e -1  8.38215e -1

the residual is
   4.76837e -7  4.76837e -7  3.57628e -7
```

As a final embellishment to our program we have included code to calculate the residual vector

$$r = b - A\alpha$$

(and really we should be doing this in double precision). In Algol 68 it is possible to define new operators and to define new meanings for existing operators. In order to calculate $b - A\alpha$ it seems a good idea to define two operators, one to multiply a vector by a matrix and the other to subtract one vector from another. The Algol 68 program for Case Study 4.1 contains operator declarations for $*$ and $-$. The first declaration expects the lower bound of the vector to be one whereas the second imposes no such restriction. Note that the bounds of the operands of both operators must be compatible. Having declared these operators the code to output the values of the residual vector is simply

    print (saveb [, p] − savea∗alpha [, p]) ;

As it is not possible to define operators in Pascal, the Pascal program of Case Study 4.1 uses the following code to output the values of the residual vector

```
FOR i : = 1 TO n DO
BEGIN
   residual: = saveb[i, p];
   FOR j : = 1 TO n DO
      residual : = residual − savea[i, j]∗alpha[j, p];
   write(residual)
END
```

Since the operator declarations take up 30 lines of code, the Algol 68 code seems verbose compared with the Pascal code. However, the Algol 68 code is superior since the expression

saveb [, p] − savea∗alpha [, p]

closely resembles the mathematical formula $\mathbf{b} - A\boldsymbol{\alpha}$. And the additional code that is required to produce this effect is unimportant since the operator declarations will be useful in other programs.

It is often useful to give names to constant values. In Algol 68 we can do this by an identity declaration

INT m = 5 ;   REAL e = exp (1.0) ;

and we can do a similar thing in Pascal by using a constant identifier

CONST m = 5;

However, in Pascal, the object to the right of = must be a literal whereas in Algol 68 any expression may be used. Thus, in Pascal, it is not possible to do:

CONST e = exp(1.0);

and, instead, a variable has to be introduced and assigned the value exp(1.0).

4.3.4 Ill-conditioning of the normal equations

We now look at Example 4.4 and the program of the previous subsection in more detail. For polynomial approximation on [0,1] the general form of the coefficient matrix is given by (4.22). This matrix is the nth principal minor of the infinite Hilbert matrix and it is well known that an equation of the form (4.14) involving this matrix is likely to be very ill-conditioned.

If we look at the sample output of Case Study 4.1 we see that if the accuracy to which b_1 is computed is decreased by just one decimal place the solution vector changes by a much greater factor. The residual vector indicates that the elimination and back-substitution processes have worked quite well; the observed behaviour of the solution vector is due to the problem itself, not the method of solution. Investigating the matter further, the program of Case Study 4.1 was executed for values of $n = 3, 4,..., 7$ and Table 4.2 lists the results for the case in which b_1 is rounded to five decimal places.

For n equal to three and four the polynomial coefficients start to look something like the first few terms in the Taylor series expansion of e^x about the point $x = 0$ (see Example 4.1). However, from $n = 5$ onwards there is no

Table 4.2

	n				
	3	4	5	6	7
α_1	1.012 79	0.996 249	0.949 770	−0.188 766	27.732 4
α_2	0.852 254	1.050 74	1.980 15	36.000 4	−1009.63
α_3	0.838 089	0.341 889	−3.840 13	−241.345	9354.44
α_4		0.330 798	6.835 84	638.958	−35252.9
α_5			−3.252 41	−713.348	63042.8
α_6				283.712	−53376.4
α_7					17229.8

consistency at all in the results and, in fact, they appear to be getting significantly worse. What we are in effect saying is that the ill-conditioning property of the problem becomes more pronounced as n increases.

To overcome these difficulties we must reformulate the problem and we now consider ways of doing this.

4.4 USE OF ORTHOGONAL POLYNOMIALS

4.4.1 Legendre Polynomials

In the previous section we saw that polynomial approximation of a function on the finite interval $[a,b]$ leads to the derivation of a system of equations which may well be ill-conditioned. We now look at ways of getting round this difficulty and observe first of all that the situation would be considerably healthier if the coefficient matrix were not full but diagonal. The solution vector could then be read off directly. Not only would we avoid the pitfalls of ill-conditioning but we would also reduce considerably the operations count involved in solving the system of equations. We shall see shortly that such a situation can be readily achieved by employing a fairly simple transformation of the problem.

In order to proceed further we introduce a sequence of polynomials, $P_n(x) : n = 0, 1,...$, known as the *Legendre* polynomials. These polynomials can be defined via the *three term recurrence relation*

$$P_n(x) = \frac{2n-1}{n} x \, P_{n-1}(x) - \frac{n-1}{n} P_{n-2}(x) \qquad n = 2, 3,... \quad (4.24)$$

with the starting values

$$P_0(x) = 1; \quad P_1(x) = x \qquad (4.25)$$

and it is fairly easy to see that $P_n(x)$ is a polynomial of degree n. Legendre polynomials possess a property which is very important in the present context, namely they are *orthogonal* on the interval $[-1,1]$. By this we mean that the integral over $[-1,1]$ of the product of two Legendre polynomials is zero unless they happen to be the same. Formally, we have

$$\int_{-1}^{1} P_n(x) P_m(x) \, dx = 0 \qquad n \neq m$$

and also that

$$\int_{-1}^{1} (P_n(x))^2 \, dx = \frac{2}{2n+1} \qquad n = 0, 1, \ldots. \tag{4.26}$$

This definition of the orthogonality of functions should be compared with the corresponding definition for vectors. Recall that two vectors are orthogonal if their *scalar product* is zero. Here the scalar, or *inner*, product is the integral of the product of two polynomials and we write

$$(P_n(x), P_m(x)) = \int_{-1}^{1} P_n(x) P_m(x) \, dx \qquad n, m = 0, 1, \ldots.$$

EXAMPLE 4.5
From (4.25) we have that

$$\int_{-1}^{1} P_0(x) P_1(x) \, dx = \int_{-1}^{1} x \, dx = \left[\frac{x^2}{2}\right]_{-1}^{1} = \tfrac{1}{2} - \tfrac{1}{2} = 0$$

and so $P_0(x)$ and $P_1(x)$ are orthogonal. From (4.24)

$$P_2(x) = \tfrac{3}{2} x.x - \tfrac{1}{2} = \tfrac{1}{2}(3x^2 - 1) . \tag{4.27}$$

Now

$$\int_{-1}^{1} P_0(x) P_2(x) \, dx = \tfrac{1}{2} \int_{-1}^{1} (3x^2 - 1) \, dx = \tfrac{1}{2} [x^3 - x]_{-1}^{1} = 0$$

and

$$\int_{-1}^{1} P_1(x) P_2(x) \, dx = \tfrac{1}{2} \int_{-1}^{1} (3x^3 - x) \, dx = \tfrac{1}{2} \left[\frac{3x^4}{4} - \frac{x^2}{2}\right]_{-1}^{1} = 0$$

and hence $P_2(x)$ is orthogonal to both $P_0(x)$ and $P_1(x)$.

Now, suppose that we wish to find the least squares approximation of the form

$$p_2(x) = \alpha_1 + \alpha_2 x + \alpha_3 x^2$$

to the continuous function $f(x)$ on the interval $[-1,1]$. Then rearranging (4.27) we have

$$x^2 = \tfrac{1}{3}(2P_2(x) + 1) = \tfrac{1}{3}(2P_2(x) + P_0(x))$$

so that

$$\begin{aligned} p_2(x) &= \alpha_1 P_0(x) + \alpha_2 P_1(x) + \alpha_3 \tfrac{1}{3}(2P_2(x) + P_0(x)) \\ &= (\alpha_1 + \tfrac{1}{3}\alpha_3) P_0(x) + \alpha_2 P_1(x) + \tfrac{2}{3}\alpha_3 P_2(x) \\ &= \beta_1 P_0(x) + \beta_2 P_1(x) + \beta_3 P_2(x) \end{aligned} \tag{4.28}$$

where

$$\beta_1 = \alpha_1 + \tfrac{1}{3}\alpha_3, \quad \beta_2 = \alpha_2, \quad \beta_3 = \tfrac{2}{3}\alpha_3 .$$

We have, therefore, transformed $p_2(x)$ from a linear combination of the monomials $\{1, x, x^2\}$ to a linear combination of the first three Legendre polynomials.

From the material of section 4.3, the least squares quadratic approximation of the form (4.28) to the function $f(x)$ on the interval $[-1,1]$ is given as the solution to a minimisation problem. We need to find the values of β_1, β_2 and β_3 which minimise the quantity

$$\int_{-1}^{1} \left(f(x) - \sum_{j=1}^{3} \beta_j P_{j-1}(x) \right)^2 dx .$$

Proceeding in exactly the same way as before we end up with the matrix equation

$$A\boldsymbol{\beta} = \mathbf{b} \tag{4.29}$$

where

$$A_{ij} = \int_{-1}^{1} P_{i-1}(x) P_{j-1}(x)\, dx = (P_{i-1}(x), P_{j-1}(x)) \qquad i,j = 1, 2, 3$$

Sec. 4.4] USE OF ORTHOGONAL POLYNOMIALS 163

$$b_i = \int_{-1}^{1} f(x) P_{i-1}(x) \, dx = (f(x), P_{i-1}(x)) \quad i = 1, 2, 3.$$

But, from the orthogonality property of the Legendre polynomials, we have that the coefficient matrix has non-zero entries only on the diagonal so that, using the result (4.26),

$$\beta_i = \frac{(f(x), P_{i-1}(x))}{(P_{i-1}(x), P_{i-1}(x))} = \frac{2i-1}{2} (f(x), P_{i-1}(x)) \quad i = 1, 2, 3.$$

The solution to our least squares approximation problem is now in terms of $P_0(x)$, $P_1(x)$ and $P_2(x)$. In order to find out what the solution looks like as a straightforward quadratic we need to solve the system

$$\alpha_1 \qquad\qquad +\tfrac{1}{3}\alpha_3 = \beta_1$$
$$\alpha_2 \qquad\qquad = \beta_2$$
$$\qquad\qquad \tfrac{2}{3}\alpha_3 = \beta_3$$

which gives

$$\alpha_3 = \tfrac{3}{2}\beta_3, \quad \alpha_2 = \beta_2, \quad \alpha_1 = \beta_1 - \tfrac{1}{2}\beta_3.$$

EXAMPLE 4.6

Suppose that we wish to find the least squares quadratic approximation to e^x on $[-1,1]$. Then the polynomial coefficients in $p_2(x) = \beta_1 P_0(x) + \beta_2 P_1(x) + \beta_3 P_2(x)$ are given by

$$\beta_1 = \tfrac{1}{2} \int_{-1}^{1} e^x P_0(x) \, dx = \tfrac{1}{2} \int_{-1}^{1} e^x \, dx = \tfrac{1}{2}(e - e^{-1})$$

$$\beta_2 = \tfrac{3}{2} \int_{-1}^{1} e^x P_1(x) \, dx = \tfrac{3}{2} \int_{-1}^{1} e^x x \, dx = 3e^{-1}$$

$$\beta_3 = \tfrac{5}{2} \int_{-1}^{1} e^x P_2(x) \, dx = \tfrac{5}{2} \int_{-1}^{1} e^x \tfrac{1}{2}(3x^2 - 1) \, dx = \tfrac{5}{2}(e - 7e^{-1})$$

using integration by parts. Hence the quadratic approximation is

$$p_2(x) = \tfrac{1}{2}(e - e^{-1}) + 3e^{-1}x + \tfrac{5}{2}(e - 7e^{-1})\tfrac{1}{2}(3x^2 - 1).$$

The coefficients in $p_2(x) = \alpha_1 + \alpha_2 x + \alpha_3 x^2$ are given by

$$\alpha_3 = \tfrac{3}{2}\beta_3 = \tfrac{15}{4}(e - 7e^{-1})$$
$$\alpha_2 = \beta_2 = 3e^{-1}$$
$$\alpha_1 = \beta_1 - \tfrac{1}{2}\beta_3 = \tfrac{1}{2}(e - e^{-1}) - \tfrac{5}{4}(e - 7e^{-1}) = \tfrac{1}{4}(-3e + 33e^{-1})$$

so that

$$p_2(x) = \tfrac{1}{4}(-3e + 33e^{-1}) + 3e^{-1}x + \tfrac{15}{4}(e - 7e^{-1})x^2.$$

The generalisation of the above is obvious and it is clearly advantageous to be able to express an approximating polynomial in terms of orthogonal polynomials. We no longer need to compute all of the entries in the coefficient matrix but just those on the diagonal. This has important consequences for the storage requirements of a least squares approximation program; the two-dimensional $n \times n$ array of coefficients is no longer required. Although in Example 4.6 we converted the polynomial to standard form this is not really necessary and in subsection 4.4.3 we give code for evaluating a polynomial expressed as a linear combination of Legendre polynomials at a point.

Unless we are very fortunate the domain of interest is unlikely to be $[-1,1]$ and we need to take account of this. One simple way is to map the given range $[a,b]$, say, onto $[-1,1]$ using the mapping

$$z = \frac{2x - a - b}{b - a}.$$

Alternatively we could use polynomials which are orthogonal on the interval $[a,b]$. This does not mean that we need to know a large set of orthogonal polynomials; it is possible to generate from the monomials a sequence of polynomials which are orthogonal on a given interval using a technique known as the *Gram–Schmidt orthogonalisation process*.

4.4.2 Polynomial orthogonalisation

Suppose that we wish to generate the sequence of polynomials $\{Q_i(x) : i = 0, 1, 2\}$ with $Q_i(x)$ of degree i such that the $Q_i(x)$s are orthogonal on $[a,b]$. Define

$$Q_0(x) = c_{00}$$
$$Q_1(x) = c_{10} + c_{11}x$$
$$Q_2(x) = c_{20} + c_{21}x + c_{22}x^2.$$

Now, orthogonality implies that

$$(Q_0(x), Q_1(x)) = \int_a^b Q_0(x) Q_1(x) \, dx = 0$$

$$(Q_0(x), Q_2(x)) = \int_a^b Q_0(x) Q_2(x) \, dx = 0 \quad (4.30)$$

$$(Q_1(x), Q_2(x)) = \int_a^b Q_1(x) Q_2(x) \, dx = 0$$

and this gives a system of three equations for the six unknowns $\{c_{ij}\}$. For the time being we choose $c_{00} = c_{10} = c_{20} = 1$ arbitrarily and then the system (4.30) defines c_{11}, c_{21} and c_{22} uniquely.

EXAMPLE 4.7
To find a sequence of three polynomials which are orthogonal on [0,1] we have

$$0 = (Q_0(x), Q_1(x)) = \int_0^1 (1 + c_{11}x) \, dx = \left[x + c_{11} \frac{x^2}{2} \right]_0^1 = 1 + \frac{c_{11}}{2}$$

so that $c_{11} = -2$. Then

$$0 = (Q_0(x), Q_2(x)) = \int_0^1 \left[1 + c_{21}x + c_{22}x^2 \right] dx$$

$$= \left[x + c_{21} \frac{x^2}{2} + c_{22} \frac{x^3}{3} \right]_0^1 = 1 + \frac{c_{21}}{2} + \frac{c_{22}}{3} \quad (4.31)$$

and

$$0 = (Q_1(x), Q_2(x)) = \int_0^1 (1 - 2x) \left[1 + c_{21}x + c_{22}x^2 \right] dx$$

$$= \int_0^1 (1 + (c_{21} - 2)x + (c_{22} - 2c_{21})x^2 - 2c_{22}x^3) \, dx$$

$$= \left[x + (c_{21} - 2) \frac{x^2}{2} + (c_{22} - 2c_{21}) \frac{x^3}{3} - c_{22} \frac{x^4}{2} \right]_0^1$$

$$= 1 + \tfrac{1}{2}(c_{21} - 2) + \tfrac{1}{3}(c_{22} - 2c_{21}) - \tfrac{1}{2} c_{22}$$

$$= -\tfrac{1}{6} c_{21} - \tfrac{1}{6} c_{22} \, .$$

Hence $c_{21} = -c_{22}$ and therefore, from (4.31) $c_{22} = 6$ and $c_{21} = -6$. We conclude that the polynomials are

$$Q_0(x) = 1$$

$$Q_1(x) = 1 - 2x$$

$$Q_2(x) = 1 - 6x + 6x^2 .$$

Having generated a sequence of polynomials which are orthogonal on a given interval the solution to the corresponding least squares approximation problem may be obtained using a straightforward extension of the material of the previous subsection. All that we need to do is to replace the Legendre polynomials with the orthogonal polynomials we have just generated and integrate over $[a,b]$ instead of $[-1,1]$. In the matrix equation (4.29) which defines the coefficients in the polynomial $p_2(x) = \beta_1 + \beta_2 x + \beta_3 x^2$ the matrix A has components

$$A_{ij} = (Q_{i-1}(x), Q_{j-1}(x)) = \int_a^b Q_{i-1}(x) \, Q_{j-1}(x) \, \mathrm{d}x \qquad i,j = 1, 2, 3$$

and is diagonal. The entries in the right-hand side vector are given by

$$b_i = (f(x), Q_{i-1}(x)) = \int_a^b f(x) \, Q_{i-1}(x) \, \mathrm{d}x \qquad i = 1, 2, 3$$

and hence the solution vector may be found directly since its components are

$$\beta_i = \frac{(f(x), Q_{i-1}(x))}{(Q_{i-1}(x), Q_{i-1}(x))} \qquad i = 1, 2, 3 . \tag{4.32}$$

Note that in the last three equations the definition of the inner product has been modified to take account of the fact that the domain of interest is now the general interval $[a,b]$.

EXAMPLE 4.8
From Example 4.7 we know that the polynomials $Q_0(x) = 1$, $Q_1(x) = 1 - 2x$ and $Q_2(x) = 1 - 6x + 6x^2$ are orthogonal on $[0,1]$. Therefore the coefficients in the least squares polynomial approximation of the form $p_2(x) = \beta_1 Q_0(x) + \beta_2 Q_1(x) + \beta_3 Q_2(x)$ to e^x on $[0,1]$ are given by

$$\beta_1 = \int_0^1 e^x \, dx / \int_0^1 dx = e - 1$$

$$\beta_2 = \int_0^1 e^x(1-2x) \, dx / \int_0^1 (1-2x)^2 \, dx = 3(e-3)$$

$$\beta_3 = \int_0^1 e^x(1-6x+6x^2) \, dx / \int_0^1 (1-6x+6x^2)^2 \, dx = 5(7e-19) \, .$$

Hence

$$p_2(x) = e - 1 + 3(e-3)(1-2x) + 5(7e-19)(1-6x+6x^2).$$

This solution should now be compared with that of Example 4.4.

The generation of higher order polynomials which are orthogonal on $[a,b]$ follows in an obvious manner. Suppose that we have already obtained the polynomials $Q_0(x), Q_1(x),\ldots, Q_{n-2}(x)$ which are such that

$$(Q_i(x), Q_j(x)) = 0 \qquad i, j = 0, 1,\ldots, n-2.$$

Then to ensure that $Q_{n-1}(x)$ is orthogonal to $Q_0(x), Q_1(x),\ldots, Q_{n-2}(x)$ we let

$$Q_{n-1}(x) = c_{n-1,0} + c_{n-1,1}x + \ldots + c_{n-1,n-1}x^{n-1}$$

with $c_{n-1,0} = 1$ and find values for $c_{n-1,1}, c_{n-1,2},\ldots, c_{n-1,n-1}$ such that

$$(Q_i(x), Q_{n-1}(x)) = 0 \qquad i = 0, 1,\ldots, n-2 \, .$$

Equation (4.32) is now valid for all values of i and hence the least squares polynomial approximation to the function $f(x)$ on $[a,b]$ may be found immediately.

At the start of this subsection we chose to set the first term of each orthogonal polynomial to be one. We could set the coefficient of x^i in $Q_i(x)$ to be one instead and then find the remaining terms. Other variations of the basic method are of course possible. An alternative approach however is to impose the additional constraints

$$(Q_i(x), Q_i(x)) = 1 \qquad i = 0, 1,\ldots \, . \tag{4.33}$$

The original orthogonality conditions plus the *normalising* conditions (4.33) give a system of equations which uniquely defines the coefficients of the

168 APPROXIMATION OF CONTINUOUS FUNCTIONS [Ch. 4

$Q_i(x)$s. The polynomials so generated are now said to be *orthonormal* on the interval $[a,b]$.

EXAMPLE 4.9
Using (4.30) and (4.33) it can be shown that the polynomials

$$Q_0(x) = 1$$
$$Q_1(x) = \sqrt{3} - 2\sqrt{3}x$$
$$Q_2(x) = \sqrt{5} - 6\sqrt{5}x + 6\sqrt{5}x^2$$

are orthonormal on [0,1]. We note that these polynomials are the same as those derived in Example 4.7 apart from the presence of a multiplying factor in $Q_1(x)$ and $Q_2(x)$. Indeed a straightforward way of generating orthonormal polynomials is to generate a sequence of orthogonal polynomials $\{Q_i(x)\}$ and then divide $Q_i(x)$ by the normalising factor $(Q_i(x), Q_i(x))^{1/2}$. The reader should now check that the polynomials $Q_1(x)$ and $Q_2(x)$ of Example 4.7 satisfy $(Q_1(x), Q_1(x)) = \frac{1}{3}$ and $(Q_2(x), Q_2(x)) = \frac{1}{5}$ (these values were used in Example 4.8).

If we express a least squares approximation polynomial in terms of orthonormal polynomials the matrix in the equation defining the coefficients is now the identity matrix and so the solution vector is just the right hand side vector.

EXAMPLE 4.10
The coefficients in the least squares polynomial approximation of the form $p_2(x) = \beta_1 Q_0(x) + \beta_2 Q_1(x) + \beta_3 Q_2(x)$ (with $Q_0(x)$, $Q_1(x)$ and $Q_2(x)$ defined as in Example 4.9) to e^x on [0,1] are given by

$$\beta_1 = \int_0^1 e^x \, dx = e - 1$$

$$\beta_2 = \int_0^1 e^x(\sqrt{3} - 2\sqrt{3}x) \, dx = \sqrt{3}e - 3\sqrt{3}$$

$$\beta_3 = \int_0^1 e^x(\sqrt{5} - 6\sqrt{5}x + 6\sqrt{5}x^2) = \sqrt{5}(7e - 19)$$

so that

$$p_2(x) = e - 1 + (\sqrt{3}e - 3\sqrt{3})(\sqrt{3} - 2\sqrt{3}x)$$
$$+ \sqrt{5}(7e - 19)(\sqrt{5} - 6\sqrt{5}x + 6\sqrt{5}x^2) \ .$$

Sec. 4.4] USE OF ORTHOGONAL POLYNOMIALS

In the previous subsection we considered the advantages of expressing a least squares approximating polynomial valid on $[-1,1]$ as a linear combination of Legendre polynomials. This approach assumes that it is always possible to find coefficients $\{\beta_i: i = 1, 2,..., n\}$ such that

$$p_{n-1}(x) = \sum_{i=1}^{n} \alpha_i x^{i-1} = \sum_{i=1}^{n} \beta_i P_{i-1}(x) .$$

In fact this assumption is perfectly valid since the two sets of functions $\{x^{i-1}: i = 1, 2,..., n\}$ and $\{P_{i-1}(x): i = 1, 2,..., n\}$ *span* the same *function space*, namely the space of polynomials of degree at most $n - 1$. By this we mean any element of the space may be written as a linear combination of the monomials or the Legendre polynomials. In this context it is natural to refer to the two sets as being sets of *basis* functions. We say that the Legendre polynomials form an *orthogonal basis* on $[-1,1]$. The polynomials $Q_0(x)$, $Q_1(x)$ and $Q_2(x)$ of Example 4.7 form an orthogonal basis on $[0,1]$ whilst the polynomials defined in Example 4.9 form an *orthonormal basis* on $[0,1]$. As a corollary any polynomial expressed in terms of orthogonal, or orthonormal, polynomials may be written as a linear combination of the monomials. However, in solving a least squares approximation problem using orthogonal polynomials it is not really necessary to convert the solution to standard form. It can be shown (see Ralston and Rabinowitz, 1978) that all orthogonal polynomials satisfy a three term recurrence relation. We now show how this result may be used to evaluate at a point a polynomial expressed as a linear combination of orthogonal polynomials in a straightforward manner.

4.4.3 A program for polynomial evaluation

In Case Study 4.2 we give programs for evaluating at a point a polynomial expressed as a linear combination of Legendre polynomials. It is fairly easy to modify these codes for the case where the orthogonal polynomials are other than Legendre.

The programs are quite straightforward and make use of the three term recurrence relation (4.24) and the initial values (4.25). In the Algol 68 program, an operator has, once again, been used but here the name of the operator is not one of the standard operator symbols; its priority is, therefore, by default, equal to 1.

CASE STUDY 4.2

Algol 68 Program

```
BEGIN
  COMMENT
    a program to evaluate a polynomial expressed as a linear combination of
    legendre polynomials at a point using the three term recurrence relation
  COMMENT
```

```
        OP EVALUATEAT = ( REF [ ] REAL coeffs , REAL x ) REAL :
        BEGIN
          COMMENT
          this operator does the evaluation part
          COMMENT
          REAL pn := x , pnminus1 := 1.0 , pnminus2 ,
               sum := coeffs [ 0 ] + coeffs [ 1 ] * x ;
          FOR n FROM 2 TO UPB coeffs
          DO
            pnminus2 := pnminus1 ;
            pnminus1 := pn ;
            pn := ( ( 2.0 * n - 1.0 ) * x * pnminus1 - ( n - 1.0 ) * pnminus2 ) / n ;
            sum +:= coeffs [ n ] * pn
          OD ;
          sum
        END ;
        INT n ; read ( n ) ;
        [ 0 : n ] REAL beta ; read ( beta ) ;
        REAL x ; read ( x ) ;
        print ( ( newline , "the coefficients of the legendre polynomials are" ,
                 " ( constant term first )" , newline ) ) ;
        FOR i FROM 0 TO n
        DO
          print ( ( newline , beta [ i ] ) )
        OD ;
        print ( ( newline , newline , "and when evaluated at the point   " , x ,
                 "    this gives   " , beta EVALUATEAT x , newline ) )
END
```

Pascal Program

```
PROGRAM legendrepolynomials(input, output);

    { a program to evaluate a polynomial expressed as a linear
      combination of legendre polynomials at a point using the
      three term recurrence relation }

    CONST
       maxn=50;

    TYPE
       indextype=0..maxn;
       coeffsvector=ARRAY[ indextype ] OF real;

    VAR
       i, n:indextype;
       beta:coeffsvector;
       x:real;

    FUNCTION evaluate(n:indextype; VAR coeffs:coeffsvector; x:real):real;
       VAR
           pn, pnminus1, pnminus2, sum:real;
           i:indextype;
       BEGIN
          pn:= x;
          pnminus1:= 1.0;
          sum:= coeffs[0] + coeffs[1]*x;
          FOR i:= 2 TO n DO
          BEGIN
             pnminus2:= pnminus1;
             pnminus1:= pn;
             pn:= ((2.0*i - 1.0)*x*pnminus1 - (i - 1.0)*pnminus2)/i;
             sum:= sum + coeffs[i]*pn
          END ;
          evaluate:= sum
       END { evaluate } ;
```

```
BEGIN
   read(n);
   FOR i:= 0 TO n DO
      read(beta[i]);
   read(x);
   writeln;
   writeln('the coefficients of the legendre polynomials are',
           ' ( constant term first )');
   writeln;
   FOR i:= 0 TO n DO
      writeln(beta[i]);
   writeln;
   writeln('and when evaluated at the point   ', x,
           '   this gives   ', evaluate(n, beta, x))
END { legendrepolynomials } .
```

Sample Data

```
4
   0.25    0.75   -0.25   -0.75    0.50
   0.5
```

Sample Output

```
the coefficients of the legendre polynomials are ( constant term first )

   2.50000e -1
   7.50000e -1
  -2.50000e -1
  -7.50000e -1
   5.00000e -1

and when evaluated at the point    5.00000e -1   this gives    8.39844e -1
```

4.5 GENERALISED LEAST SQUARES METHODS

4.5.1 Weighted least squares

Whilst in practice an approximating function is usually a polynomial there is no reason at all why we should be restricted in this way. In certain circumstances it may prove profitable to use a linear combination of functions other than monomials (or, equivalently, orthogonal polynomials). Just what form these other functions should take we return to later and for the moment we consider in an abstract sense a more general approach to least squares approximation.

Let $\{\phi_i(x): i = 1, 2, \ldots, n\}$ be a set of known functions which we refer to as *expansion* functions (or *basis* functions). Let

$$L(\mathbf{a}, x) = \sum_{j=1}^{n} \alpha_j \phi_j(x) \tag{4.34}$$

be a *linear* approximating function. By linear we mean that the parameters α_j: $j = 1, 2, \ldots, n$ (the *expansion coefficients*) appear linearly in (4.34). Polynomial approximation is therefore a special case of the more general form being considered here and corresponds to the choice

$$\phi_j(x) = x^{j-1} \qquad j = 1, 2, \ldots, n$$

if the polynomial is expressed in the 'normal' form.

In section 4.4 we saw that if the domain of interest is $[-1, 1]$ the choice

$$\phi_j(x) = P_{j-1}(x) \qquad j = 1, 2, \ldots, n$$

(where $P_{j-1}(x)$ is a Legendre polynomial) leads to the polynomial coefficients being found as the solution to a system of equations in which the coefficient matrix is diagonal. The aim here is to achieve a similar situation for the more general case. For the moment the only condition that we impose on the basis functions is that they be *linearly independent*, that is, it must not be possible to express one of the basis functions as a linear combination of the others. (Without this restriction the coefficient matrix will be singular).

The least squares approximation of the form (4.34) to the function $f(x)$ on the finite interval $[a, b]$ minimises with respect to the α_js the quantity

$$\int_a^b (f(x) - L(\boldsymbol{\alpha}, x))^2 \, dx. \tag{4.35}$$

This basic definition may be modified by introducing a *weighting function* $w(x) > 0$ so that (4.35) becomes

$$\int_a^b (f(x) - L(\boldsymbol{\alpha}, x))^2 w(x) \, dx.$$

This weight function allows us to have some control over the approximating function; it can be used to emphasise (or reduce) the square error $(f(x) - L(\boldsymbol{\alpha}, x))^2$ over certain sections of $[a, b]$. In addition it allows us to use polynomials which are orthogonal on $[-1, 1]$ other than the Legendre polynomials (see subsection 4.5.2).

Corresponding to (4.11) we introduce the function

$$I(\boldsymbol{\alpha}) = \int_a^b (f(x) - L(\boldsymbol{\alpha}, x))^2 w(x) \, dx$$

which we wish to minimise with respect to $\alpha_1, \alpha_2, \ldots, \alpha_n$. As before, at a minimum

$$0 = \frac{\partial I(\boldsymbol{\alpha})}{\partial \alpha_i}$$

$$= -2 \int_a^b \left(f(x) - \sum_{j=1}^n \alpha_j \phi_j(x) \right) \phi_i(x) w(x) \, dx \qquad i = 1, 2, \ldots, n$$

or, rearranging,

$$\sum_{j=1}^n \alpha_j \int_a^b \phi_i(x) \phi_j(x) w(x) \, dx = \int_a^b f(x) \phi_i(x) w(x) \, dx$$

$$i = 1, 2, \ldots, n$$

and these are the normal equations corresponding to (4.13). To keep the notation as simple as possible we introduce the inner product

$$(g(x), h(x)) = \int_a^b g(x) h(x) w(x) \, dx$$

for any two functions $g(x)$ and $h(x)$. Then, the optimal values of the α_js are given by

$$\sum_{j=1}^n \alpha_j (\phi_i(x), \phi_j(x)) = (f(x), \phi_i(x)) \qquad i = 1, 2, \ldots, n$$

or in matrix form as

$$A\boldsymbol{\alpha} = \mathbf{b} \tag{4.36}$$

where

$$A_{ij} = (\phi_i(x), \phi_j(x)) \qquad i, j = 1, 2, \ldots, n$$
$$b_i = (f(x), \phi_i(x)) \qquad i = 1, 2, \ldots, n \ .$$

The ideas of orthogonality can be extended to this more general case if we say that a set of basis functions $\{\phi_i(x): i = 1, 2, \ldots, n\}$ is orthogonal on the interval $[a,b]$ with respect to $w(x)$ if

$$\int_a^b \phi_i(x)\,\phi_j(x)w(x)\,dx = 0 \qquad i \neq j.$$

Further, we say that the set is orthonormal if

$$\int_a^b \phi_i^2(x)w(x)\,dx = 1.$$

If we use orthogonal basis functions the coefficient matrix in (4.36) will be diagonal and the solution vector can be obtained immediately. In subsection 4.4.2 we saw that it is possible to generate a sequence of polynomials orthogonal on $[a,b]$ with respect to the weight $w(x) = 1$ from the monomials. The ideas presented there are now applied to the more general case.

4.5.2 Gram–Schmidt orthogonalisation process

Let $\psi_i(x): i = 1, 2, \ldots, n$ be a set of basis functions defined by

$$\psi_1(x) = \phi_1(x)$$

$$\psi_i(x) = \phi_i(x) - \sum_{j=1}^{i-1} c_{ij}\psi_j(x) \qquad i = 2, 3, \ldots, n$$

so that $\psi_i(x)$ is a linear combination of $\{\phi_j(x): j = 1, 2, \ldots, i\}$. For the functions $\psi_i(x)$ to be orthogonal we need to find values for the c_{ij}s such that

$$(\psi_k(x), \psi_i(x)) = 0 \qquad k = 1, 2, \ldots, i-1.$$

Now, suppose that we construct the $\psi_i(x)$s one by one in the order $\psi_1(x)$, $\psi_2(x),\ldots$ so that at the $i-1$th stage the functions $\{\psi_k(x): k = 1, 2, \ldots, i-1\}$ form an orthogonal set. Then

$$(\psi_k(x), \psi_i(x)) = \int_a^b \psi_k(x)\psi_i(x)w(x)\,dx$$

$$= \int_a^b \psi_k(x)\left(\phi_i(x) - \sum_{j=1}^{i-1} c_{ij}\psi_j(x)\right)w(x)\,dx$$

$$= \int_a^b \psi_k(x)\phi_i(x)w(x)\,dx$$

$$- c_{ik}\int_a^b \psi_k(x)\psi_k(x)w(x)\,dx \qquad k = 1, 2, \ldots, i-1$$

remembering that $(\psi_k(x), \psi_j(x)) = 0: j = 1, 2, \ldots, k-1, k+1, \ldots, i-1$.

Hence $\psi_i(x)$ will be orthogonal to $\psi_1(x), \psi_2(x),\ldots, \psi_{i-1}(x)$ if

$$c_{ik} = \int_a^b \psi_k(x)\, \phi_i(x) w(x)\, dx \Big/ \int_a^b \psi_k(x)\, \psi_k(x) w(x)\, dx$$

and so the process

$$\psi_1(x) = \phi_1(x)$$

$$\psi_i(x) = \phi_i(x) - \sum_{j=1}^{i-1} \frac{(\psi_j(x),\, \phi_i(x))}{(\psi_j(x),\, \psi_j(x))} \psi_j(x) \quad i = 2, 3,\ldots, n \quad (4.37)$$

can be used to generate a sequence of orthogonal functions from the basis $\{\phi_i(x): i = 1, 2,\ldots, n\}$.

The process (4.37) is known as the Gram–Schmidt orthogonalisation process and it can be used to generate a sequence of orthogonal functions starting with any linearly independent basis. If we normalise at each stage, that is divide $\psi_i(x)$ by $(\psi_i(x), \psi_i(x))^{1/2}$, then the new functions will form an orthonormal basis.

EXAMPLE 4.11
Let $[a,b] = [-1,1]$ and choose $w(x) = (1-x^2)^{-1/2}$. Then, starting with the monomials $\phi_i(x) = x^{i-1}$: $i = 1, 2,\ldots$ we have

$$\psi_1(x) = 1$$

$$\psi_2(x) = x - \frac{(1,\, x)}{(1,\, 1)}\, 1\ .$$

Now

$$(1,\, x) = \int_{-1}^{1} \frac{x}{\sqrt{1-x^2}}\, dx\ .$$

Make the change of variable $x = \cos(u)$ so that $dx = -\sin(u)du$ and $\sqrt{1-x^2} = \sqrt{1-\cos^2(u)} = \sin(u)$. Then

$$(1,\, x) = -\int_\pi^0 \cos(u)\, du = \int_0^\pi \cos(u)\, du = [\sin(u)]_0^\pi = 0\ .$$

Hence

$\psi_2(x) = x.$

The next polynomial in the sequence, $\psi_3(x)$, is given by

$$\psi_3(x) = x^2 - \frac{(1, x^2)}{(1, 1)} 1 - \frac{(x, x^2)}{(x, x)} x$$

and we have

$$(x, x) = (1, x^2) = \int_{-1}^{1} \frac{x^2}{\sqrt{1-x^2}} \, dx = \frac{\pi}{2}$$

$$(1, 1) = \int_{-1}^{1} \frac{1}{\sqrt{1-x^2}} \, dx = \pi$$

$$(x, x^2) = \int_{-1}^{1} \frac{x^3}{\sqrt{1-x^2}} \, dx = 0 \ .$$

Hence

$$\psi_3(x) = x^2 - \tfrac{1}{2} \ .$$

4.5.3 Chebyshev polynomials

In Example 4.11 we generated the sequence of polynomials 1, x, $x^2 - \tfrac{1}{2}$ which are orthogonal on $[-1,1]$ with respect to the weight function $(1-x^2)^{-1/2}$. Now consider the polynomials

$$T_i(x) = \cos(i \cos^{-1}(x)) \qquad i = 0, 1, \ldots \tag{4.38}$$

so that we have

$$T_0(x) = \cos(0) = 1$$

$$T_1(x) = \cos(\cos^{-1}(x)) = x$$

$$T_2(x) = \cos(2 \cos^{-1}(x)) = 2 \cos^2(\cos^{-1}(x)) - 1 = 2x^2 - 1$$

and so on. Then these polynomials are also orthogonal on $[-1,1]$ with respect to the weight function $(1-x^2)^{-1/2}$ since

$$(T_i(x), T_j(x)) = \int_{-1}^{1} \cos(i \cos^{-1}(x)) \cos(j \cos^{-1}(x)) \frac{1}{\sqrt{1-x^2}} \, dx$$

$$= \int_0^{\pi} \cos(iu) \cos(ju) \, du = 0 \quad i \neq j \quad i, j = 0, 1, \ldots.$$

Now, using a standard result, we have

$$\cos(i \cos^{-1}(x)) + \cos((i-2)\cos^{-1}(x))$$
$$= 2\cos((i-1)\cos^{-1}(x)) \cos(\cos^{-1}(x)) \quad i = 2, 3, \ldots$$

so that

$$T_i(x) + T_{i-2}(x) = 2T_{i-1}(x)x$$

or

$$T_i(x) = 2xT_{i-1}(x) - T_{i-2}(x) \tag{4.39}$$

and so these polynomials satisfy a three term recurrence relation. From (4.39) it is clear that $T_i(x)$ will be of degree i and the next two polynomials in the sequence are

$$T_3(x) = 4x^3 - 3x$$
$$T_4(x) = 8x^4 - 8x^2 + 1.$$

The polynomials $T_i(x)$ are known as Chebyshev polynomials and apart from a multiplicative constant they are the same as those generated by the Gram–Schmidt orthogonalisation process. The next example shows how Chebyshev polynomials may be used in least squares approximation.

EXAMPLE 4.12
Suppose that we wish to find a least squares approximation to e^x on $[-1, 1]$ of the form $p_3(x) = \alpha_1 T_0(x) + \alpha_2 T_1(x) + \alpha_3 T_2(x) + \alpha_4 T_3(x)$ where the $T_i(x)$s are Chebyshev polynomials. Now

$$\int_{-1}^{1} T_i(x) T_i(x) \frac{1}{\sqrt{1-x^2}} \, dx = \int_0^{\pi} \cos^2(iu) \, du = \begin{cases} \pi & i = 0 \\ \pi/2 & i = 1, 2, \ldots \end{cases} \tag{4.40}$$

so that if we use a weighted inner product with $w(x) = (1-x^2)^{-1/2}$ the coefficient matrix in (4.36) is diagonal. We have immediately that

$$\alpha_1 = \frac{1}{\pi} \int_{-1}^{1} \frac{e^x T_0(x)}{\sqrt{1-x^2}} \, dx = 1.26607$$

$$\alpha_2 = \frac{2}{\pi} \int_{-1}^{1} \frac{e^x T_1(x)}{\sqrt{1-x^2}} \, dx = 1.13032$$

$$\alpha_3 = \frac{2}{\pi} \int_{-1}^{1} \frac{e^x T_2(x)}{\sqrt{1-x^2}} \, dx = 0.271495$$

$$\alpha_4 = \frac{2}{\pi} \int_{-1}^{1} \frac{e^x T_3(x)}{\sqrt{1-x^2}} \, dx = 0.0443368$$

so that

$$p_3(x) = 1.26607 + 1.13032x + 0.271495\,(2x^2 - 1) + 0.0443368\,(4x^3 - 3x)\ .$$

The integrals which define α_1, α_2, α_3 and α_4 have been evaluated not analytically but numerically to six significant figures. The subject of numerical integration is not discussed until Chapter 6 but the reader may be interested to know that an eight-point Gauss Chebyshev rule was employed.

4.5.4 Choosing a basis

Before leaving least squares approximation it is important that something be said about the choice of basis functions. It is advisable to obtain as much information as possible about the overall form of the function to be approximated and then choose a basis which reflects any known behaviour. For example it may be known that the function is even (odd) and hence there is no point in including odd (even) functions in the basis set. Thus a suitable choice of basis functions for the approximation of $\cos(x)$ over $[-\pi/2, \pi/2]$ would be $1, x^2, x^4,\ldots$ or, better still, the polynomials produced by the Gram–Schmidt orthogonalisation process from this set. Another situation that might arise is that in which the function is periodic on $[-\pi, \pi]$. In such a case it is natural to use as an approximating function a linear combination of periodic functions.

4.6 APPROXIMATION ON NON-FINITE INTERVALS

We now look at the problem of approximating a function $f(x)$ over an interval for which one, or possibly both, of the end points is not finite. To find a least squares approximation the approach is exactly the same as that outlined elsewhere in this chapter but we must employ an appropriate weight function. We simply derive the normal equations and then solve for the polynomial coefficients. Once again orthogonal polynomials can be used to keep the amount of computational effort involved to a minimum.

Suppose that we wish to approximate $f(x)$ over the interval $]-\infty, \infty[$. Then with respect to the inner product

$$(g(x), h(x)) = \int_{-\infty}^{\infty} g(x)h(x)e^{-x^2}\,dx \qquad (4.41)$$

the polynomials defined by the three term recurrence relation

$$H_n(x) = 2xH_{n-1}(x) - 2(n-1)H_{n-2}(x) \qquad n = 2, 3,\ldots$$

with

$$H_0(x) = 1,\ H_1(x) = 2x$$

are orthogonal and are known as the *Hermite polynomials*. If we seek an approximation of the form $p_{n-1}(x) = \alpha_1 H_0(x) + \alpha_2 H_1(x) + \ldots + \alpha_n H_{n-1}(x)$ and use the inner product (4.41) the coefficient matrix in the normal equations is diagonal and so the α_is can be found as

$$\alpha_i = \int_{-\infty}^{\infty} f(x)H_{i-1}(x)e^{-x^2}\,dx \bigg/ \int_{-\infty}^{\infty} H_{i-1}^2(x)e^{-x^2}\,dx\ .$$

It can be shown that

$$\alpha_i = \int_{-\infty}^{\infty} f(x)H_{i-1}(x)e^{-x^2}\,dx \bigg/ \sqrt{\pi}\,2^{i-1}(i-1)!$$

$$i = 1, 2,\ldots, n\ .$$

For semi-infinite intervals we consider without loss of generality $[0, \infty[$ since any other semi-infinite range can easily be mapped onto this interval. With respect to the inner product

$$(g(x), h(x)) = \int_0^{\infty} g(x)h(x)e^{-x}\,dx \qquad (4.42)$$

the polynomials defined by the three term recurrence relation

$$L_n(x) = (2n-1-x)L_{n-1}(x) - (n-1)^2 L_{n-2}(x) \qquad n = 2, 3,\ldots$$

with

$$L_0(x) = 1, \; L_1(x) = 1 - x$$

are orthogonal and are known as the *Laguerre* polynomials. Using (4.42) the coefficients in the least squares approximation of the form $p_{n-1}(x) = \alpha_1 L_0(x) + \alpha_2 L_1(x) + \ldots + \alpha_n L_{n-1}(x)$ to $f(x)$ on $[0, \infty[$ are given by

$$\alpha_i = \int_0^\infty f(x) L_{i-1}(x) e^{-x} \, dx \Big/ \int_0^\infty L_{i-1}^2(x) e^{-x} \, dx \; .$$

It can be shown that

$$\alpha_i = \int_0^\infty f(x) L_{i-1}(x) e^{-x} \, dx / ((i-1)!)^2 \qquad i = 1, 2, \ldots, n \; .$$

4.7 ALTERNATIVE CRITERIA FOR BEST APPROXIMATION

4.7.1 L_p approximation

In the last four sections we have looked at least squares approximations in which the measure (4.10) (or a generalisation of it) has been used to determine the optimal values of the coefficients in the approximating function. As we remarked in subsection 4.3.1 other measures (such as (4.9)) are possible and we now investigate the class of measures which define the L_p *approximations*.

Recall that the (unweighted) least squares approximation to a given function on the finite interval $[a,b]$ gives

$$\underset{\alpha_1, \alpha_2, \ldots, \alpha_n}{\text{minimum}} \int_a^b (f(x) - L(\mathbf{\alpha}, x))^2 \, dx$$

that is, the integral of the square of the residual function $r(x) = f(x) - L(\mathbf{\alpha}, x)$ is minimised with respect to the coefficients $\alpha_1, \alpha_2, \ldots, \alpha_n$. The *least first power* solution minimises not the integral of the square but the integral of the absolute value of the residual, that is, it gives

$$\underset{\alpha_1, \alpha_2, \ldots, \alpha_n}{\text{minimum}} \int_a^b |f(x) - L(\mathbf{\alpha}, x)| \, dx$$

whilst the *Chebyshev* (or *minimax*) solution gives

$$\underset{\alpha_1,\alpha_2,\ldots,\alpha_n}{\text{minimum}} \underset{a\leq x\leq b}{\text{maximum}} |f(x) - L(\mathbf{\alpha}, x)| \tag{4.43}$$

and here the maximum absolute value of the residual is minimised. (The quantity (4.43) is a generalisation of the measure (4.9).)

Fig. 4.1 illustrates the difference between least squares, least first power and minimax approximation. The function $f(x)$ is to be approximated by the straight line $p_1(x) = \alpha_1 + \alpha_2 x$. Fig. 4.1(a) shows an arbitrary straight line approximation and 4.1(b) the corresponding residual function. To obtain the least squares solution we need to find those values of α_1 and α_2 which minimise the area under the curve $r^2(x)$ of Fig. 4.1(c) between a and b. The least first power solution gives the minimum area under the curve $|r(x)|$ of Fig. 4.1(d) whilst for the minimax solution the maximum deviation of the curve $r(x)$ in Fig. 4.1(b) from the x-axis within $[a,b]$ needs to be minimised. Note that in the last three diagrams of Fig. 4.1 the given curve crosses, or touches, the x-axis at the points at which the curves $y = f(x)$ and $y = \alpha_1 + \alpha_2 x$ intersect in Fig. 4.1(a).

The three approximations considered so far in this subsection are members of the L_p family of approximations. Formally we have that the L_p approximation to $f(x)$ on $[a,b]$ gives

$$\underset{\alpha_1,\alpha_2,\ldots,\alpha_n}{\text{minimum}} \int_a^b |f(x) - L(\mathbf{\alpha}, x)|^p \, dx \tag{4.44}$$

and this definition may be generalised further by the insertion of a weight function, $w(x)$, inside the integral. It can be shown that as $p \to \infty$ the approximation which gives (4.44) tends to the minimax (or L_∞) approximation which we now consider in detail.

4.7.2 Minimax approximation

In Example 4.11 and subsection 4.5.3 we introduced a sequence of polynomials, the Chebyshev polynomials, which are orthogonal on $[-1,1]$ with respect to the weight $w(x) = (1 - x^2)^{-1/2}$. These polynomials play an important role in the computation of minimax polynomial approximations and we begin this subsection by stating a very important property that they possess.

THEOREM 4.1 (Minimum deviation)
Of all polynomials of degree n with leading coefficient one, $T_n(x)/2^{n-1}$ has minimum maximum amplitude (that is, smallest deviation from zero) on $[-1,1]$.

A proof of this theorem may be found in Ralston and Rabinowitz (1978). It is fairly easy to see from the three term recurrence relation (4.39) that $T_n(x)/2^{n-1}$ is a polynomial of degree n with leading coefficient 1. Now

$$\underset{-1\leq x\leq 1}{\text{max}} |T_n(x)| = 1$$

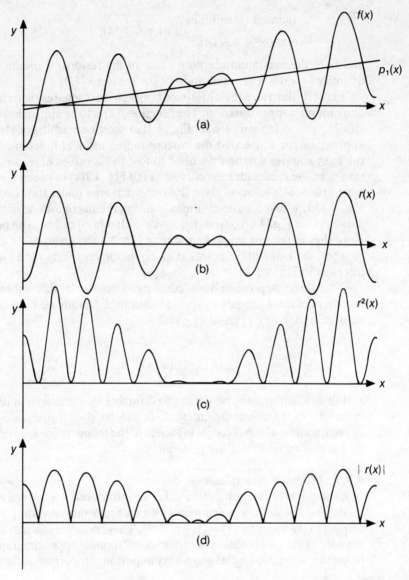

Fig. 4.1

and $T_n(x)$ takes on this maximum value at the points $x_j = \cos(j\pi/n)$: $j = 0, 1, \ldots, n$, these results following from the definition (4.38). Moreover $T_n(x_j) = -T_n(x_{j+1})$: $j = 0, 1, \ldots, n-1$ and $T_n(x)$ is said to *equioscillate* with amplitude one on $n+1$ points of $[-1, 1]$. (See Fig. 4.2 for the case $n = 4$.)

Now, Theorem 4.1 says that if $p_n(x)$ is any other polynomial with leading

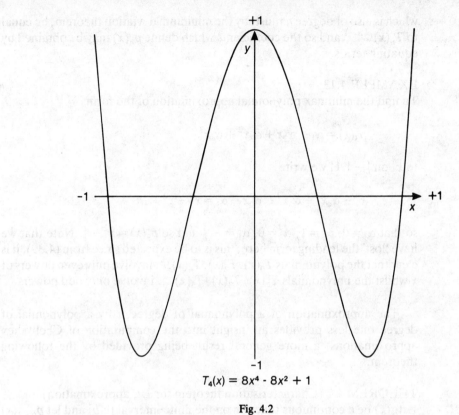

$T_4(x) = 8x^4 - 8x^2 + 1$

Fig. 4.2

coefficient 1 then

$$\max_{-1 \leq x \leq 1} |T_n(x)| \leq \max_{-1 \leq x \leq 1} |p_n(x)|.$$

Consider the problem of computing the minimax polynomial approximation of degree $n-1$ to the polynomial $q_n(x)$ of degree n and let $p_{n-1}(x)$ be that polynomial which gives

$$\min_{\alpha_1, \alpha_2, \ldots \alpha_n} \max_{-1 \leq x \leq 1} \left| q_n(x) - \sum_{i=1}^{n} \alpha_i x^{i-1} \right|.$$

Then if the leading coefficient of $q_n(x)$ is 1

$$r_n(x) = q_n(x) - p_{n-1}(x)$$

which is also of degree n must, by the minimum deviation theorem, be equal to $T_n(x)/2^{n-1}$ and so the coefficients which define $p_n(x)$ may be obtained by equating terms.

EXAMPLE 4.13
To find the minimax polynomial approximation of the form

$$p_3(x) = \alpha_1 + \alpha_2 x + \alpha_3 x^2 + \alpha_4 x^3$$

to x^4 on $[-1,1]$ we write

$$x^4 - \alpha_4 x^3 - \alpha_3 x^2 - \alpha_2 x - \alpha_1 = x^4 - x^2 + 1/8$$

so that $\alpha_4 = 0$, $\alpha_3 = 1$, $\alpha_2 = 0$, $\alpha_1 = -\frac{1}{8}$ and so $p_3(x) = x^2 - \frac{1}{8}$. Note that we have 'lost' the leading term here; this is to be expected since from (4.39), it is clear that the polynomials $T_0(x), T_2(x), T_4(x),\ldots$ involve only even powers of x whilst the polynomials $T_1(x), T_3(x), T_5(x),\ldots$ involve only odd powers.

The approximation of a polynomial of degree n by a polynomial of degree one less provides an insight into the computation of Chebyshev approximations, a more general result being provided by the following theorem.

THEOREM 4.2 (Characterisation theorem for L_∞ approximation)
Let $f(x)$ be a continuous function on the finite interval $[a,b]$ and let $p_{n-1}(x)$ be the corresponding minimax polynomial approximation of degree at most $n-1$. Then the residual function $r(x) = f(x) - p_{n-1}(x)$ equioscillates on $n+1$ distinct points in $[a,b]$. (See Fig. 4.3.)

This theorem (see Phillips and Taylor (1973) for a proof) can be extended to a much wider class of basis functions subject only to a few minor constraints — see Powell (1981) for further details. Moreover it can be shown (see Powell, 1981) that if a polynomial can be found such that the residual function equioscillates on $n+1$ points of $[-1,1]$ then that polynomial must be the minimax approximation. Algorithms which use the equioscillation property to find a minimax approximation are rather complex in nature. However we now show that it is possible to find a polynomial approximation which almost satisfies the Chebyshev criterion without difficulty.

4.7.3 Chebyshev series approximation
In section 4.2 a polynomial approximation to $f(x)$ was obtained by truncating its Taylor series expansion. Clearly we can also express the Taylor expansion as a linear combination of Legendre or Chebyshev polynomials and truncate that instead.

Sec. 4.7] ALTERNATIVE CRITERIA FOR BEST APPROXIMATION 185

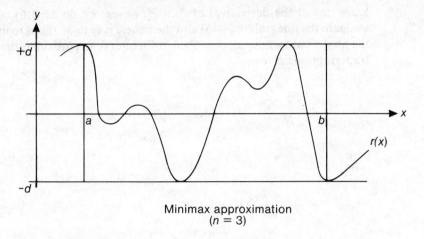

Minimax approximation $(n = 3)$

Fig. 4.3

Let

$$f(x) = \sum_{i=0}^{\infty}{}' a_i T_i(x) \ .$$

(The prime on the summation sign indicates that the first term in the series is halved. This is a conventional nicety only and merely serves to make life easy for us later on.) Now since the Chebyshev polynomials are orthogonal on $[-1,1]$ with respect to the weight $(1-x^2)^{-1/2}$ we have

$$\int_{-1}^{1} \frac{f(x)\, T_j(x)}{\sqrt{1-x^2}} \, dx = \int_{-1}^{1} \sum_{i=0}^{\infty}{}' a_i T_i(x)\, T_j(x) \frac{1}{\sqrt{1-x^2}} \, dx$$

$$= \sum_{i=0}^{\infty}{}' a_i \int_{-1}^{1} T_i(x)\, T_j(x) \frac{1}{\sqrt{1-x^2}} \, dx$$

$$= \frac{\pi}{2} a_j \quad j = 0, 1, \dots \ .$$

(Note the effect of the prime in relation to the result (4.40).) Making the usual change of variable we have

$$a_j = \frac{2}{\pi} \int_0^{\pi} f(\cos(\theta))\, \cos(j\theta) d\theta. \tag{4.45}$$

To obtain the Chebyshev series coefficients $a_j : j = 0, 1, \dots$ we do not need to know any of the derivatives of $f(x)$. However, we do need to be able to evaluate the integrals in (4.45) and the easiest way to do this is to use one of the numerical techniques described in Chapter 6. In particular, the N-panel trapezium rule gives

$$a_j \approx \frac{2}{N} \sum_{k=0}^{N} {}'' f(x_k) T_j(x_k) \qquad j = 0, 1, \ldots$$

with

$$x_k = \cos\left(\frac{\pi k}{N}\right) \qquad k = 0, 1, \ldots, N$$

where the primes on the summation sign indicate that both the first and the last terms are to be halved.

Suppose now that we truncate the Chebyshev series expansion so that

$$p_{n-1}(x) = \sum_{i=0}^{n-1} {}' a_i T_i(x)$$

may be regarded as a polynomial approximation of degree at most $n-1$ to $f(x)$. Then the residual function is given as

$$r(x) = f(x) - p_{n-1}(x) = \sum_{i=0}^{\infty} {}' a_i T_i(x) - \sum_{i=0}^{n-1} {}' a_i T_i(x) = \sum_{i=n}^{\infty} a_i T_i(x).$$

If the series converges rapidly then the form of $r(x)$ will be dominated by the term $a_n T_n(x)$ provided that a suitably large value of n has been taken. This means that the residual function behaves like a Chebyshev polynomial so that, effectively, $p_{n-1}(x)$ is the minimax approximation to $f(x)$.

Chebyshev series approximation on intervals other than $[-1, 1]$ is quite straightforward as the next example shows.

EXAMPLE 4.14

Suppose that we wish to find a Chebyshev series approximation to e^x valid on $[0, 1]$. Then we need to make the change of variable

$$z = 2x - 1 \ . \tag{4.46}$$

Now $x = (z+1)/2$ so we need to compute the Chebyshev coefficients for $e^{(z+1)/2}$ and, using the program of the next subsection, the first five terms are

$a_0/2 = 3.50677$

$a_1 = 0.850391$

$a_2 = 0.105208$

$a_3 = 0.00872238$

$a_4 = 0.000542785$.

The series is converging at a reasonably fast rate and if we truncate after the term involving a_3, to give

$$p_3(x) = \frac{a_0}{2} T_0(2x-1) + a_1 T_1(2x-1) + a_2 T_2(2x-1) + a_3 T_3(2x-1)$$

the residual will be dominated by the term $a_4 T_4(2x-1)$. Note that the change of variable (4.46) must be used when evaluating the Chebyshev polynomials. Now, from the definition (4.38), $|T_4(2x-1)| < 1$ so that

$$\underset{x \in [0,1]}{\text{maximum}} |e^x - p_3(x)| \simeq |a_4| = 0.000542785 .$$

Looking at the output from the program of Case Study 4.3 we see that $|e^x - p_3(x)|$ takes a value which is close to $|a_4|$ at five points and that the sign of the residual at these points is alternately positive then negative.

4.7.4 A program for Chebyshev series approximation

The programs of Case Study 4.3 find a Chebyshev series approximation to e^x valid on [0,1]. These programs make use of some of the code of Case Study 6.1 to evaluate the Chebyshev series coefficients numerically using the trapezium rule. Full details of the procedure *trapeziumrule* are given in subsection 6.2.3.

In Case Study 4.2 we declared an operator in the Algol 68 program and a function in the Pascal program in order to evaluate a polynomial expressed as a linear combination of Legendre polynomials. These declarations were derived from the three term recurrence relation (4.24) and the initial values (4.25). By using the corresponding recurrence relation (4.39) and initial values for the Chebyshev polynomials, it is possible to produce an operator/function for evaluating a polynomial expressed as a linear combination of Chebyshev polynomials. Such an operator/function is required in the programs of Case Study 4.3. We have left it as an exercise for the reader to provide the code for the operator/function declaration.

Although the program computes the coefficients $a_i : i = 0, 1, ..., $ *highest-coefficient* in the Chebyshev series approximation to e^x, the final term is only used to check the equioscillation property. Hence when evaluating our approximation we omit this term. In the Algol 68 program this is achieved by using a *trim* of the array *a* as the first operand of *evaluateat*. Note that the @ operator is used to ensure that the lower bound of this trim is 0 and not, by default, 1.

In the Pascal program the value of π is calculated from *arctan* using a rearrangement of the identity

$$\tan\left(\frac{\pi}{4}\right) = 1$$

and this value is assigned to the variable *pi*. This ensures that the value of π used in the program is correct to machine precision. For the reasons given in subsection 4.3.3, in Pascal, a variable has to be used to represent π even though we do not intend to alter the variable's value; in Algol 68, we can use the predefined constant *pi*.

We noted earlier that there is no need to declare an Algol 68 loop control parameter. A consequence of this is that the scope of a loop control parameter is limited to the body of the *DO* clause. In the program of Case Study 4.3 we need to refer to the value of the loop control parameter *i* within the function *g*. We do this by copying its value into the variable *term* and using *term* instead of *i* in the body of *g*. This problem does not occur in the Pascal program because the control variable has a scope which encloses the function *g*.

The program approximates e^x on [0,1] using the mapping (4.46). Changing the function to be approximated is easy but changing the domain of interest is not. The reader is now encouraged to modify the program so that the lower and upper limits of the range [*a*,*b*] are read in as data.

CASE STUDY 4.3

```
Algol 68 Program
BEGIN
  COMMENT
  a program to evaluate chebyshev series coefficients and to check the
  approximate equioscillation property of a truncated series
  COMMENT
  REAL tol = 0.000005 ;
  PROC f = ( REAL x ) REAL :
    COMMENT
    the function whose chebyshev coefficients are required
    COMMENT
    exp ( x ) ;
  INT term ;
  PROC g = ( REAL theta ) REAL :
    COMMENT
    the function f ( x ) must undergo two transformations ; the first of these
    maps the range [ 0 , 1 ] onto [ - 1 , + 1 ] ( using equation (4.46) ) whilst
    the second uses the change of variable x = cos ( u ) to ensure that the
    integral to be approximated is in the form given in equation (4.45) ;
    g ( theta ) defines the integrand in this latter equation
    COMMENT
    f ( ( cos ( theta ) + 1.0 ) * 0.5 ) * cos ( term * theta ) ;
```

```
PROC trapeziumrule = ( PROC ( REAL ) REAL integrand ,
                       REAL lower , upper , tolerance ,
                       INT maximumdepth , BOOL monitorprogress ,
                       REF BOOL converged , REF REAL newintegral ) VOID :
BEGIN
  COMMENT
  a procedure for approximating a definite integral using the trapezium rule
  we omit the details
  COMMENT
END ;
OP EVALUATEAT = ( REF [ ] REAL coeffs , REAL x ) REAL :
BEGIN
  COMMENT
  an operator to evaluate a polynomial expressed as a linear combination of
  chebyshev polynomials
  we omit the details
  COMMENT
END ;
INT highestcoefficient , numberofevaluationpoints ;
read ( ( highestcoefficient , numberofevaluationpoints ) ) ;
BOOL converged ;
[ 0 : highestcoefficient ] REAL coefficients ;
REAL halfpi = pi * 0.5 ;
COMMENT
compute the chebyshev series coefficients
COMMENT
FOR i FROM 0 TO highestcoefficient
DO
  term := i ;
  trapeziumrule ( g , 0.0 , pi , tol , 10 , FALSE , converged ,
                                                    coefficients [ i ] ) ;
  coefficients [ i ] /:= halfpi ;
  IF NOT converged
  THEN
    print ( ( newline , "the computed approximation to the  " , term ,
              " th coefficient is not correct to the tolerance  " , tol ) )
  FI
OD ;
print ( ( newline , "the chebyshev polynomial coefficients are" ,
          " ( constant term first )" , newline ) ) ;
FOR i FROM 0 TO highestcoefficient
DO
  print ( ( newline , coefficients [ i ] ) )
OD ;
COMMENT
do not forget to halve the first term
COMMENT
coefficients [ 0 ] *:= 0.5 ;
COMMENT
check the equioscillation property
COMMENT
REAL step = 1.0 / ( numberofevaluationpoints - 1 ) ;
REAL xvalue := - step , approximatevalue ;
print ( ( newline , newline , newline ,
          "table of approximate function values" , newline , newline ,
          "     x              approximate value          error",
          newline ) ) ;
FOR j TO numberofevaluationpoints
DO
  xvalue +:= step ;
  approximatevalue := coefficients [ 0 : highestcoefficient - 1 @ 0 ]
                      EVALUATEAT ( 2.0 * xvalue - 1.0 ) ;
  print ( ( xvalue , "         " , approximatevalue , "            " ,
            f ( xvalue ) - approximatevalue , newline ) )
OD
END
```

Pascal Program

```
PROGRAM chebyshevseriesapproximation(input, output);

    {  a program to evaluate chebyshev series coefficients and to check
       the approximate equioscillation property of a truncated series  }

    CONST
       maxhighestcoefficient=50;
       tol=0.000005;

    TYPE
       indextype=0..maxhighestcoefficient;
       coeffsvector=ARRAY[ indextype ] OF real;

    VAR
       halfpi, pi, step, xvalue, approximatevalue:real;
       i, highestcoefficient:indextype;
       j, numberofevaluationpoints:1..maxint;
       converged:boolean;
       coefficients:coeffsvector;

    FUNCTION f(x:real):real;
       {  the function whose chebyshev coefficients are required  }
       BEGIN
          f:= exp(x)
       END { f } ;

    FUNCTION g(theta:real):real;
       {  the function f(x) must undergo two transformations;
          the first of these maps the range [0,1] onto [-1,+1] (using
          equation (4.46)) whilst the second uses the change of variable
          x = cos(u) to ensure that the integral to be approximated is
          in the form given in equation (4.45);
          g(theta) defines the integrand in this latter equation  }
       BEGIN
          g:= f((cos(theta) + 1.0)*0.5)*cos(i*theta)
       END { g } ;

    PROCEDURE trapeziumrule(FUNCTION integrand(x:real):real;
                            lower, upper, tolerance:real;
                            maximumdepth:integer; monitorprogress:boolean;
                            VAR converged:boolean; VAR newintegral:real);
       {  a procedure for approximating a definite integral using
          the trapezium rule - we omit the details  }
       END { trapeziumrule } ;

    FUNCTION evaluate(n:indextype; VAR coeffs:coeffsvector; x:real):real;
       {  a function to evaluate a polynomial expressed as a linear
          combination of chebyshev polynomials
          we omit the details  }
       END { evaluate } ;

BEGIN
    pi:= 4.0*arctan(1.0);
    halfpi:= pi*0.5;
    read(highestcoefficient, numberofevaluationpoints);
    writeln;
    FOR i:= 0 TO highestcoefficient DO
    BEGIN
       trapeziumrule(g, 0.0, pi, tol, 10, false, converged, coefficients[i]);
       coefficients[i]:= coefficients[i]/halfpi;
       IF NOT converged
       THEN
          writeln('the computed approximation to the   ', i,
                  ' th coefficient is not correct to the tolerance',
                  tol)
    END ;
```

Sec. 4.7] ALTERNATIVE CRITERIA FOR BEST APPROXIMATION 191

```
         writeln('the chebyshev polynomial coefficients are',
                '( constant term first )');
         writeln;
         FOR i:= 0 TO highestcoefficient DO
            writeln(coefficients[i]);
         { do not forget to halve the first term }
         coefficients[0]:= coefficients[0]*0.5;
         { check the equioscilation property }
         writeln;
         writeln;
         writeln('table of approximate function values');
         writeln;
         writeln('      x              approximate value        error');
         step:= 1.0/(numberofevaluationpoints - 1);
         xvalue:= -step;
         FOR j:= 1 TO numberofevaluationpoints DO
         BEGIN
            xvalue:= xvalue + step;
            approximatevalue:= evaluate(highestcoefficient - 1,
                               coefficients, 2.0*xvalue - 1.0);
            writeln(xvalue, ' ':9, approximatevalue,
                            ' ':9, f(xvalue) - approximatevalue)
         END
   END { chebyshevseriesapproximation } .
```

Sample Data

4   9

Sample Output

the chebyshev polynomial coefficients are (constant term first)

```
   3.50677e  0
   8.50391e -1
   1.05208e -1
   8.72246e -3
   5.42886e -4
```

table of approximate function values

x	approximate value	error
0.00000e 0	9.99481e -1	5.18557e -4
1.25000e -1	1.13365e 0	-5.01618e -4
2.50000e -1	1.28431e 0	-2.83301e -4
3.75000e -1	1.45473e 0	2.63886e -4
5.00000e -1	1.64818e 0	5.43621e -4
6.25000e -1	1.86793e 0	3.16132e -4
7.50000e -1	2.11725e 0	-2.53918e -4
8.75000e -1	2.39942e 0	-5.47871e -4
1.00000e 0	2.71771e 0	5.74272e -4

4.7.5 Least first power approximation

Finally in this section we look at the least first power approximation problem. The solution to this problem is, in certain circumstances, surprisingly easy to calculate as shown by the following theorem.

THEOREM 4.3

Let $f(x)$ be a continuous function on the finite interval $[-1,1]$ and let $p_{n-1}(x)$ be the least first power polynomial approximation to $f(x)$ of degree at most $n-1$. Suppose that the residual function $r(x) = f(x) - p_{n-1}(x)$ has precisely n zeros. Then the position of each zero is independent of $f(x)$ and is given by

$$\theta_i = \cos\left(\frac{i\pi}{n+1}\right) \qquad i = 1, 2, \ldots, n.$$

This theorem (proof in Powell, 1981) says that, provided certain conditions are satisfied, the computation of an L_1 approximation reduces to finding values for the polynomial coefficients such that

$$r(\theta_i) = f(\theta_i) - p_{n-1}(\theta_i) = 0 \qquad i = 1, 2, \ldots, n$$

that is, we need to find values for the coefficients which ensure that the approximation passes through the coordinates $\{(\theta_i, f(\theta_i)) : i = 1, 2, \ldots, n\}$. This type of problem is known as an *interpolation* problem and is discussed in the next chapter. The following example indicates the way in which the process works.

EXAMPLE 4.15

Consider the approximation of e^x on $[0,1]$. For the general interval $[a,b]$ the zeros of the residual function are given by

$$\theta_i = \tfrac{1}{2}(a+b) + \tfrac{1}{2}(b-a)\cos\left(\frac{i\pi}{n+1}\right) \qquad i = 1, 2, \ldots, n$$

so that for $[0,1]$

$$\theta_i = \tfrac{1}{2} + \tfrac{1}{2}\cos\left(\frac{i\pi}{n+1}\right) \qquad i = 1, 2, \ldots n.$$

Choosing $n = 3$ it is fairly easy to see that the polynomial

$$p_2(x) = \frac{(x-\theta_2)(x-\theta_3)}{(\theta_1-\theta_2)(\theta_1-\theta_3)} e^{\theta_1} + \frac{(x-\theta_1)(x-\theta_3)}{(\theta_2-\theta_1)(\theta_2-\theta_3)} e^{\theta_2}$$

$$+ \frac{(x-\theta_1)(x-\theta_2)}{(\theta_3-\theta_1)(\theta_3-\theta_2)} e^{\theta_3}$$

passes through the coordinates $\{(\theta_1, e^{\theta_1}), (\theta_2, e^{\theta_2}), (\theta_3, e^{\theta_3})\}$.
Rearranging we get

$$p_n(x) = e^{\theta_1} \frac{\theta_2 \theta_3}{(\theta_1 - \theta_2)(\theta_1 - \theta_3)} + e^{\theta_2} \frac{\theta_1 \theta_3}{(\theta_2 - \theta_1)(\theta_2 - \theta_3)}$$

$$+ e^{\theta_3} \frac{\theta_1 \theta_2}{(\theta_3 - \theta_1)(\theta_3 - \theta_2)} - x \left\{ e^{\theta_1} \frac{\theta_2 + \theta_3}{(\theta_1 - \theta_2)(\theta_1 - \theta_3)} \right.$$

$$+ e^{\theta_2} \frac{\theta_1 + \theta_3}{(\theta_2 - \theta_1)(\theta_2 - \theta_3)} + e^{\theta_3} \frac{\theta_1 + \theta_2}{(\theta_3 - \theta_1)(\theta_3 - \theta_2)} \right\}$$

$$+ x^2 \left\{ \frac{e^{\theta_1}}{(\theta_1 - \theta_2)(\theta_1 - \theta_3)} + \frac{e^{\theta_2}}{(\theta_2 - \theta_1)(\theta_2 - \theta_3)} \right.$$

$$\left. + \frac{e^{\theta_3}}{(\theta_3 - \theta_1)(\theta_3 - \theta_2)} \right\} \simeq 1.0153 + 0.85030x + 0.83298x^2.$$

Note that Theorem 4.3 says that $p_2(x)$ is the L_1 quadratic approximation on [0,1] to any function which satisfies the conditions of that theorem and which passes through the given coordinates.

The result of Theorem 4.3 holds provided that the residual function has precisely n zeros. This is not as stringent a condition as it might seem at first sight since the condition holds if the first n derivatives of $f(x)$ are continuous on $[-1,1]$ and $f^{(n)}(x)$ is everywhere positive in $[-1,1]$ (and this means that the solution obtained in Example 4.15 is valid). The points $\{\theta_i : i = 1, 2,..., n\}$ (the interpolation points) are the interior points of $[-1,1]$ at which $T_{n+1}(x)$ takes on its maximum value. They are also the zeros of $U_n(x)$, the Chebyshev polynomial of the second kind of degree n. The polynomials $\{U_n(x) : n = 0, 1,...\}$ are orthogonal on $[-1,1]$ with respect to the weight function $w(x) = \sqrt{(1 - x^2)}$ and may be defined by

$$U_0(x) = 1; \quad U_1(x) = 2x;$$
$$U_n(x) = 2x\, U_{n-1}(x) - U_{n-2}(x) \quad n \geq 2.$$

4.8 FURTHER COMMENTS

Chapter 's' of the NAG library contains routines for evaluating a number of standard mathematical functions at a point. Most of these routines make use of truncated Chebyshev series expansions. The only other routines which deal directly with function approximation solve the discrete problem in which a set of coordinates defines the function to be approximated. We shall see that many of the techniques for the approximation of a discrete function are very similar indeed to the methods outlined in this chapter.

So far we have looked at linear approximating functions only. Lifting the restriction that the unknowns in $L(\alpha, x)$ appear linearly can complicate matters considerably. However, rational polynomials, that is polynomials of the form

$$p_{n-1,m-1}(x) = \sum_{i=1}^{n} \alpha_i x^{i-1} / \sum_{j=1}^{m} \beta_j x^{j-1}$$

often possess desirable properties when used as approximating functions. It is normal to choose $\beta_1 = 1$ in order to define $p_{n-1,m-1}(x)$ uniquely. The remaining parameters can be defined by ensuring that $f(x)$ and $p_{n-1,m-1}(x)$ agree at $x = 0$ and have as many derivatives as possible equal at 0. Such approximations, known as Padé approximations, possess the characteristic that they become less accurate away from the origin and hence it is important that the domain of interest be transformed so that zero is at its centre. Note that if $m = 1$ this method gives the first $n - 1$ terms in the Taylor series expansion for $f(x)$ expanded about $x = 0$ (see section 4.2).

Another way of finding values for the parameters in a rational approximation is to solve the associated minimax (or any other L_p) approximation problem. It can be shown (see Ralston and Rabinowitz, 1978) that the residal fnction associated with the L_∞ approximation satisfies an equioscillation property and therefore any standard numerical technique for minimax approximation which uses this property may be used. (One of the most well known techniques for minimax approximation is the *exchange algorithm* (see Powell, 1981).) Rational approximating polynomials have one serious disadvantage in that it is possible for the denominator to become zero and great care must be exercised to ensure that if this happens it is recognised.

EXERCISES

4.1 Expand abot the point $x = 0$ the following functions using Taylor's theorem:

(i) $\sin(x)$;
(ii) $\cos(x)$;   and
(iii) $\ln(1 + x)$.

Use the series expansions to find a polynomial approximation of degree at most five to each function. Find a bound for the error in each approximation if it is to be valid on $[-\pi/4, \pi/4]$ for $\sin(x)$ and $\cos(x)$, and

$[-\frac{1}{2},\frac{1}{2}]$ for $\ln(1+x)$.

4.2 Find the least squares straight line approximation to $\sin(x)$ on $[0,\pi/2]$

(i) directly;
(ii) using polynomials orthogonal on $[0,\pi/2]$; and
(iii) by mapping the interval onto $[-1,1]$ and using Legendre polynomials.

4.3 Expand e^x about the point $x=0$ using Taylor's theorem and use this to find a polynomial approximation of degree five to e^x valid on $[0,1]$. Express this polynomial as a linear combination of Chebyshev polynomials. Show that if the term involving $T_5(x)$ is ignored the bound on the error for this further approximation is smaller than the bound on the error for the original series truncation. (This process is known as *economisation*.)

4.4 A least squares approximation to e^x on $[-\pi,\pi]$ of the form $L(\alpha, x) = \alpha_1 + \alpha_2 \sin(x) + \alpha_3 \cos(x)$ is sought. Construct the coefficient matrix for the system of equations which defines the solution to this problem. Determine values for α_1, α_2 and α_3.

4.5 Find the minimax polynomial approximation of degree 6 to x^7 on $[-1,1]$.

4.6 Show that if the function $f(x)$ is approximated on the interval $[-1,1]$ by $p_{n-1}(x) = \beta_1 P_0(x) + \beta_2 P_1(x) + \ldots + \beta_n P_{n-1}(x)$ where the $P_i(x)$s are Legendre polynomials, the least squares error is given by

$$\int_{-1}^{1} (f(x) - p_{n-1}(x))^2 \, dx = \int_{-1}^{1} f^2(x) \, dx - 2 \sum_{i=1}^{n} \frac{\beta_i^2}{2i-1}.$$

4.7 Write a program which converts a polynomial expressed in the form $\alpha_1 + \alpha_2 x + \ldots + \alpha_n x^{n-1}$ to one expressed as a linear combination of

(i) Legendre polynomials; and
(ii) Chebyshev polynomials.

In both cases write a program to perform the reverse process.

4.8 Write a program which produces a minimal degree least squares polynomial approximation to e^x on $[-1,1]$. To do this you should find the polynomial approximation of degree $0, 1, 2, \theta$ and so on until a point is reached at which the effect of increasing the order of the polynomial becomes negligible. The polynomial approximation should be in the form of a linear combination of Legendre polynomials so that the expression for the least squares error derived in Exercise 4.6 may be used. All integrals should be performed analytically.

4.9 Modify the code of Case Study 4.3 so that it may be used to find the Chebyshev series approximation to a function on the general interval $[a,b]$. Test your program using the function $f(x) = e^x$ and the interval $[0.5, 1.5]$.

5

Approximation of Numerically Defined Functions

5.1 INTRODUCTION

In Chapter 4 we looked at the approximation of a continuous function over a given interval. In many cases of interest, however, the only information available consists of a set of data points $\{(x_i, f_i) : i = 1, 2, \ldots, N\}$ (where f_i is used to represent $f(x_i)$). We are therefore interested with the problem of representing these data in some way, for example by a polynomial. We can demand that this polynomial agrees with the given function at all the given points. We call this an *interpolating polynomial*. However, it is possible to modify the ideas of least squares and minimax approximation introduced in Chapter 4 in order to obtain a polynomial which in some sense 'best' represents the data. Having obtained an approximating function, it can be evaluated at points other than the x_is, differentiated, integrated and so on. Indeed the integration of approximating polynomials is of such importance that we devote an entire chapter to it (Chapter 6).

Interpolating polynomials can be easy to construct as the following example shows.

EXAMPLE 5.1

To find the polynomial which passes through the coordinates

x_i	1	2	3
f_i	1	2	7

we need to find values for a, b and c in $p_2(x) = a + bx + cx^2$ such that $p_2(1) = 1$, $p_2(2) = 2$ and $p_2(3) = 7$. Hence we have

$$a + b + c = 1$$
$$a + 2b + 4c = 2$$
$$a + 3b + 9c = 7.$$

The solution to this system of equations is $a = 4$, $b = -5$ and $c = 2$ so that $p_2(x) = 4 - 5x + 2x^2$.

Polynomials of best approximation will not necessarily pass through any of the given coordinates. We look for a measure to quantify the discrepancy between the known values and those given by the approximating function and attempt to minimise it. An example of the sort of thing that we are trying to do here is to fit a straight line to eleven data points (see Fig. 5.1). The

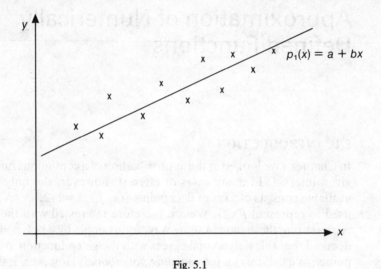

Fig. 5.1

slope (b) and the intercept (a) of the polynomial approximation $p_1(x) = a + bx$ need to be optimised in some way.

5.2 LAGRANGE INTERPOLATION

5.2.1 Interpolation of function values

With Lagrange interpolation we are trying to find a polynomial which exactly fits a given set of data points. Suppose that we have available two coordinates $\{(x_1, f_1), (x_2, f_2)\}$ (with $x_1 \neq x_2$). Let

$$p_1(x) = a + bx$$

be the straight line which links these two points. Then we wish to find values for a and b such that

Sec. 5.2] LAGRANGE INTERPOLATION

$$p_1(x_1) = a + bx_1 = f_1 \tag{5.1}$$

$$p_1(x_2) = a + bx_2 = f_2 \tag{5.2}$$

that is, such that the interpolating polynomial takes on the values f_1 and f_2 at the points x_1 and x_2 respectively. Subtracting (5.1) from (5.2) gives

$$b(x_2 - x_1) = f_2 - f_1$$

or

$$b = (f_2 - f_1)/(x_2 - x_1). \tag{5.3}$$

In fact this result is fairly obvious from a close inspection of Fig. 5.2.

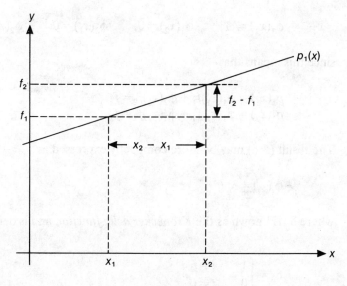

Fig. 5.2

Substituting (5.3) into (5.1) yields the result

$$a = f_1 - \frac{f_2 - f_1}{x_2 - x_1} x_1 = \frac{f_1 x_2 - f_2 x_1}{x_2 - x_1}$$

so that

$$p_1(x) = \frac{1}{x_2 - x_1} \{f_1 x_2 - f_2 x_1 + (f_2 - f_1)x\}$$

$$= \frac{x - x_2}{x_1 - x_2} f_1 + \frac{x - x_1}{x_2 - x_1} f_2$$

$$= \phi_1(x) f_1 + \phi_2(x) f_2 \qquad (5.4)$$

where

$$\phi_1(x) = \frac{x - x_2}{x_1 - x_2}; \quad \phi_2(x) = \frac{x - x_1}{x_2 - x_1}. \qquad (5.5)$$

The fact that $p_1(x)$ passes through the given data points may be easily deduced by noting that

$$\phi_1(x_1) = 1; \quad \phi_1(x_2) = 0; \quad \phi_2(x_1) = 0; \quad \phi_2(x_2) = 1 \qquad (5.6)$$

since this means that

$$p_1(x_1) = \phi_1(x_1) f_1 + \phi_2(x_1) f_2 = f_1$$
$$p_1(x_2) = \phi_1(x_2) f_1 + \phi_2(x_2) f_2 = f_2 .$$

The result (5.6) may be more compactly expressed as

$$\phi_i(x_j) = \delta_{ij}$$

where δ_{ij} is known as the *Kronecker delta function* and is defined by

$$\delta_{ij} = \begin{cases} 1 & i = j \\ 0 & i \neq j \end{cases}.$$

When expressed in the form (5.4), $p_1(x)$ is referred to as the *Lagrange interpolating polynomial* (*LIP*) for the given data points.

The ideas outlined above can now be extended in a fairly obvious manner. Given N coordinates (with the x_is distinct) it is always possible to find a polynomial of degree at most $N - 1$ which passes through them. (We say 'at most' since, for example, it may just happen that all the coordinates lie on a straight line.) Consider the polynomial

$$p_{N-1}(x) = \sum_{i=1}^{N} \phi_i(x)f_i \tag{5.7}$$

with

$$\phi_i(x) = \frac{(x-x_1)(x-x_2)\cdots(x-x_{i-1})(x-x_{i+1})\cdots(x-x_N)}{(x_i-x_1)(x_i-x_2)\cdots(x_i-x_{i-1})(x_i-x_{i+1})\cdots(x_i-x_N)}$$

$$= \prod_{\substack{k=1 \\ k \neq i}}^{N} \frac{x-x_k}{x_i-x_k} \qquad i = 1, 2, \ldots, N \tag{5.8}$$

(so that if $N=2$ we obtain the polynomials defined by (5.5)). Now each $\phi_i(x)$ is a polynomial of degree $N-1$. Further

$$\phi_i(x_j) = \delta_{ij} \tag{5.9}$$

since if $i \neq j$ the term $x_j - x_j = 0$ appears in the numerator whilst if $i = j$ each term in the numerator appears in the denominator. Hence

$$p_{N-1}(x_j) = \sum_{i=1}^{N} \delta_{ij} f_i = f_j$$

so that $p_{N-1}(x)$ passes through the given coordinates.

EXAMPLE 5.2
The polynomial which passes through the points

x_i	1	2	3	4
f_i	6	5	2	−9

is given by

$$p_3(x) = \frac{(x-2)(x-3)(x-4)}{(1-2)(1-3)(1-4)} 6 + \frac{(x-1)(x-3)(x-4)}{(2-1)(2-3)(2-4)} 5$$
$$+ \frac{(x-1)(x-2)(x-4)}{(3-1)(3-2)(3-4)} 2 - \frac{(x-1)(x-2)(x-3)}{(4-1)(4-2)(4-3)} 9.$$

We now express this polynomial in 'standard' form (but note that in practice we would rarely want to do this). We have

$$p_3(x) = -\frac{6}{6}(x^3 - 9x^2 + 26x - 24) + \frac{5}{2}(x^3 - 8x^2 + 19x - 12)$$
$$-\frac{2}{2}(x^3 - 7x^2 + 14x - 8) - \frac{9}{6}(x^3 - 6x^2 + 11x - 6)$$
$$= \frac{1}{2}(-2x^3 + 18x^2 - 52x + 48 + 5x^3 - 40x^2 + 95x - 60 -$$
$$- 2x^3 + 14x^2 - 28x + 16 - 3x^3 + 18x^2 - 33x + 18)$$
$$= \frac{1}{2}(-2x^3 + 10x^2 - 18x + 22)$$
$$= -x^3 + 5x^2 - 9x + 11.$$

Check

$p_3(1) = -1 + 5 - 9 + 11 = 6;\quad p_3(2) = -8 + 20 - 18 + 11 = 5;$
$p_3(3) = -27 + 45 - 27 + 11 = 2;\ p_3(4) = -64 + 80 - 36 + 11 = -9.$

We may now use $p_3(x)$ to obtain an estimate of $f(x)$ at a point. For example, $f(2.5) \simeq p_3(2.5) = 4.125$.

When expressed in the form (5.7), $p_{N-1}(x)$ is known as the Lagrange interpolating polynomial of degree $N-1$ for the points $\{(x_i, f_i) : i = 1, 2, \ldots, N\}$. There is, in fact, only one polynomial of this degree which passes through the given points as shown by the following theorem.

THEOREM 5.1
Let $p_{N-1}(x)$ be a polynomial of degree at most $N-1$ which passes through the coordinates $\{(x_i, f_i) : i = 1, 2, \ldots, N\}$ (with the x_is being distinct). Then $p_{N-1}(x)$ is unique.

Proof
Suppose that $q_{N-1}(x) \not\equiv p_{N-1}(x)$ is another polynomial of degree at most $N-1$ which passes through the given coordinates. Then

$$r_{N-1}(x) \equiv p_{N-1}(x) - q_{N-1}(x)$$

is a polynomial of degree at most $N-1$ and is such that

$$r_{N-1}(x_i) = 0 \quad i = 1, 2, \ldots, N.$$

Thus, $r_{N-1}(x)$ has N zeros and therefore it must be identically equal to zero.

The proof of Theorem 5.1 is quite straightforward and presents a further example of proof-by-contradiction; we assume the required result to be false and then prove that this leads to a contradiction.

EXAMPLE 5.3

We include this example to indicate that the use of an LIP can lead to a very poor approximation. Suppose that we evaluate the function $f(x) = e^{-x} \sin(4x)$ at the points $x = 0.1, 0.6$ and 2 and derive

$$p_2(x) = \frac{(x-0.6)(x-2)}{(-0.5)(-1.9)} 0.352360$$
$$+ \frac{(x-0.1)(x-2)}{(0.5)(-1.4)} 0.370702$$
$$+ \frac{(x-0.1)(x-0.6)}{(1.9)(1.4)} 0.133895.$$

Then we have the situation shown in Fig. 5.3. Clearly $p_2(x)$ is a poor

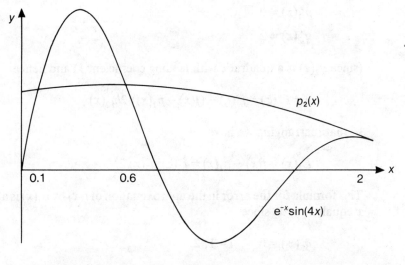

Fig. 5.3

approximation to $f(x)$ and in order to obtain a better result we would have to consider the possibility of evaluating $f(x)$ at a number of other points and using a higher order approximation. Other techniques which we may care to consider are discussed in section 5.5.

We now attempt to obtain an expression for the error (which is obviously zero at the x_is) in an LIP. Let $f(x)$ be the continuous function which gives the values $\{f_i : i = 1, 2, \ldots, N\}$ and let $E_{N-1}(x)$ be defined by

$$E_{N-1}(x) = f(x) - p_{N-1}(x)$$

that is, $E_{N-1}(x)$ is the difference between the true and approximating functions. Return to the case $N = 2$ and define the function

$$F(z) = f(z) - p_1(z) - (f(x) - p_1(x))q_2(z)/q_2(x)$$

where

$$q_2(x) = (x - x_1)(x - x_2) = \prod_{i=1}^{2} (x - x_i)$$

and x is some chosen point other than x_1 or x_2. Then $F(z)$ has three zeros, namely x_1, x_2 and x. (It may have others but this does not concern us.) Hence by Rolle's theorem (Theorem 1.3), $F'(z)$ (where the prime denotes differentiation with respect to z) has at least two zeros in the smallest interval containing $\{x_1, x_2, x\}$. A further application of Rolle's theorem says that $F''(z)$ has at least one zero, call it ξ, in the same interval. Now

$$p''_1(z) = 0$$
$$q''_2(z) = 2$$

(since $q_2(z)$ is a quadratic with leading coefficient 1) and hence

$$0 = F''(\xi) = f''(\xi) - (f(x) - p_1(x)) \, 2/q_2(x)$$

so that rearranging we have

$$E_1(x) = f(x) - p_1(x) = f''(\xi) \, q_2(x)/2.$$

This formula for the error in the approximation of $f(x)$ by $p_1(x)$ is also valid if x equals x_1 or x_2 since

$$q_2(x_i) = 0 \qquad i = 1, 2.$$

The generalisation of the result is straightforward and leads to the expression

$$E_{N-1}(x) = \frac{f^{(N)}(\xi)}{N!} \prod_{i=1}^{N} (x - x_i)$$

where ξ is some point in the smallest interval containing the x_is and x. We have no way of knowing precisely what the value of ξ is and hence we cannot get an exact measure of the error in Lagrange interpolation. However, if an upper bound on the magnitude of $f^{(N)}(x)$ is available it will be possible to obtain an upper bound on $|E_{N-1}(x)|$.

5.2.2 Function minimisation using LIP's

As an example of the use of the LIP, consider the problem of approximating the position of the *minimum* of a real-valued function (whose form is known analytically) of one real variable. Suppose that we wish to minimise the function $f(x)$ and that this function is such that the usual approach for finding a value of x for which $f'(x) = 0$ is far from easy to accomplish. We aim to replace the function with a low order approximating polynomial and minimise that instead.

We recall that at any extremum (that is, a *maximum* or minimum) $\bar{x}$ of $f(x)$

$$f'(\bar{x}) = 0.$$

Suppose that we can find an interval $[x_1, x_3]$ containing the minimum $\bar{x}$. Then by finding the Lagrange interpolating polynomial based on the point set $\{x_1, x_2, x_3\}$ where x_2 is some interior point of $[x_1, x_3]$ the point $\hat{x}$ which satisfies

$$p'(\hat{x}) = 0$$

will be an estimate of $\bar{x}$.

Let us consider this process in more detail. First we need to bracket the minimum $\bar{x}$. The following algorithm attempts to achieve the desired result.

```
choose an initial approximation, z, to x̄ and a step length h;
set x₂ := z; x₃ := z + h;
— minimum has not yet been bracketed
if f(x₃) > f(x₂)
then
  set x₁ := z − h;
  if f(x₁) > f(x₂)
  then
    — minimum has been bracketed
  else
    — minimum must lie in the opposite direction
    set h := − h; x₃ := x₁
  fi
fi;
while minimum has not yet been bracketed
do
  double the step length, that is, set h := 2*h;
  set x₁ := x₂; x₂ := x₃; x₃ := z + h;
  if f(x₃) > f(x₂)
  then
    — minimum has been bracketed
  fi
```

od;
— minimum lies within the interval $[x_1, x_3]$

In this algorithm the *step length*, h, represents a movement in the direction of decreasing $f(x)$. Now at the start of the process if $f(z+h) > f(z)$ and $f(z-h) > f(z)$ we already have a bracket for the minimum (see Fig. 5.4(a)). If $f(z+h) > f(z)$ and $f(z-h) \leq f(z)$ we are going in the wrong direction and need to negate h (see Fig. 5.4(b)). If a minimum has not yet been bracketed then it lies in the direction of increasing h. We may be a long way from it so to avoid taking a large number of small steps keep doubling the step length until a situation is reached at which $f(x)$ starts to increase (see Fig. 5.4(c)). The smallest interval in which $\bar{x}$ is known to lie is $[x_1, x_3]$.

Having bracketed the minimum we consider the interpolating polynomial which passes through the coordinates $\{(x_1, f_1), (x_2, f_2), (x_3, f_3)\}$. From (5.7) this can be written as

$$p_2(x) = \frac{(x-x_2)(x-x_3)}{(x_1-x_2)(x_1-x_3)} f_1 + \frac{(x-x_1)(x-x_3)}{(x_2-x_1)(x_2-x_3)} f_2$$
$$+ \frac{(x-x_1)(x-x_2)}{(x_3-x_1)(x_3-x_2)} f_3 = ax^2 + bx + c$$

where a, b and c are rather complicated coefficients. Differentiating, we find that

$$p_2'(x) = 2ax + b.$$

Thus, the point that satisfies $p_2'(\hat{x}) = 0$ is

$$\hat{x} = -\frac{b}{2a}.$$

If we now substitute the actual values of a and b we obtain

$$\hat{x} = \frac{(x_2^2 - x_3^2)f_1 + (x_3^2 - x_1^2)f_2 + (x_1^2 - x_2^2)f_3}{2((x_2 - x_3)f_1 + (x_3 - x_1)f_2 + (x_1 - x_2)f_3)}$$

and this value may be used as an approximation to the minimum of $f(x)$. The method described above (due originally to Davies, Swann and Campey, 1964) finds an initial estimate of the minimum. It is natural to attempt to refine this first approximation in some way and this suggests that we should employ an iterative process which, hopefully, converges to the minimum. A process based on the repeated use of quadratic interpolation is described by Wolfe (1978) who discusses the provision of convergence criteria.

Sec. 5.2]   LAGRANGE INTERPOLATION   207

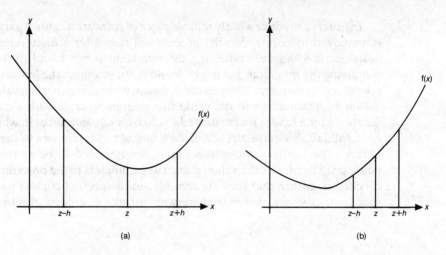

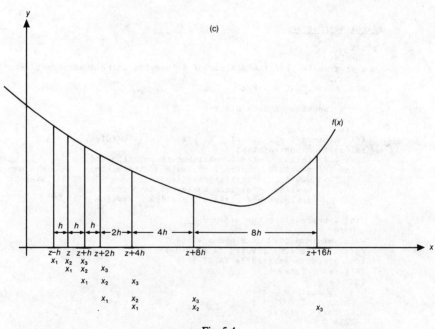

Fig. 5.4

5.2.3 A program for function minimisation based on quadratic interpolation

The programs of Case Study 5.1 implement the search technique of the previous subsection to estimate the minimum of $\sin(x)/x$.

In order to produce a fairly reliable piece of software a control parameter is employed to ensure that the process will stop after a finite number of iterations if a bracket containing the minimum is not found. A bracket containing the minimum will not be found if, for example, the function does not possess a minimum. Despite the inclusion of this check certain values for the input parameters will still cause the program to crash with a run-time failure and the reader is encouraged to discover circumstances in which the code will fail. We have already made a number of references to the NAG library. The routines within this library are intended to be as robust as possible; in particular the values passed as parameters to the procedures are checked to ensure that they are sensible. In the present context we could check, for example, that *maximumnumberofsteps* is, at least, positive.

CASE STUDY 5.1

Algol 68 Program

```
BEGIN
  COMMENT
  a program to find the minimum of a function using quadratic interpolation
  COMMENT
  PROC f = ( REAL x ) REAL :
    COMMENT
    the function to be minimised
    COMMENT
    sin ( x ) / x ;
  REAL minimum , step ,  x1 , x2 , x3 , fofx1 , fofx2 , fofx3 ;
  INT maximumnumberofsteps ;
  read ( ( minimum , step , maximumnumberofsteps ) ) ;
  print ( ( newline , "initial estimate of the minimum    " , minimum ,
            newline , "initial step length                " , step ,
            newline , "maximum number of step doublings   " ,
            maximumnumberofsteps , newline , newline ) ) ;
  COMMENT
  find a bracket for the minimum
  COMMENT
  x2 := minimum ; x3 := minimum + step ;
  fofx2 := f ( x2 ) ; fofx3 := f ( x3 ) ;
  BOOL minimumbracketed := FALSE ;
  IF fofx3 > fofx2
  THEN
    x1 := minimum - step ;
    fofx1 := f ( x1 ) ;
    IF fofx1 > fofx2
    THEN
      minimumbracketed := TRUE
    ELSE
      COMMENT
      reverse the direction of search
      COMMENT
      step := - step ; x3 := x1 ; fofx3 := fofx1
    FI
  FI ;
```

```
    INT numberofsteps := 0 ;
    WHILE NOT minimumbracketed AND numberofsteps < maximumnumberofsteps
    DO
       numberofsteps +:= 1 ;
       step *:= 2.0 ;
       x1 := x2 ; x2 := x3 ; x3 := minimum + step ;
       fofx1 := fofx2 ; fofx2 := fofx3 ; fofx3 := f ( x3 ) ;
       IF fofx3 > fofx2
       THEN
          minimumbracketed := TRUE
       FI
    OD ;
    IF minimumbracketed
    THEN
       [ 1 : 2 ] REAL bracket = IF x1 < x3 THEN ( x1 , x3 ) ELSE ( x3 , x1 ) FI ;
       COMMENT
       find the minimum of the quadratic interpolating polynomial
       COMMENT
       REAL x1sqrd = x1 * x1 , x2sqrd = x2 * x2 , x3sqrd = x3 * x3 ;
       minimum := ( ( ( x2sqrd - x3sqrd ) * fofx1
                    + ( x3sqrd - x1sqrd ) * fofx2
                    + ( x1sqrd - x2sqrd ) * fofx3 ) * 0.5
                  / ( ( x2 - x3 ) * fofx1 + ( x3 - x1 ) * fofx2
                    + ( x1 - x2 ) * fofx3 ) ;
       REAL fofminimum = f ( minimum ) ;
       print ( ( "the minimum of the given function lies in the range" , newline ,
                 bracket , newline ,
                 "and the estimate given by quadratic interpolation" , newline ,
                 "is a minimum value for the function of" , newline , newline ,
                 "          " , fofminimum , newline , newline , "at the point" ,
                 newline , newline , "          " , minimum , newline ) )
    ELSE
       print ( ( newline , newline ,
                 "failure in an attempt to bracket the minimum" , newline ,
                 "try again with different input parameters" , newline ) )
    FI
END
```

Pascal Program

```
PROGRAM quadraticinterpolation(input, output);

    { a program to find the minimum of a function using
      quadratic interpolation }

    VAR
       step, x1, x2, x3, min, x1sqrd, x2sqrd, x3sqrd,
                       fofx1, fofx2, fofx3, fofmin:real;
       numberofsteps, maximumnumberofsteps:0..maxint;
       minimumbracketed:boolean;

    FUNCTION f(x:real):real;
       { the function to be minimised  }
    BEGIN
       f:= sin(x)/x
    END { f } ;

BEGIN
    { read and print the input data }
    read(min, step, maximumnumberofsteps);
    writeln;
    writeln('initial estimate of the minimum     ', min);
    writeln('initial step length                 ', step);
```

```
            writeln('maximum number of step doublings   ', maximumnumberofsteps);
            writeln;
            { find a bracket for the minimum   }
            x2:= min;   x3:= min + step;
            fofx2:= f(x2);   fofx3:= f(x3);
            minimumbracketed:= false;
            IF fofx3>fofx2
            THEN
                BEGIN
                    x1:= min - step;
                    fofx1:= f(x1);
                    IF fofx1>fofx2
                    THEN
                        minimumbracketed:= true
                    ELSE
                        BEGIN
                            { reverse the direction of search   }
                            step:= -step;   x3:= x1;   fofx3:= fofx1
                        END
                END ;
            numberofsteps:= 0;
            WHILE NOT minimumbracketed AND (numberofsteps<>maximumnumberofsteps) DO
            BEGIN
                numberofsteps:= numberofsteps + 1;
                step:= step*2.0;
                x1:= x2;   x2:= x3;   x3:= min + step;
                fofx1:= fofx2;   fofx2:= fofx3;   fofx3:= f(x3);
                IF fofx3>fofx2
                THEN
                    minimumbracketed:= true
            END { WHILE } ;
            { minimumbracketed OR (numberofsteps=maximumnumberofsteps)   }
            IF minimumbracketed
            THEN
                BEGIN
                    { find the minimum of the quadratic interpolating polynomial   }
                    x1sqrd:= sqr(x1);   x2sqrd:= sqr(x2);   x3sqrd:= sqr(x3);
                    min:= ((x2sqrd - x3sqrd)*fofx1
                              + (x3sqrd - x1sqrd)*fofx2
                              + (x1sqrd - x2sqrd)*fofx3)*0.5
                        / ((x2 - x3)*fofx1 + (x3 - x1)*fofx2 + (x1 - x2)*fofx3);
                    fofmin:= f(min);
                    writeln('the minimum of the given function lies in the range');
                    IF x1<x3
                    THEN
                        write(x1, x3)
                    ELSE
                        write(x3, x1);
                    writeln;
                    writeln('and the estimate given by quadratic interpolation');
                    writeln('is a minimum value for the function of');
                    writeln;
                    writeln(' ':12, fofmin);
                    writeln;
                    writeln('at the point');
                    writeln;
                    writeln(' ':12, min)
                END
            ELSE
                BEGIN
                    writeln;
                    writeln;
                    writeln('failure in an attempt to bracket the minimum');
                    writeln('try again with different input parameters')
                END
END { quadraticinterpolation } .
```

Sample Data

    4.0   0.025   20

Sample Output

    initial estimate of the minimum   4.00000e 0
    initial step length   2.50000e -2
    maximum number of step doublings   20

    the minimum of the given function lies in the range
      4.20000e 0   4.80000e 0
    and the estimate given by quadratic interpolation
    is a minimum value for the function of

              -2.17229e -1

    at the point

              4.50012e 0

5.2.4 The case of equally spaced points

It sometimes happens that the points at which the function $f(x)$ is given (or is assumed to be known) are equally spaced; that is, the N points $x_1, x_2, \ldots, x_N$ are

$$x_i = x_1 + (i-1)h \qquad i = 1, 2, \ldots, N$$

where h is the *spacing* between two adjacent points. We can construct the LIP (5.7) based on the information $\{(x_i, f_i) : i = 1, 2, \ldots, N\}$ in the usual way. However, the factors $x_i - x_k$ appearing in the denominator of $\phi_i(x)$ defined by (5.8) will be integral multiples of h. Further, if we express the continuous variable x as $x_1 + \theta h$, where θ is now the continuous variable, the typical factor $x - x_k$ in the numerator of $\phi_i(x)$ will be $x_1 + \theta h - (x_1 + (k-1)h)$ $= (\theta - k + 1)h$. It is clear then that a considerable simplification of all the $\phi_i(x)$ appearing in (5.8) and hence of the interpolating polynomial itself is possible, although we shall not pursue the algebraic details formally. However, if we were to do this for $N = 3$, we would find that the Lagrange interpolating polynomial could be written

$$p_2(\theta) = f_1 + \theta(f_2 - f_1) + \frac{\theta(\theta - 1)}{2}(f_3 - 2f_2 + f_1) \qquad (5.10)$$

(where, for reasons to be explained below, the terms in f_2 and f_1 have not been collected together). That this is the LIP is easily verified. If we give θ the values 0, 1 and 2 in turn (so that x is x_1, x_2 and x_3), the expression (5.10) has the values f_1, f_2 and f_3 respectively. Remembering uniqueness (Theorem 5.1), (5.10) is indeed the LIP.

In order to be able to indicate briefly the generalisation of (5.10) to the case of N points, it is necessary to introduce the idea of a *difference*. The quantity $f_{i+1} - f_i$ is known as a *first difference* and, in one of the conventional notations, it is written Δf_i. It is clear that from the given data $N-1$ first differences can be formed. From these first differences we can form $N-2$ *second differences* defined by

$$\Delta^2 f_i = \Delta f_{i+1} - \Delta f_i \quad i = 1, 2, \ldots, N-2$$

and, if we wish, we can use the definition of a first difference to express $\Delta^2 f_i$ in terms of function values. We have

$$\Delta^2 f_i = f_{i+2} - f_{i+1} - (f_{i+1} - f_i)$$
$$= f_{i+2} - 2f_{i+1} + f_i.$$

Putting $i = 1$ in these definitions of first and second differences, we see that (5.10) can be written as

$$p_2(\theta) = f_1 + \theta \Delta f_1 + \frac{\theta(\theta-1)}{2} \Delta^2 f_1.$$

From the second differences, we can construct *third differences* and so on. The rth order difference is defined as

$$\Delta^r f_i = \Delta^{r-1} f_{i+1} - \Delta^{r-1} f_i$$

and, if data at N points are available, it is possible to construct differences up to order $N-1$. The generalisation of (5.10) can now be given:

$$p_{N-1}(\theta) = f_1 + \theta \Delta f_1 + \frac{\theta(\theta-1)}{2!} \Delta^2 f_1 + \ldots$$
$$+ \frac{\theta(\theta-1) \ldots (\theta-N+2)}{(N-1)!} \Delta^{N-1} f_1.$$

This can be shown to be the LIP based on the data $\{(x_i, f_i): i = 1, 2, \ldots, N\}$ and it is known as *Newton's forward difference formula*. The adjective 'forward' appears in this description because strictly speaking the quantity $\Delta f_i = f_{i+1} - f_i$ should have been called a first forward difference and similarly for the higher differences. In another notation, $f_{i+1} - f_i$ is denoted by ∇f_{i+1} and is then called a first backward difference and higher orders of backward difference can be defined; in yet another notation, it is denoted by $\delta f_{i+\frac{1}{2}}$ and is then called a first central difference and higher orders of central difference can be defined. It is possible to express the Lagrange interpolating polynomial in terms of the backward differences of $f(x)$ and in terms of the central differences of $f(x)$ but we shall not give any details. The interested reader is referred to Ralston and Rabinowitz (1978).

5.3 NUMERICAL DIFFERENTIATION AND THE APPROXIMATION OF DERIVATIVES

Suppose we have available the information $\{(x_i, f_i) : i = 1, 2, \ldots, N\}$ about a function $f(x)$. We have shown in section 5.2 how to construct the LIP $p_{N-1}(x)$ and this can be used to make a (possibly poor) estimate of $f(\alpha)$ for some given α by saying $f(\alpha) \simeq p_{N-1}(\alpha)$. It may be that we would like to estimate $f'(\alpha)$ (and perhaps higher derivatives of $f(x)$ at $x = \alpha$). We can use the LIP again and estimate $f'(\alpha)$ as

$$f'(\alpha) \simeq p'_{N-1}(\alpha)$$

where we would first have to differentiate $p_{N-1}(x)$ with respect to x and then set $x = \alpha$. This can lead to an inaccurate estimate and, in the extreme case, it can be worthless. (The reader should now return to Fig. 5.3 and conclude that the estimate provided by $p'_2(x)$ is unacceptable for any value of x.) This situation is more likely to occur if the given values of $f(x)$ are subject to uncertainty.

Consider the case $N = 2$ (that is, straight line approximation). Taylor's theorem (Theorem 1.1) gives the result

$$f(c + h) = f(c) + hf'(c) + \frac{h^2}{2} f''(\xi)$$

where ξ is some point in the smallest interval containing c and $c + h$ so that

$$f'(c) = \frac{f(c+h) - f(c)}{h} - \frac{h}{2} f''(\xi) \ .$$

By ignoring the term $-(h/2)f''(\xi)$ we can obtain an approximation to $f'(c)$ involving the function values $f(c+h)$ and $f(c)$, namely

$$f'(c) \simeq \frac{f(c+h) - f(c)}{h} \qquad (5.11)$$

and this is just the derivative of the LIP for the points $\{(c, f(c)), (c+h, f(c+h))\}$. Recalling that the formal definition of the derivative of a continuous function at the point $x = c$ is

$$f'(c) = \lim_{h \to 0} \frac{f(c+h) - f(c)}{h}$$

we would expect (5.11) to give increasingly more accurate estimates of $f'(c)$ as the value of h decreases. Unfortunately, in practice, this is not so.

Consider two consecutive points x_{i+1} and x_i, distance h apart. Assume that x_{i+1} and x_i are exactly represented and that $\bar{f}(x_{i+1})$ and $\bar{f}(x_i)$ are computed values for $f(x_{i+1})$ and $f(x_i)$ with errors δ_1 and δ_2 respectively. Suppose we use (5.11) to approximate $f'(x_i)$. Then the error in what we actually compute is

$$E = \frac{f(x_{i+1}) - f(x_i)}{h} - \frac{\bar{f}(x_{i+1}) - \bar{f}(x_i)}{h}$$
$$= \frac{\delta_1 - \delta_2}{h}$$

so that

$$|E| \leq \frac{1}{h}[|\delta_1| + |\delta_2|].$$

Clearly, if h is not big the estimate we obtain of the derivative is likely to be very poor. However, if h is very small there will be a tendency to magnify the errors in the computed function values and a poor estimate will again be derived. Clearly there will be an optimal value of h somewhere in between these two extremes. We conclude that numerical differentiation can be fraught with danger and should be treated with considerable caution.

Before leaving this section we return to the case of equally spaced points. The finite difference approximation

$$\frac{\Delta f_i}{h} = \frac{f_{i+1} - f_i}{h}$$

to f'_i is just the slope (that is, derivative) at the point $x = x_i$ of the LIP for the coordinates $\{(x_i, f_i), (x_{i+1}, f_{i+1})\}$. Further, forming the second derivative at the point $x = x_i$ of the LIP for the coordinates $\{(x_i, f_i), (x_{i+1}, f_{i+1}), (x_{i+2}, f_{i+2})\}$ yields the finite difference approximation

$$\frac{\Delta^2 f_i}{h^2} = \frac{f_{i+2} - 2f_{i+1} + f_i}{h^2}$$

5.4 HERMITE INTERPOLATION

5.4.1 Interpolation of function and derivative values

A Lagrange interpolating polynomial, by construction, passes through a prescribed number of coordinates. It may be that we have also available derivative values $f'(x_i)$ at some, possibly all, points x_i and we would like to make use of this additional information when constructing a polynomial approximation. Fitting a polynomial to both function and derivative values is known as *Hermite interpolation* and we consider in this subsection the case in which we have a value for $f'(x_i), f'_i$, at each point $x_i : i = 1, 2, \ldots, N$.

Consider the polynomial of degree $2N - 1$

$$p_{2N-1}(x) = \sum_{i=1}^{N} \alpha_i(x) f_i + \sum_{i=1}^{N} \beta_i(x) f'_i \tag{5.12}$$

with

$$\left. \begin{array}{l} \alpha_i(x) = (1 - 2(x - x_i)\phi'_i(x_i))\phi_i^2(x) \\ \\ \beta_i(x) = (x - x_i)\phi_i^2(x) \end{array} \right\} \quad i = 1, 2, \ldots, N$$

where the polynomials $\phi_i(x)$ (of degree $N - 1$) are those appearing in the LIP (see equation (5.8)). Note that in the definition of $\alpha_i(x)$

$$\phi'_i(x_i) = \left. \left(\frac{\mathrm{d}}{\mathrm{d}x} \phi_i(x) \right) \right|_{x=x_i} \quad i = 1, 2, \ldots, N.$$

Now, from the result (5.9) we have

$$\left. \begin{array}{l} \alpha_i(x_j) = (1 - 2(x_j - x_i)\phi'_i(x_i))\delta_{ij}^2 = \delta_{ij} \\ \\ \beta_i(x_j) = (x_j - x_i)\delta_{ij}^2 = 0 \end{array} \right\} \quad i,j = 1, 2, \ldots, N$$

and hence

$$p_{2N-1}(x_j) = f_j \quad j = 1, 2, \ldots, N$$

that is, $p_{2N-1}(x)$ passes through the coordinates $\{(x_j, f_j) : j = 1, 2, \ldots, N\}$. Further

$$\left.\begin{array}{l}\alpha_i'(x) = (1 - 2(x - x_i)\phi_i'(x_i))2\phi_i(x)\phi_i'(x) - 2\phi_i'(x_i)\phi_i^2(x) \\ \beta_i'(x) = (x - x_i)2\phi_i(x)\phi_i'(x) + \phi_i^2(x)\end{array}\right\} \quad i = 1, 2, \ldots, N$$

using the formula for the differentiation of the product of two functions. Therefore

$$\left.\begin{array}{l}\alpha_i'(x_j) = (1 - 2(x_j - x_i)\phi_i'(x_i))2\delta_{ij}\phi_i'(x_j) - 2\phi_i'(x_i)\delta_{ij}^2 = 0 \\ \beta_i'(x_j) = (x_j - x_i)2\delta_{ij}\phi_i'(x_j) + \delta_{ij}^2 = \delta_{ij}\end{array}\right\} \quad i,j = 1, 2, \ldots, N$$

which means that

$$p_{2N-1}'(x_j) = f_j' \quad j = 1, 2, \ldots, N$$

that is, the polynomial $p_{2N-1}(x)$, when differentiated, takes on the given derivative values at the points $\{x_i : i = 1, 2, \ldots, N\}$.

EXAMPLE 5.4
Consider the function defined by the following values

x_i	0	1	2
f_i	5	3	27
f_i'	−7	8	41

Then

$$\phi_1(x) = \tfrac{1}{2}(x-1)(x-2) = \tfrac{1}{2}(x^2 - 3x + 2)$$
$$\phi_2(x) = -x(x-2) = -x^2 + 2x$$
$$\phi_3(x) = \tfrac{1}{2}x(x-1) = \tfrac{1}{2}(x^2 - x)$$

and so

$$\phi_1'(x) = \tfrac{1}{2}(2x - 3); \quad \phi_1'(0) = -\tfrac{3}{2}$$

$$\phi_2'(x) = -2x + 2; \quad \phi_2'(1) = 0$$

$$\phi_3'(x) = \tfrac{1}{2}(2x - 1); \quad \phi_3'(2) = \tfrac{3}{2}.$$

Hence the polynomial which fits the given function and derivative values is

$$p_5(x) = \left(1 - 2x\left(-\frac{3}{2}\right)\right)\frac{1}{4}(x^2 - 3x + 2)^2 5 + (1 - 2(x-1)0)(-x^2 + 2x)^2 3$$

$$+ \left(1 - 2(x-2)\frac{3}{2}\right)\frac{1}{4}(x^2 - x)^2 27 + x\frac{1}{4}(x^2 - 3x + 2)^2(-7)$$

$$+ (x-1)(-x^2 + 2x)^2 8 + (x-2)\frac{1}{4}(x^2 - x)^2 41.$$

Expressing this in 'standard' form (although, in general, we would not wish to do this) we have

$$p_5(x) = (x^4 - 6x^3 + 13x^2 - 12x + 4)\left[\frac{5}{4}(1 + 3x) - \frac{7}{4}x\right]$$
$$+ (x^4 - 4x^3 + 4x^2)[3 + 8(x-1)]$$
$$+ (x^4 - 2x^3 + x^2)\left[\frac{27}{4}(7 - 3x) + \frac{41}{4}(x-2)\right]$$
$$= \frac{1}{4}\{8x^5 - 43x^4 + 74x^3 - 31x^2 - 28x + 20 + 32x^5 - 148x^4$$
$$\quad + 208x^3 - 80x^2 - 40x^5 + 187x^4 - 254x^3 + 107x^2\}$$
$$= \frac{1}{4}\{-4x^4 + 28x^3 - 4x^2 - 28x + 20\}$$
$$= -x^4 + 7x^3 - x^2 - 7x + 5.$$

Check

$$p_5(0) = 5; \quad p_5(1) = 3; \quad p_5(2) = 27$$
$$p_5'(x) = -4x^3 + 21x^2 - 2x - 7$$
$$p_5'(0) = -7; \quad p_5'(1) = 8; \quad p_5'(2) = 41.$$

When expressed in the form (5.12) $p_{2N-1}(x)$ is known as the *Hermite interpolating polynomial* (*HIP*) for the given values. A straightforward extension of the proof of Theorem 5.1 will show that it is unique.

An expression for the error in Hermite interpolation may be derived in a similar way to that for Lagrange interpolation using the repeated application of Rolle's theorem (Theorem 1.3). In fact we have

$$E_{2N-1}(x) = f(x) - p_{2N-1}(x) = \frac{f^{(2N)}(\xi)}{(2N)!} \prod_{i=1}^{N} (x - x_i)^2$$

where ξ is some point in the interval spanned by $\{x, x_1, x_2, \ldots, x_N\}$.

5.4.2 Function minimisation using HIP's

Like Lagrange interpolation, Hermite interpolation is of use in the field of function minimisation although we do, of course, have to compute derivative values. The process starts by bracketing the minimum with two points at which both function and derivative values are known. By fitting a cubic Hermite interpolating polynomial to the available information and finding its minimum we should be able to find a good approximation to the minimum of the original function.

The following algorithm attempts to locate a bracket $[x_1, x_2]$ in which the minimum, $\bar{x}$, of the function $f(x)$ lies.

> choose an initial approximation, z, to $\bar{x}$ and a step length h;
> if $f'(z) > 0$
> then
>   set $h := -h$
> fi;
> while $f'(z)$ and $f'(z+h)$ are of the same sign
> do
>   set $z := z+h$; $h := 2*h$
> od;
> set $x_1 := z$; $x_2 := z+h$

Having established the direction of decreasing $f(x)$ the algorithm searches for an interval for which the derivative of $f(x)$ is of opposite sign at the end points (see Fig. 5.5). Again the tactic of doubling the step length is employed in case the initial guess at the position of the minimum is poor.

Having bracketed the minimum, $f(x)$ is evaluated at x_1 and x_2 and a cubic HIP is fitted to the data $f(x_1)$, $f'(x_1)$, $f(x_2)$ and $f'(x_2)$. If we express this polynomial in the form

$$p_3(x) = ax^3 + bx^2 + cx + d$$

then it has turning points at the solutions of

$$p_3'(x) = 3ax^2 + 2bx + c = 0 \ .$$

The value which satisfies

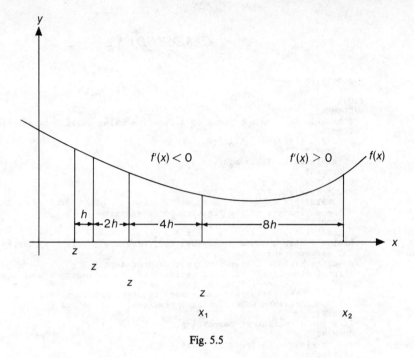

Fig. 5.5

$$p_3''(x) = 6ax + 2b > 0$$

gives a minimum. This value is

$$x = (-b + \sqrt{(b^2 - 3ac)})/3a \qquad (5.13)$$

(the negative square root gives the maximum) if $a \neq 0$ and

$$x = -c/2b$$

if $a = 0$. Precisely what we mean by '$a \neq 0$' is a matter for some discussion. Further, the formula (5.13) will lead to loss of significance if $b^2 \gg 3ac$. Alternative formulae have been proposed which get round these problems and details may be found in Wolfe (1978).

5.4.3 A program for function minimisation based on cubic interpolation

The programs of Case Study 5.2 implement the minimum location technique of the previous subsection. The algorithm employed to find the required root of the cubic Hermite interpolating polynomial is that given by Wolfe (1978, p. 267) and is valid in all cases.

220 APPROXIMATION OF NUMERICALLY DEFINED FUNCTIONS   [Ch. 5

CASE STUDY 5.2

<u>Algol 68 Program</u>

```
BEGIN
  COMMENT
  a program to find a minimum of a function using cubic interpolation
  COMMENT
  PROC f = ( REAL x ) REAL :
    COMMENT
    the function to be minimised
    COMMENT
    sin ( x ) / x ;
  PROC fdash = ( REAL x ) REAL :
    COMMENT
    the derivative of the function to be minimised
    COMMENT
    cos ( x ) / x - sin ( x ) / x ↑ 2 ;
  REAL z , step , minimum , zplusstep , fdashofz , fdashofzplusstep ;
  INT maximumnumberofsteps ;
  read ( ( minimum , step , maximumnumberofsteps ) ) ;
  print ( ( newline , "initial estimate of the minimum   " , minimum ,
            newline , "initial step length               " , step ,
            newline , "maximum number of step doublings  " ,
            maximumnumberofsteps , newline , newline ) ) ;
  COMMENT
  find the direction of decreasing f
  COMMENT
  z := minimum ; fdashofz := fdash ( z ) ;
  IF fdashofz > 0.0
  THEN
    step := - step
  FI ;
  zplusstep := z + step ; fdashofzplusstep := fdash ( zplusstep ) ;
  COMMENT
  search for a bracket containing the minimum
  COMMENT
  BOOL minimumbracketed := fdashofz * fdashofzplusstep < 0.0 ;
  INT numberofsteps := 0 ;
  WHILE NOT minimumbracketed AND numberofsteps < maximumnumberofsteps
  DO
    numberofsteps +:= 1 ;
    step *:= 2.0 ;
    z := zplusstep ; zplusstep +:= step ;
    fdashofz := fdashofzplusstep ; fdashofzplusstep := fdash ( zplusstep ) ;
    minimumbracketed := fdashofz * fdashofzplusstep < 0.0
  OD ;
  IF minimumbracketed
  THEN
    REAL x1 , x2 , fdashofx1 , fdashofx2 ;
    IF z < zplusstep
    THEN
      x1 := z ; x2 := zplusstep ;
      fdashofx1 := fdashofz ; fdashofx2 := fdashofzplusstep
    ELSE
      x1 := zplusstep ; x2 := z ;
      fdashofx1 := fdashofzplusstep ; fdashofx2 := fdashofz
    FI ;
    COMMENT
    find the minimum of the cubic interpolating polynomial
    COMMENT
    REAL mu = 3.0 * ( f ( x1 ) - f ( x2 ) )
              / ( x2 - x1 ) + fdashofx1 + fdashofx2 ;
    REAL lambda = sqrt ( mu ↑ 2 - fdashofx1 * fdashofx2 ) ;
    minimum := x1 + ( x2 - x1 ) * ( 1.0 - ( fdashofx2 + lambda - mu )
                                    / ( fdashofx2 - fdashofx1
                                        + 2.0 * lambda ) ) ;
```

```
            REAL fofminimum = f ( minimum ) ;
            print ( ( "the minimum of the given function lies in the range" , newline ,
                      x1 , x2 , newline ,
                      "and the estimate given by cubic interpolation" , newline ,
                      "is a minimum value for the function of" , newline , newline ,
                      "           " , fofminimum , newline , newline , "at the point" ,
                      newline , newline , "              " , minimum , newline ) )
    ELSE
        print ( ( newline , newline ,
                  "failure in an attempt to bracket the minimum" , newline ,
                  "try again with different input parameters" , newline ) )
    FI
END
```

Pascal Program

```
PROGRAM cubicinterpolation(input, output);

    {  a program to find a minimum of a function using
       cubic interpolation  }

    VAR
        step, mu, lambda, min, fofmin, z, zplusstep, x1, x2,
                    fdashz, fdashzplusstep, fdashx1, fdashx2:real;
        numberofsteps, maximumnumberofsteps:0..maxint;
        minimumbracketed:boolean;

    FUNCTION f(x:real):real;
        {  the function to be minimised  }
    BEGIN
        f:= sin(x)/x
    END { f } ;

    FUNCTION fdash(x:real):real;
        { the derivative of the function to be minimised  }
    BEGIN
        fdash:= cos(x)/x - sin(x)/sqr(x)
    END { fdash } ;

BEGIN
    {  read and print the input data  }
    read(min, step, maximumnumberofsteps);
    writeln;
    writeln('initial estimate of the minimum    ', min);
    writeln('initial step length                ', step);
    writeln('maximum number of step doublings   ', maximumnumberofsteps);
    writeln;
    {  find the direction of decreasing f  }
    z:= min;   fdashz:= fdash(z);
    IF fdashz>0.0
    THEN
        step:= -step;
    zplusstep:= z + step;    fdashzplusstep:= fdash(zplusstep);
    {  search for a bracket containing the minimum  }
    minimumbracketed:= fdashz*fdashzplusstep<0.0;
    numberofsteps:= 0;
    WHILE NOT minimumbracketed AND (numberofsteps<>maximumnumberofsteps) DO
    BEGIN
        numberofsteps:= numberofsteps + 1;
        step:= step*2.0;
        z:= zplusstep;    zplusstep:= zplusstep + step;
        fdashz:= fdashzplusstep;   fdashzplusstep:= fdash(zplusstep);
        minimumbracketed:= fdashz*fdashzplusstep<0.0
    END ;
```

```
        { minimumbracketed OR (numberofsteps=maximumnumberofsteps) }
        IF minimumbracketed
        THEN
            BEGIN
                IF z<zplusstep
                THEN
                    BEGIN
                        x1:= z;  x2:= zplusstep;
                        fdashx1:= fdashz;  fdashx2:= fdashzplusstep
                    END
                ELSE
                    BEGIN
                        x1:= zplusstep;  x2:= z;
                        fdashx1:= fdashzplusstep;  fdashx2:= fdashz
                    END ;
                { find the minimum of the cubic interpolating polynomial }
                mu:= 3.0*(f(x1) - f(x2))/(x2 - x1) + fdashx1 + fdashx2;
                lambda:= sqrt(mu*mu - fdashx1*fdashx2);
                min:= x1 + (x2 - x1)*(1.0 - (fdashx2 + lambda - mu)
                                   /(fdashx2 - fdashx1 + 2.0*lambda));
                fofmin:= f(min);
                writeln('the minimum of the given function lies in the range');
                writeln(x1, x2);
                writeln('and the estimate given by cubic interpolation');
                writeln('is a minimum value for the function of');
                writeln;
                writeln(' ':12, fofmin);
                writeln;
                writeln('at the point');
                writeln;
                writeln(' ':12, min)
            END
        ELSE
            BEGIN
                writeln;
                writeln;
                writeln('failure in an attempt to bracket the minimum');
                writeln('try again with different input parameters')
            END
END { cubicinterpolation } .
```

Sample Data

    4.0   0.025   20

Sample Output

```
initial estimate of the minimum      4.00000e  0
initial step length                  2.50000e -2
maximum number of step doublings        20
```

the minimum of the given function lies in the range
  4.37500e 0   4.77500e 0
and the estimate given by cubic interpolation
is a minimum value for the function of

          -2.17234e -1

at the point

          4.49315e 0

5.5 SPLINE APPROXIMATION

5.5.1 Approximation using piecewise polynomials

At first sight it would appear that, using Lagrange interpolation, by increasing the number of points, and hence the degree of the polynomial, successively better approximations to the underlying continuous function will be obtained (provided, of course, that a sufficient number of function values are, or can be, made available). However, there is no guarantee that this will always happen. Moreover, in between the interpolation points the approximating polynomial may oscillate quite violently and not truly reflect the behaviour of the function which it claims to represent.

An alternative approach is to use *piecewise polynomial approximation*. Instead of looking for an overall continuous approximation we aim to construct a polynomial that is composed of a sequence of low degree polynomials which, for the time being, we consider as being only locally valid. We do this by defining a different polynomial approximation within each sub-interval $[x_i, x_{i+1}]$, keeping the overall degree of approximation constant (although this is not essential).

The simplest form of piecewise polynomial approximation is provided by the *broken line* (or *piecewise linear*) approximation (see Fig. 5.6). Over the

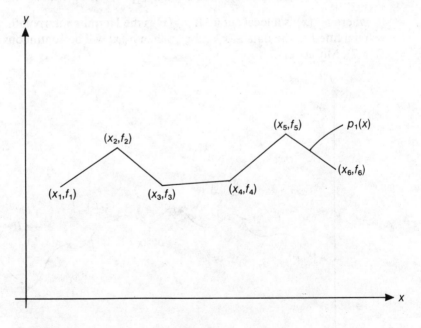

Fig. 5.6

sub-interval $[x_i, x_{i+1}]$ the approximating polynomial $p_1(x)$ is defined to be the straight line passing through the coordinates $\{(x_i, f_i), (x_{i+1}, f_{i+1})\}$ and we denote this local approximation as $p_{1,i}(x)$. We have

$$p_1(x) \equiv p_{1,i}(x) = \frac{x - x_{i+1}}{x_i - x_{i+1}} f_i + \frac{x - x_i}{x_{i+1} - x_i} f_{i+1} \quad x \in [x_i, x_{i+1}]$$
$$i = 1, 2, \ldots, N-1.$$

Clearly, the overall approximation $p_1(x)$ will be continuous but its derivative will not be (unless all the points happen to lie on the same straight line).

The concept of piecewise linear approximation may be extended in a fairly obvious manner. Suppose, for example, we have available an odd number, $2N-1$, say, of function values. Then over each sub-interval $[x_{2i-1}, x_{2i+1}] : i = 1, 2, \ldots, N-1$ we could fit a quadratic passing through the coordinates $\{(x_{2i-1}, f_{2i-1}), (x_{2i}, f_{2i}), (x_{2i+1}, f_{2i+1})\}$. Again we will end up with a continuous piecewise polynomial with discontinuous derivative. This may mean that the approximating polynomial is less smooth than we would like and we seek ways of improving this situation.

Consider the case in which we have both function and derivative values available at each *grid* point $x_i : i = 1, 2, \ldots, N$ and let $p_3(x)$ be defined by

$$p_3(x) \equiv p_{3,i}(x) \quad x \in [x_i, x_{i+1}] \quad i = 1, 2, \ldots, N-1 \quad (5.14)$$

where $p_{3,i}(x)$ is a local cubic. If $p_{3,i}(x)$ is the Hermite interpolating polynomial fitted to the data at x_i and x_{i+1} then $p_3(x)$ will be continuous (see Fig. 5.7). Moreover

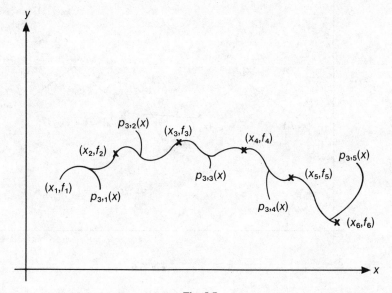

Fig. 5.7

$$p_3'(x_i) = f_i' = p_{3,i}'(x_i) = p_{3,i-1}'(x_i) \qquad i = 2, 3, \ldots, N-1$$

and hence the first derivative of $p_3(x)$ is continuous also.

In general, derivative values will not be available and so we need some technique for forcing continuity of first (and higher) derivatives of the approximating polynomial at the interior points x_i: $i = 2, 3, \ldots, N-1$ (which are often referred to as *knots*). We again consider piecewise cubic polynomial approximation and attempt to find values for the polynomial coefficients in

$$p_{3,i}(x) = a_i + b_i x + c_i x^2 + d_i x^3 \qquad i = 1, 2, \ldots, N-1$$

which ensure that the polynomial $p_3(x)$ of the form (5.14) has a continuous first and second derivative everywhere, making use of function values only. The equations

$$\left. \begin{array}{l} p_{3,i}(x_i) = a_i + b_i x_i + c_i x_i^2 + d_i x_i^3 = f_i \\ p_{3,i}(x_{i+1}) = a_i + b_i x_{i+1} + c_i x_{i+1}^2 + d_i x_{i+1}^3 = f_{i+1} \end{array} \right\} \quad i = 1, 2, \ldots, N-1 \qquad (5.15)$$

need to be satisfied if each $p_{3,i}(x)$ is to pass through the coordinates $\{(x_i, f_i), (x_{i+1}, f_{i+1})\}$. This gives $2N - 2$ equations involving the $4N - 4$ coefficients $\{a_i, b_i, c_i, d_i : i = 1, 2, \ldots, N-1\}$ and so we have a further $2N - 2$ conditions to find. We now force continuity of the first derivative of $p_3(x)$ by ensuring that the derivatives of adjacent polynomials are equal at the common point; that is, we impose the conditions

$$p_{3,i}'(x_i) = b_i + 2c_i x_i + 3d_i x_i^2 = p_{3,i-1}'(x_i) = b_{i-1} + 2c_{i-1} x_i + 3d_{i-1} x_i^2$$
$$i = 2, 3, \ldots, N-1 \qquad (5.16)$$

which gives a further $N - 2$ equations. Similarly, we impose continuity of the second derivative of $p_3(x)$ by specifying the conditions

$$p_{3,i}''(x_i) = 2c_i + 6d_i x_i = p_{3,i-1}''(x_i) = 2c_{i-1} + 6d_{i-1} x_i \qquad i = 2, 3, \ldots, N-1 \qquad (5.17)$$

which yields another $N - 2$ equations, making $4N - 6$ in all. (Note that we are not making use of any first or second derivative values of the function being approximated; all we are doing here is imposing continuity conditions on $p_3(x)$.) We are still two conditions short and one way of resolving this situation is to make the approximating polynomial 'smooth' at the two end points x_1 and x_N using the *end conditions*

$$p_{3,1}''(x_1) = 2c_1 + 6d_1 x_1 = 0$$
$$p_{3,N-1}''(x_N) = 2c_{N-1} + 6d_{N-1} x_N = 0 \ .$$
(5.18)

EXAMPLE 5.5
Suppose we wish to find a piecewise cubic polynomial approximation to the coordinates $\{(0, 1), (\tfrac{1}{2}, e^{\tfrac{1}{2}}), (1, e)\}$. Then

$$p_3(x) = \begin{cases} p_{3,1}(x) & x \in [0, \tfrac{1}{2}] \\ p_{3,2}(x) & x \in [\tfrac{1}{2}, 1] \end{cases}.$$

The conditions (5.15) yield the equations

$$\begin{aligned}
p_{3,1}(0) &= a_1 & &= 1 \\
p_{3,1}(\tfrac{1}{2}) &= a_1 + \tfrac{1}{2}b_1 + \tfrac{1}{4}c_1 + \tfrac{1}{8}d_1 & &= e^{\tfrac{1}{2}} \\
p_{3,2}(\tfrac{1}{2}) &= & a_2 + \tfrac{1}{2}b_2 + \tfrac{1}{4}c_2 + \tfrac{1}{8}d_2 &= e^{\tfrac{1}{2}} \\
p_{3,2}(1) &= & a_2 + b_2 + c_2 + d_2 &= e
\end{aligned}$$

whilst continuity of the first derivative of $p_3(x)$ at $x = \tfrac{1}{2}$ (equation (5.16)) gives

$$p_{3,1}'(\tfrac{1}{2}) - p_{3,2}'(\tfrac{1}{2}) = b_1 + c_1 + \tfrac{3}{4}d_1 - b_2 - c_2 - \tfrac{3}{4}d_2 = 0.$$

Continuity of the second derivative of $p_3(x)$ at $x = \tfrac{1}{2}$ (equation (5.17)) gives

$$p_{3,1}''(\tfrac{1}{2}) - p_{3,2}''(\tfrac{1}{2}) = 2c_1 + 3d_1 - 2c_2 - 3d_2 = 0.$$

and, finally, the end conditions (5.18) give

$$\begin{aligned}
p_{3,1}''(0) &= & 2c_1 & &= 0. \\
p_{3,2}''(1) &= & & 2c_2 + 6d_2 &= 0.
\end{aligned}$$

Solving these eight equations using Gauss elimination (with row interchanges) we have

$$\begin{aligned}
a_1 &= 1 \\
b_1 &= -\tfrac{1}{2}e + 3e^{\tfrac{1}{2}} - 5/2 & &\approx 1.08702 \\
c_1 &= 0 \\
d_1 &= 2e - 4e^{\tfrac{1}{2}} + 2 & &\approx 0.84168 \\
a_2 &= \tfrac{1}{2}e - e^{\tfrac{1}{2}} + 3/2 & &\approx 1.21042 \\
b_2 &= -7e/2 + 9e^{\tfrac{1}{2}} - 11/2 & &\approx -0.17549 \\
c_2 &= 6e - 12e^{\tfrac{1}{2}} + 6 & &\approx 2.52504 \\
d_2 &= -2e + 4e^{\tfrac{1}{2}} - 2 & &\approx -0.84168.
\end{aligned}$$

A piecewise polynomial of the form just given is known as a *spline* and higher order splines may be derived in an analogous manner. We now look at an alternative way of expressing spline functions.

5.5.2 B-splines

In recent years much attention has been given to an alternative representation of splines. In order to demonstrate the approach we restrict ourselves to the case

$$x_i = a + (i-1)h \qquad i = 1, 2, \ldots, N$$

where

$$h = (b-a)/(N-1)$$

is a constant step length and a and b are the first and last grid points. Piecewise linear approximation may be regarded as the simplest form of spline approximation and we consider this case in detail. The extension to higher order spline approximation will be clear.

Within the sub-interval $[x_i, x_{i+1}]$ the piecewise linear approximation $p_1(x)$ takes the form

$$p_1(x) = p_{1,i}(x) = \frac{x - (a + ih)}{-h} f_i + \frac{x - (a + (i-1)h)}{h} f_{i+1}$$

as a Lagrange interpolating polynomial. $p_{1,i}(x)$ may also be expressed as

$$p_{1,i}(x) = \frac{f_{i+1} - f_i}{h} x + \frac{1}{h} [(a + ih)f_i - (a + (i-1)h)f_{i+1}] \tag{5.19}$$

which means that it is a straight line with slope $(f_{i+1} - f_i)/h$ and intercept $((a + ih)f_i - (a + (i-1)h)f_{i+1})/h$.

Consider the function

$$\phi_i(x) = \begin{cases} x - x_{i-1} \\ 2h - (x - x_{i-1}) = \\ 0 \end{cases} \begin{cases} x - (a + i - 2)h & x \in [x_{i-1}, x_i] \\ 2h - (x - (a + i - 2)h)) & x \in [x_i, x_{i+1}] \\ 0 & x \notin [x_{i-1}, x_{i+1}] \end{cases}$$

$$i = 2, 3, \ldots, N-1$$

(see Fig. 5.8). Then it is clear that $\phi_i(x)$ (sometimes known as a *hat function*) satisfies the condition

$$\phi_i(x_j) = h\delta_{ij} \qquad i = 2, 3, \ldots, N-1; j = 1, 2, \ldots, N. \tag{5.20}$$

We now form the function $\Phi_i(x)$ as a linear combination of two of these functions (which we refer to as *basis functions* for reasons which will shortly

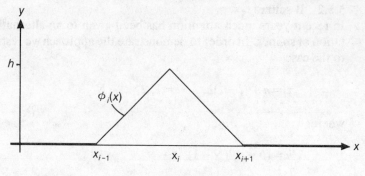

Fig. 5.8

become apparent); in detail we have

$$\Phi_i(x) = \frac{f_i}{h} \phi_i(x) + \frac{f_{i+1}}{h} \phi_{i+1}(x) \tag{5.21}$$

where $\phi_{i+1}(x)$ has the same form as $\phi_i(x)$ but shifted right by h so that it is non-zero in $]x_i, x_{i+2}[$.

Expanding we find that, for $x \in [x_i, x_{i+1}]$,

$$\Phi_i(x) = \frac{f_i}{h}(2h - (x - (a + (i-2)h))) + \frac{f_{i+1}}{h}(x - (a + (i-1)h))$$

$$= \frac{f_{i+1} - f_i}{h} x + \frac{1}{h}[(a + ih)f_i - (a + (i-1)h)f_{i+1}]$$

and we note that this form for $\Phi_i(x)$ is exactly the same as the form of the polynomial $p_{1,i}(x)$ given in equation (5.19). We conclude that it is always possible to write a piecewise linear approximating polynomial in terms of the $\phi_i(x)$s (which are independent of the function values f_j); that is, the $\phi_i(x)$s form a basis for the space of broken line functions. Using the representation (5.21) for $p_{1,i}(x)$ we can write

$$p_1(x) = \frac{1}{h} \sum_{i=1}^{N} f_i \phi_i(x) \qquad x \in [a, b]$$

provided that we introduce the additional grid points $x_0 \, (= a - h)$ and x_{N+1} ($= b + h$) so that we can define $\phi_1(x)$ and $\phi_N(x)$.

As with piecewise linears, splines of higher order may be written as a linear combination of continuous basis functions (or *B-splines*) and, in fact, they should always be written in such a form since this is both computationally (it reduces the overall operations count) and mathematically (it eases the convergence analysis) desirable. The higher the order of the spline approximation, the greater the number of additional grid points which will have to be introduced. B-splines do not necessarily satisfy a condition of the form (5.20) but if they do they are known as *cardinal splines*. Further details may be found in Powell (1981), Schumaker (1981) and de Boor (1978).

5.6 DISCRETE APPROXIMATION

5.6.1 Introduction to discrete best approximation

Most of us have, at some time, performed an experiment which attempts to verify the existence of a linear relationship between two quantities. Having taken a number of readings, these are plotted and a straight line drawn which is considered to be a reasonable fit to the data. Inevitably, errors creep in and we seek a straight line of best approximation which may, in fact, not pass through any of the observation points.

In section 5.2 we considered the construction of a polynomial approximation of degree $N-1$ which exactly fits a prescribed set of N coordinates. It was convenient for us to express this as a Lagrange interpolating polynomial although we could also have approached the problem by aiming to find the values of the coefficients of

$$p_{N-1}(x) = \sum_{i=1}^{N} \alpha_i x^{i-1}$$

by solving the system of N equations

$$p_{N-1}(x_k) = f_k \quad k = 1, 2, \ldots, N.$$

The situation we now wish to consider is that in which the number of function values, M, is greater than (and, in some cases, considerably greater than) N. The system

$$p_{N-1}(x_k) = f_k \quad k = 1, 2, \ldots, M.$$

with $M > N$, is known as an *overdetermined system* (that is, there are more equations than unknowns) and in this section we consider ways of solving such a system.

What we do is to consider the function

$$L(\mathbf{a}, x) = \sum_{j=1}^{N} \alpha_j \phi_j(x)$$

where the $\phi_j(x)$s are some chosen basis functions and choose the coefficients $\alpha_1, \alpha_2, \ldots, \alpha_N$ in some 'best' way. In order to measure the accuracy of such an approximation we introduce the residual vector **r** whose components are given by

$$r_k = f_k - L(\mathbf{a}, x_k) \qquad k = 1, 2, \ldots, M.$$

The aim is, then, to keep the 'size' of **r** as small as possible.

5.6.2 Discrete least squares approximation

The most commonly used criterion for determining a best approximation to a set of coordinates is that the sum of the squares of each of the residuals, r_k, should be as small as possible; that is, we seek values for the α_is which ensure that

$$S(\mathbf{a}) = \sum_{k=1}^{M} (f_k - L(\mathbf{a}, x_k))^2$$

is a minimum. The situation is, in fact, almost identical to the continuous case. Here, in place of an integral we have a summation over a set of points. Before proceeding further the reader is encouraged to adapt the material of the early part of subsection 4.3.2 to the present case and to derive a system of equations equivalent to (4.14).

Formally we have that at the minimum

$$\begin{aligned}
0 &= \frac{\partial}{\partial \alpha_i} S(\mathbf{a}) \qquad i = 1, 2, \ldots, N \\
&= \frac{\partial}{\partial \alpha_i} \sum_{k=1}^{M} \left(f_k - \sum_{j=1}^{N} \alpha_j \phi_j(x_k) \right)^2 \\
&= -2 \sum_{k=1}^{M} \left(f_k - \sum_{j=1}^{N} \alpha_j \phi_j(x_k) \right) \phi_i(x_k).
\end{aligned}$$

Rearranging, we have

$$\sum_{j=1}^{N} \alpha_j \sum_{k=1}^{M} \phi_j(x_k) \phi_i(x_k) = \sum_{k=1}^{M} f_k \phi_i(x_k) \qquad i = 1, 2, \ldots, N. \tag{5.22}$$

If we introduce the vector inner product

$$\langle \mathbf{g}, \mathbf{h} \rangle = \sum_{k=1}^{M} g_k h_k \tag{5.23}$$

(which should be compared with the inner product introduced in subsection

4.4.1) for any two M-vectors $\mathbf{g}$ and $\mathbf{h}$ then the system (5.22) may be rewritten as the system of N equations in the N unknowns $\alpha_1, \alpha_2, \ldots, \alpha_N$

$$\Phi\mathbf{a} = \mathbf{b}$$

where

$$\Phi_{ij} = \langle \phi_i, \phi_j \rangle$$
$$b_i = \langle \mathbf{f}, \phi_i \rangle$$

and

$$\phi_i^T = (\phi_i(x_1), \phi_i(x_2), \ldots, \phi_i(x_M))$$
$$\mathbf{f}^T = (f_1, f_2, \ldots, f_M).$$

(Compare this with equation (4.36).)

EXAMPLE 5.6
To find the least squares straight line approximation to the data points

x_i	1	2	3	4	5
f_i	3	4	7	6	7

we choose $\phi_1(x) = 1$ and $\phi_2(x) = x$ and form the inner products

$\langle \phi_1, \phi_1 \rangle = 1 \times 1 + 1 \times 1 + 1 \times 1 + 1 \times 1 + 1 \times 1 = 5$
$\langle \phi_1, \phi_2 \rangle = 1 \times 1 + 1 \times 2 + 1 \times 3 + 1 \times 4 + 1 \times 5 = 15 = \langle \phi_2, \phi_1 \rangle$
$\langle \phi_2, \phi_2 \rangle = 1 \times 1 + 2 \times 2 + 3 \times 3 + 4 \times 4 + 5 \times 5 = 55$
$\langle \mathbf{f}, \phi_1 \rangle = 3 \times 1 + 4 \times 1 + 7 \times 1 + 6 \times 1 + 7 \times 1 = 27$
$\langle \mathbf{f}, \phi_2 \rangle = 3 \times 1 + 4 \times 2 + 7 \times 3 + 6 \times 4 + 7 \times 5 = 91.$

Then the slope (α_2) and the intercept (α_1) of the polynomial approximation $p_1(x) = \alpha_1 + \alpha_2 x$ are given as the solution of the pair of simultaneous equations

$$\begin{pmatrix} 5 & 15 \\ 15 & 55 \end{pmatrix} \begin{pmatrix} \alpha_1 \\ \alpha_2 \end{pmatrix} = \begin{pmatrix} 27 \\ 91 \end{pmatrix}.$$

Using Gauss elimination we have

$$\begin{pmatrix} 5 & 15 \\ 0 & 10 \end{pmatrix} \begin{pmatrix} \alpha_1 \\ \alpha_2 \end{pmatrix} = \begin{pmatrix} 27 \\ 10 \end{pmatrix}$$

that is

$$\alpha_2 = 1$$
$$\alpha_1 = \frac{1}{5}(27 - 15\alpha_2) = \frac{12}{5} = 2\frac{2}{5}$$

so that

$$p_1(x) = 2\frac{2}{5} + x$$

(see Fig. 5.9).

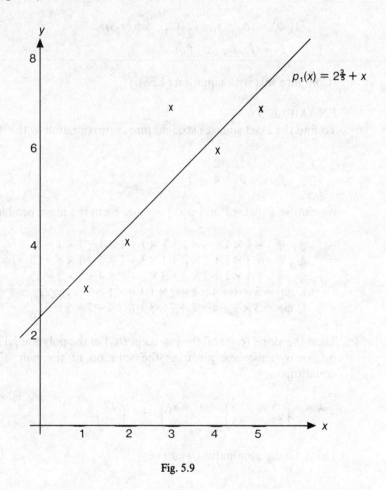

Fig. 5.9

As in the continuous case it is often useful to introduce the concept of a weighted least squares approximation and we do this by redefining the inner product (5.23) as

$$\langle g, h \rangle = \sum_{k=1}^{M} w_k g_k h_k$$

where $\{w_k : k = 1, 2, \ldots, M : w_k > 0\}$ is some chosen set of weights. Using this definition we can reduce the influence of any value known to be susceptible to error by selecting a small value for the weight at that point. Conversely, we might want the approximating function to fit one of the data points very closely and this result can be achieved by employing a very large weight at that point.

5.6.3 Discrete Gram–Schmidt orthogonalisation

In section 4.4 we saw that by choosing an appropriate set of basis functions (Legendre polynomials if the interval of approximation is $[-1,1]$) the coefficient matrix for the system of equations defining the least squares approximation to a continuous function can be made diagonal. Clearly we would like to be able to do the same sort of thing here. In addition, if

$$x_k = (k-1)/(M-1) \qquad k = 1, 2, \ldots, M$$

and

$$\phi_j(x) = x^{j-1} \qquad j = 1, 2, \ldots, N$$

then for large M

$$\langle \phi_i, \phi_j \rangle = \sum_{k=1}^{M} x_k^{i-1} x_k^{j-1} = \sum_{k=1}^{M} x_k^{i+j-2} \simeq M \int_0^1 x^{i+j-2} \, dx = \frac{M}{i+j-1}$$

and, once again, we encounter problems of ill-conditioning.

To obtain a diagonal matrix we need to ensure that the basis functions are *orthogonal*. By this, we mean that they must satisfy

$$\langle \phi_i, \phi_j \rangle \begin{cases} = 0 & i \neq j \\ \neq 0 & i = j \end{cases}$$

and any set of functions which satisfies this condition is said to form an orthogonal set on the points $\{x_k : k = 1, 2, \ldots, M\}$ (and with respect to the weights $\{w_k : k = 1, 2, \ldots, M\}$ if a weighted inner product is employed). If the basis functions are orthogonal then the solution to the discrete least squares approximation problem is given as

$$\alpha_j = \langle \mathbf{f}, \boldsymbol{\phi}_j \rangle / \langle \boldsymbol{\phi}_j, \boldsymbol{\phi}_j \rangle \qquad j = 1, 2, \ldots, N.$$

In subsection 4.5.3 we introduced the Chebyshev polynomials and noted that they are orthogonal on $[-1,1]$ with respect to the weight function $(1-x^2)^{-\frac{1}{2}}$. These polynomials have the additional property

$$\sum_{k=1}^{M}{}'' T_i(x_k) T_j(x_k) = \begin{cases} 0 & i \neq j \\ M & i = j = 0 \text{ or } M \\ M/2 & i = j \neq 0 \text{ or } M \end{cases}$$

(the double prime indicating that the first and last terms are to be halved) where

$$x_k = \cos(\pi(k-1)/(M-1)) \qquad k = 1, 2, \ldots, M \tag{5.24}$$

that is, they are orthogonal on the point set (5.24) with respect to the weights

$$w_1 = w_M = \tfrac{1}{2}$$
$$w_k = 1 \qquad k = 2, 3, \ldots, M-1.$$

However, in the general case it is extremely unlikely that (5.24) will be the point set for our discrete approximation problem and hence we need some scheme for generating orthogonal polynomials. The obvious thing to do here is to extend the Gram–Schmidt algorithm of subsection 4.5.2. Given the set of basis functions $\{\phi_i(x) : i = 1, 2, \ldots, N\}$, the functions $\{\psi_i(x) : i = 1, 2, \ldots, N\}$ defined by

$$\psi_1(x) = \phi_1(x)$$
$$\psi_i(x) = \phi_i(x) - \sum_{j=1}^{i-1} \frac{\langle \boldsymbol{\psi}_j, \boldsymbol{\phi}_i \rangle}{\langle \boldsymbol{\psi}_j, \boldsymbol{\psi}_j \rangle} \psi_j(x) \qquad i = 2, 3, \ldots, N \tag{5.25}$$

will be orthogonal where

$$\boldsymbol{\psi}_j^T = (\psi_j(x_1), \psi_j(x_2), \ldots, \psi_j(x_M)) \qquad j = 1, 2, \ldots, N.$$

(Compare this with the system (4.37).) If we divide $\psi_i(x)$ by $\langle \boldsymbol{\psi}_i, \boldsymbol{\psi}_i \rangle^{\frac{1}{2}}$ at each stage then the $\psi_i(x)$s will be orthonormal, that is, they will satisfy

$$\langle \boldsymbol{\psi}_i, \boldsymbol{\psi}_i \rangle = 1 \qquad i = 1, 2, \ldots, N.$$

EXAMPLE 5.7
Given the data of Example 5.6, if we choose $\phi_i(x) = x^{i-1} : i = 1, 2$ then

$$\phi_1^T = (1, 1, 1, 1, 1)$$
$$\phi_2^T = (1, 2, 3, 4, 5).$$

Using the Gram–Schmidt orthogonalisation process we have

$$\psi_1(x) = \phi_1(x) = 1$$

so that

$$\psi_1^T = (1, 1, 1, 1, 1).$$

Hence

$$\langle \psi_1, \phi_2 \rangle = 1 \times 1 + 1 \times 2 + 1 \times 3 + 1 \times 4 + 1 \times 5 = 15$$
$$\langle \psi_1, \psi_1 \rangle = 1 \times 1 + 1 \times 1 + 1 \times 1 + 1 \times 1 + 1 \times 1 = 5$$

so that

$$\psi_2(x) = \phi_2(x) - \frac{\langle \psi_1, \phi_2 \rangle}{\langle \psi_1, \psi_1 \rangle} \psi_1(x)$$

$$= x - 3.$$

Hence, if we fit a curve to the data of the form

$$p_1(x) = \alpha_1 \psi_1(x) + \alpha_2 \psi_2(x)$$

then

$$\alpha_1 = \langle \mathbf{f}, \psi_1 \rangle / \langle \psi_1, \psi_1 \rangle = (3 \times 1 + 4 \times 1 + 7 \times 1 + 6 \times 1 + 7 \times 1)/5 = \frac{27}{5}$$

$$\alpha_2 = \langle \mathbf{f}, \psi_2 \rangle / \langle \psi_2, \psi_2 \rangle$$
$$= (3 \times (-2) + 4 \times (-1) + 7 \times (0) + 6 \times 1 + 7 \times 2)$$
$$/((-2)^2 + (-1)^2 + 0^2 + 1^2 + 2^2)$$
$$= 10/10 = 1.$$

Therefore

$$p_1(x) = x - 3 + \frac{27}{5} = x + 2\frac{2}{5}.$$

We note that algorithm (5.25) generates the orthogonal polynomials

$\psi_i(x) : i = 1, 2, \ldots, N$ as continuous functions. It is, however, necessary only to generate their values on the given point set since it is always possible to reformulate the solution to the least squares approximation problem in terms of the original basis functions.

5.6.4 Discrete minimax and least first power approximation

In section 4.7 we noted that the solution to the continuous least squares approximation problem is just one member of a family, the L_p approximations. Here, we have a similar situation and we refer to that solution which gives

$$\underset{\mathbf{a}}{\text{minimum}} \sum_{k=1}^{M} |f_k - L(\mathbf{a}, x_k)|^p w_k$$

as the l_p approximation to the given coordinates. Restricting ourselves to the case $w_k = 1 : k = 1, 2, \ldots, M$ the l_1 (or least first power) approximation minimises the sum of the absolute values of the residuals, that is, gives

$$\underset{\mathbf{a}}{\text{minimum}} \sum_{k=1}^{M} |f_k - L(\mathbf{a}, x_k)|$$

whilst the l_∞ (or minimax) approximation minimises the maximum absolute residual, that is, gives

$$\underset{\mathbf{a}}{\text{minimum}} \underset{1 \leq k \leq M}{\text{maximum}} |f_k - L(\mathbf{a}, x_k)|$$

(and it can be shown that as $p \to \infty$ the l_p approximations tend to the l_∞ approximation).

In subsection 4.7.2 we noted that the solution to the L_∞ polynomial approximation problem possesses a very important property (see Theorem 4.2). A similar situation arises here as shown by the following theorem.

THEOREM 5.2
Let $\{(x_k, f_k) : k = 1, 2, \ldots, M\}$ be a set of data points with $X_M = \{x_k : k = 1, 2, \ldots, M\}$ and let $p_{N-1}(x)$ be the minimax polynomial approximation to these points of degree at most $N-1$. Then the residual vector equioscillates on $N+1$ points of X_M; that is, there exists a subset $\tilde{X}_{N+1}$ of X_M such that

$$\begin{aligned} |r_k| &= d & x_k &\in \tilde{X}_{N+1} \\ |r_k| &\leq d & k &= 1, 2, \ldots, M \end{aligned}$$

and at consecutive points of $\bar{X}_{N+1}$ the residuals alternate in sign.

For a proof of this theorem see Rice (1964).

Theorem 5.2 tells us that, for an l_∞ approximation, the residual vector takes on its maximum absolute value at (at least) $N+1$ of the points $\{x_i : i = 1, 2, \ldots, M\}$ and that the sign of the error alternates from one 'maximum' point to the next. Further, if we can find an approximating polynomial such that the conditions of Theorem 5.2 are satisfied, then that polynomial is the minimax approximation.

EXAMPLE 5.8
Suppose we wish to find the minimax straight line approximation to the data of Example 5.6. Referring to Fig. 5.10 it is clear that the first two and the last

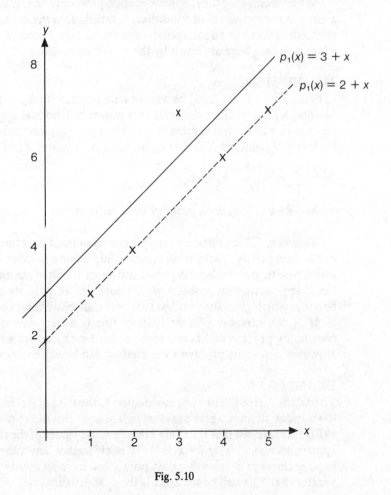

Fig. 5.10

two data points lie on the straight line $p_1(x) = 2 + x$. The residuals for this polynomial approximation would, therefore, be 0, 0, 2, 0, and 0. By raising this line a little, the residual at the point $x = 3$ will decrease but elsewhere will increase in magnitude. A cross-over point will eventually be reached at which any further attempt to reduce $|r_3|$ will have the effect of increasing the overall maximum absolute residual. This critical point is represented by the line $p_1(x) = 3 + x$ for which the maximum absolute residual is 1. For this straight line approximation to the data we observe that the residual vector is

$$\mathbf{r}^T = (-1, -1, 1, -1, -1).$$

Taking the points $x = 1$, $x = 3$ and $x = 5$ in turn we see that the sign of the residual alternates on this point set and conclude that $3 + x$ is the minimax approximation. Note that the set of equioscillation points is not unique.

We now consider the l_1 approximation problem and recall that, provided a not too restrictive set of conditions is satisfied, in the continuous case the problem reduces to an interpolation problem (see Theorem 4.3). A similar situation arises here as shown by the following theorem.

THEOREM 5.3
Let $\{(x_i, f_i) : i = 1, 2, \ldots, M\}$ be a set of data points with $X_M = \{x_i : i = 1, 2, \ldots, M\}$ and let $p_{N-1}(x)$ be the least first power polynomial approximation to these points of degree at most $N - 1$. Then $p_{N-1}(x)$ interpolates the data at (at least) N points of X_M; that is, there exists a subset $\bar{X}_N$ of X_M such that

$$r_k = 0 \qquad x_k \in \bar{X}_N.$$

See Rice (1964) for a proof of this theorem.

Theorem 5.3 says that an l_1 approximation must pass through at least N of the data points. There is, of course, only a finite number of polynomials which pass through at least N points and hence a simple algorithm for solving the l_1 approximation problem would consider all possible candidates and find that which gives the smallest sum of the absolute values of the residuals. If M is considerably greater than N this is a very laborious and time-consuming process and hence more refined methods are used in practice. However, for some problems the method can be extremely simple to use.

EXAMPLE 5.9
To find the l_1 straight line approximation to the data of Example 5.6 we need to consider all lines which pass through at least two of the points. Referring to Fig. 5.11 we see that the sum of the absolute values of the residuals for the approximation $p_1(x) = 2 + x$ is 2. It is clear that any other straight line passing through at least two data points will have an absolute residual sum greater than this and hence $2 + x$ is the l_1 approximation.

It is useful to draw a comparison between the l_1, l_2 and l_∞ approximations to the data of Example 5.6. The function values were obtained by

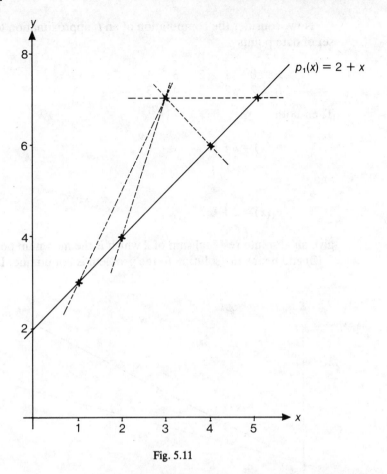

Fig. 5.11

evaluating the polynomial $2 + x$ at the points $x = 1, 2, 3, 4$ and 5, a value of 2 being added to f_3 to simulate the presence of a rogue value. We observe that the l_1 approximation is not influenced at all by this perturbation unlike the solution to the l_∞ problem. The l_2 solution is only slightly affected.

5.6.5 Non-uniqueness of l_p approximations

Before leaving l_p approximation we consider the uniqueness of the l_1, l_2 and l_∞ solutions. For the case $p = 2$ the uniqueness of the solution is governed by the form of the coefficient matrix Φ. If this matrix is not of full rank (that is, the columns of the matrix do not form a linearly independent set of vectors) then the solution is indeterminate. To avoid such a situation arising we need to impose the condition that the basis functions themselves be linearly independent; that is, it must not be possible to express one basis function as a linear combination of the others. Note that if such a condition is not satisfied the Gram–Schmidt orthogonalisation process will break down.

240 APPROXIMATION OF NUMERICALLY DEFINED FUNCTIONS [Ch. 5]

Now, consider the computation of an l_1 approximation to the following set of data points

$$\begin{array}{cccccc} x_i & 0 & 1 & 2 & 3 & 4 \\ f_i & 2 & 2 & 3 & 4 & 3 \end{array}.$$

Then both

$$p_1(x) = 2 + \tfrac{1}{4}x$$

and

$$p_1(x) = 2 + \tfrac{1}{2}x$$

give an absolute residual sum of 2 which is the minimum possible (see Fig. 5.12) and hence the solution to the problem is not unique. It is beyond the

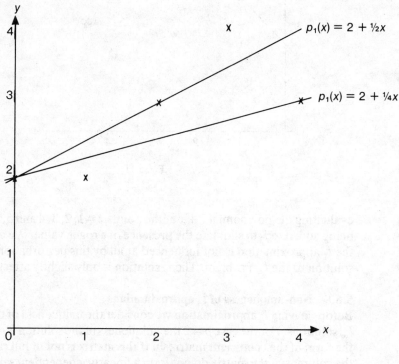

Fig. 5.12

scope of this book to embark on a discussion of under what conditions an l_1 (or l_∞) approximation is unique. However, the reader should be aware of the fact that more than one solution to a data fitting problem may exist.

5.6.6 A program for least first power straight line approximation

The programs of Case Study 5.3 implement the simple algorithm given in subsection 5.6.4 for computing the least first power straight line approximation to a given set of data points. In the event of a non-unique solution the code returns the first solution encountered. It should be noted that if supplied with the data of Example 5.6 the programs will test whether the polynomial $p_1(x) = 2 + x$ is a solution 6 times. Clearly any reasonable algorithm should be capable of avoiding such duplication of work. The variable *leastfirstpowererror* must be initialised to a suitably large enough value. In the Algol 68 program we have used the predefined constant *maxreal*, the largest representable positive real value; the value 10000.0 has been used in the Pascal program.

CASE STUDY 5.3

Algol 68 Program

```
BEGIN
  COMMENT
  a program to compute the least first power straight line fit to a given
  set of data points
  COMMENT
  REAL residualsum , intercept , slope , a , b , h ,
    leastfirstpowererror := maxreal ;
  PROC line = ( REAL x ) REAL :
    COMMENT
    a straight line approximation
    COMMENT
    a + b * x ;
  INT numberofpoints , point1 , point2 ;
  read ( numberofpoints ) ;
  [ 1 : numberofpoints ] REAL xcoords , fcoords ;
  FOR i TO numberofpoints
  DO
    read ( ( xcoords [ i ] , fcoords [ i ] ) )
  OD ;
  FOR i TO numberofpoints - 1
  DO
    FOR j FROM i + 1 TO numberofpoints
    DO
      COMMENT
      find the equation of the straight line passing through the i th and
      j th data points
      COMMENT
      h := xcoords [ j ] - xcoords [ i ] ;
      a := ( xcoords [ j ] * fcoords [ i ] - xcoords [ i ] * fcoords [ j ] )
           / h ;
      b := ( fcoords [ j ] - fcoords [ i ] ) / h ;
      COMMENT
      find the sum of the absolute value of the residuals for this approximation
      COMMENT
      residualsum := 0.0 ;
      FOR k TO numberofpoints
      DO
        IF k # i AND k # j
        THEN
          residualsum +:= ABS ( fcoords [ k ] - line ( xcoords [ k ] ) )
        FI
      OD ;
```

```
            COMMENT
            test to see if the current straight line is better than the best
            approximation found so far
            COMMENT
            IF residualsum < leastfirstpowererror
            THEN
              point1 := i ; point2 := j ;
              intercept := a ; slope := b ;
              leastfirstpowererror := residualsum
            FI
       OD
   OD ;
   print ( ( newline , newline ,
             "the least first power approximation to the data" , newline ,
             newline , "   point number    x-coordinate    function value" ) ) ;
   FOR i TO numberofpoints
   DO
      print ( ( newline , "      " , i , "         " , xcoords [ i ] ,
                         "      " , fcoords [ i ] ) )
   OD ;
   print ( ( newline , newline , newline ,
             "is the line which passes through the points numbered" , newline ,
             newline , point1 , "         and   " , point2 , newline , newline ,
             "and is" , newline , newline , intercept , "  +  " , slope ,
             " x" , newline , newline ) )
END
```

Pascal Program
──────────────

```
PROGRAM leastfirstpowerstraightline(input, output);

   { a program to compute the least first power straight line fit
     to a given set of data points }

   CONST
      maxnumberofpoints=50;

   VAR
      residualsum, intercept, slope, a, b, h, leastfirstpowererror:real;
      i, j, k, numberofpoints, point1, point2:1..maxnumberofpoints;
      xcoords, fcoords:ARRAY [ 1..maxnumberofpoints ] OF real;

   FUNCTION line(x:real):real;
     { a straight line approximation }
   BEGIN
      line:= a + b*x
   END { line } ;

BEGIN
   leastfirstpowererror:= 10000.0;
   read(numberofpoints);
   FOR i:= 1 TO numberofpoints DO
      read(xcoords[i], fcoords[i]);
   FOR i:= 1 TO numberofpoints - 1 DO
      FOR j:= i + 1 TO numberofpoints DO
      BEGIN
         { find the equation of the straight line passing through
           the i th and j th data points }
         h:= xcoords[j] - xcoords[i];
         a:= (xcoords[j]*fcoords[i] - xcoords[i]*fcoords[j])/h;
         b:= (fcoords[j] - fcoords[i])/h;
         { find the sum of the absolute value of the residuals
           for this approximation }
         residualsum:= 0.0;
         FOR k:= 1 TO numberofpoints DO
```

```
                IF (k<>i) AND (k<>j)
                THEN
                    residualsum:= residualsum +
                                    abs(fcoords[k] - line(xcoords[k]));
            { test to see if the current straight line is better than
              the best approximation found so far }
            IF residualsum<leastfirstpowererror
            THEN
                BEGIN
                    point1:= i;  point2:= j;
                    intercept:= a;  slope:= b;
                    leastfirstpowererror:= residualsum
                END
        END ;
    writeln;
    writeln;
    writeln('the least first power approximation to the data');
    writeln;
    writeln('  point number    x-coordinate    function value');
    FOR i:= 1 TO numberofpoints DO
        writeln(i:9, ' ':8, xcoords[i], ' ':4, fcoords[i]);
    writeln;
    writeln;
    writeln('is the line which passes through the points numbered');
    writeln;
    writeln(point1, '     and   ', point2);
    writeln;
    writeln('and is');
    writeln;
    writeln(intercept, '   +   ', slope, ' x');
    writeln
END { leastfirstpowerstraightline } .
```

Sample Data

```
    6
    0.0    5.0
    1.0    3.0
    2.0    7.0
    3.0    8.0
    4.0   13.0
    5.0   13.0
```

Sample Output

the least first power approximation to the data

```
    point number    x-coordinate     function value
        1            0.00000e  0      5.00000e  0
        2            1.00000e  0      3.00000e  0
        3            2.00000e  0      7.00000e  0
        4            3.00000e  0      8.00000e  0
        5            4.00000e  0      1.30000e  1
        6            5.00000e  0      1.30000e  1
```

is the line which passes through the points numbered

         3   and   6

and is

         3.00000e 0   +   2.00000e 0 x

5.7 AVAILABLE ALGORITHMS

There are a number of algorithms available which implement the methods outlined in this chapter, the most widely used computer codes being based on the least squares criterion. Some adopt the approach of looking for a *minimal degree* approximation. Here, the user indicates the maximum degree of the polynomial approximation that is acceptable. The best constant, straight line, quadratic, etc. polynomial approximations are considered in turn and a 'goodness of fit' measure, based on a weighted sum of the squares of the residuals, is used to monitor the effect of increasing the order. If adding an extra term has little effect the process is terminated.

Codes for minimax approximation are usually based on the exchange algorithm (see Powell, 1981) which exploits the equioscillation property of the Chebyshev solution. An alternative approach employs linear programming techniques (see Barrodale and Phillips, 1974). In the case of multiple solutions such codes usually return one approximation but indicate that it may not be unique. Linear programming can also be used to solve the l_1 approximation problem (see Barrodale and Roberts, 1973).

More recently attention has been given to the minimal degree approach to l_∞ approximation (see Roberts, 1976). When devising algorithms of this type it is important to ensure that, when the next term is added, the process does not start from scratch again.

The main use of Lagrange and Hermite interpolating polynomials is to provide the basis for other topics in numerical analysis such as quadrature, multi-variable function minimisation, etc. Nevertheless most computer packages contain at least one interpolating routine.

Finally, as in the continuous case we should be aware of the concept of non-linear approximation and, in fact, one of the earliest forms of function approximation is based on the use of exponentials (see Barrodale and Olesky, 1981). Clearly, methods for finding the solutions to non-linear approximation problems will tend to be rather more complex than those described in this chapter and will almost certainly be based on the use of some iterative technique.

EXERCISES

5.1 Find, in standard form, the polynomial which passes through the coordinates

x_i	0	1	2	3
f_i	−1	0	3	14

5.2 Find, in standard form, the Hermite interpolating polynomial for the following data.

x_i	-1	1	3
f_i	0	0	208
f'_i	12	4	348

5.3 Find the least squares straight line approximation to the data

x_i	1	2	3
f_i	-2	-2	2

(a) directly;
(b) using polynomials which are orthogonal on the given point set.
Calculate the sum of the squares of the residuals.

5.4 Given the data

x_i	0	1	2	3	4	5
f_i	5	3	7	8	13	13

find the least squares straight line approximation using orthogonal polynomials. Use a sketch to find the least first power approximation to this data. Verify that the residual for the approximation $p_1(x) = 3 + 2x$ equioscillates on three points and deduce that this is the minimax approximation.

5.5 Prove that for any N-dimensional vector $\mathbf{x}$

$$\underset{1 \leqslant i \leqslant N}{\text{maximum}} |x_i| \leqslant \left(\sum_{i=1}^{N} x_i^2 \right)^{\frac{1}{2}} \leqslant N^{\frac{1}{2}} \underset{1 \leqslant i \leqslant N}{\text{maximum}} |x_i|. \qquad (1)$$

For the data

x_i	1	2	3	4	5
f_i	3	4	7	6	7

the least squares straight line approximation is $p_1(x) = 2\frac{2}{5} + x$. Show that relationship (1) holds for the residual vector.

5.6 Write a program to estimate the value of a function at a point using a Lagrange interpolating polynomial. Use the data of Example 5.2 to test your program.

5.7 Write a program which sets up the matrix equation for a cubic spline approximation to the function $\sin(2x)$ using the grid points $\{0, \frac{1}{2}, 1\}$. Use the Gauss elimination code of Case Study 3.2 to compute the solution.

5.8 Write a program, based on the code of Case Study 4.1, which computes a least squares polynomial approximation to e^x using 51 equally spaced points in [0, 1] (so that $x_i = (i-1)/50 : i = 1, 2, \ldots, 51$). Produce a table of results corresponding to Table 4.2. Modify your program so that the elements of the right-hand side vector are perturbed by a small amount, say 0.00005, and observe the effect on the computed solution.

5.9 Write a program, based on the code of Case Study 5.3, which finds the least first power quadratic approximation to a set of coordinates.

6
Numerical Integration

6.1 INTRODUCTORY REMARKS

In this chapter we discuss the problem of trying to obtain the numerical value of the definite integral

$$I = \int_a^b f(x)\,dx \quad (b > a) \tag{6.1}$$

to some desired precision. The interval of integration $[a, b]$ is assumed to be finite but some consideration will be given later to integrals over semi-infinite and infinite ranges. Sometimes it will be possible to evaluate (6.1) analytically using one of the standard methods of integration. However, in many circumstances encountered in practice this will not be possible and we must use numerical techniques which evaluate an integral only approximately. Simple examples of this situation are provided by the integrals

$$I = \int_0^1 e^{-x^2}\,dx$$

and

$$I = \int_1^2 \frac{e^{-x}}{x}\,dx\;.$$

Even if an analytical technique is available, it may be difficult and tedious to apply it in practice and consequently an approach through numerical integration, or *quadrature*, may be preferable. In addition, we sometimes wish to estimate the value of an integral where the only information available about the *integrand*, $f(x)$, is its value on a discrete point set. In such a case we have little choice but to try to use one of the methods given in this chapter.

In quadrature, an approximation or *quadrature rule*

$$R = \sum_{i=1}^{n} w_i f(x_i) \qquad (6.2)$$

is used to estimate (6.1). The rule (6.2) is just a weighted sum (with *weights* $\{w_i: i = 1, 2, ..., n\}$) of values of the integrand. We shall find that the points $\{x_i: i = 1, 2, ..., n\}$ belong to $[a, b]$ and will assume that they are ordered so that $x_1 < x_2 ... < x_{n-1} < x_n$. We refer to (6.2) as an n-point rule which is

(i) *closed*   if $x_1 = a$ and $x_n = b$

(ii) *open*   if $x_1 > a$ and $x_n < b$

(iii) *half-open*   if either $x_1 = a$ and $x_n < b$ or $x_1 > a$ and $x_n = b$.

Further, we say that the rule is an *equal interval rule* if the points x_i are equally spaced.

EXAMPLE 6.1
Suppose that we wish to estimate

$$\int_{-1}^{1} f(x) dx .$$

Two rules which we might consider using are

(i) $\frac{1}{3}\{f(-1) + 4f(0) + f(1)\}$.

This is a closed, equal-interval, three-point rule (which some readers may recognise as Simpson's rule). The weights are 1/3, 4/3, and 1/3 at the points -1, 0 and 1 respectively.

(ii) $\frac{1}{9}\left\{5f\left(-\sqrt{\frac{3}{5}}\right) + 8f(0) + 5f\left(\sqrt{\frac{3}{5}}\right)\right\}$.

This is an open, three-point rule in which the weights are 5/9, 8/9 and 5/9 at the points $-\sqrt{(3/5)}$, 0 and $\sqrt{(3/5)}$.

For the integrand $f(x) = e^x$ we have the following results:

         true value   2.35040
         rule (i)   2.36205
         rule (ii)   2.35034 .

We observe that rule (i) gives an overestimate of the required integral whilst rule (ii) gives an underestimate. Further, rule (ii) is the more accurate for this particular integral.

When a quadrature rule has been derived we are interested in the *truncation error*

$$E = I - R.$$

We cannot in general say what this is (otherwise we would be able to evaluate (6.1) exactly) but we can try to find an upper limit on the magnitude of E. To do this we try to find a positive number M for which we can assert

$$|E| \leq M.$$

This is not always easy to do. It may be possible, however, to assert that E is zero when $f(x)$ is $1, x, x^2, \ldots, x^m$ but not when $f(x)$ is x^{m+1}. Such a rule is said to have *precision m*.

EXAMPLE 6.2
Consider the two rules of Example 6.1. Table 6.1 indicates the performance of these two rules on the integrands $f(x) = 1, x, x^2, \ldots$. The column headed 'True integral' has been obtained using

$$\int_{-1}^{1} x^i dx = \left[\frac{x^{i+1}}{i+1}\right]_{-1}^{1} = \begin{cases} 2/(i+1) & i \text{ even} \\ 0 & i \text{ odd} \end{cases}$$

The first monomial that rule (i) fails to integrate exactly is x^4 and hence this

Table 6.1

Integrand	True integral	Rule (i)	E for rule (i)	Rule (ii)	E for rule (ii)
1	2.00000	2.00000	0.00000	2.00000	0.00000
x	0.00000	0.00000	0.00000	0.00000	0.00000
x^2	0.66667	0.66667	0.00000	0.66667	0.00000
x^3	0.00000	0.00000	0.00000	0.00000	0.00000
x^4	0.40000	0.66667	−0.26667	0.40000	0.00000
x^5	0.00000			0.00000	0.00000
x^6	0.28571			0.24000	0.04571

rule has precision 3. Rule (ii) on the other hand has precision 5. The fact that rule (ii) has greater precision than rule (i) means that for 'suitable' functions, rule (ii) should be the more accurate of the two. The relative performance of the two rules discussed in Example 6.1 is, therefore, to be expected.

There are a number of further preliminary observations which should be made about quadrature rules. First, rules for which the weights w_i are all positive are to be preferred since they reduce the possibility of differencing troubles. However, if the integrand is not of one sign throughout the range of integration such problems cannot be totally eliminated by a rule of this form. The second point to be made is that the amount of computation is usually dominated by the number of evaluations of the integrand (which is n, unless any symmetry properties of $f(x)$ can be exploited). Thus it is useful to be able to identify, if possible, the kinds of integrand for which a particular type of rule can be expected to work well.

A third observation is that, when implementing rule (6.2), the integration points and weights will generally be represented in a computer only to some finite precision and will therefore be subject to rounding error. Moreover, even if the representation of the points is exact, $f(x)$ will be evaluated only to a finite precision. We have also to think of the error which may arise from accumulating a sum of products. Hence what we actually compute is a quantity R^* and it is necessary to try to say something about the *computational error* $R - R^*$ in terms of the particular computer and language implementation used. Further, since

$$I - R^* = (I - R) + (R - R^*)$$

we have

$$|I - R^*| \leq |I - R| + |R - R^*|$$

on using the triangle inequality and so the magnitude of the error in R^* is bounded by the sum of the bounds for the magnitude of the truncation error and for the magnitude of the computational error.

In sections 6.2, 6.3 and 6.4 we discuss in some detail the derivation of a number of different quadrature schemes although, in an elementary text book, the treatment is necessarily not mathematically complete. Sections 6.5, 6.6 and 6.7 are concerned with an outline of other important approaches, and in section 6.8 we give the basic ideas behind the concept of *adaptive quadrature* in which an attempt is made to enable a computer program to take account of the behaviour of the integrand on the interval of integration. In section 6.9 we discuss the effect of computational error as defined in the preceding paragraph. Finally, in section 6.10 we discuss some of the facilities available in certain sections of the chapter on quadrature in the NAG Algol 68 library.

6.2 NEWTON–COTES QUADRATURE RULES

6.2.1 Closed and open rules

Newton–Cotes rules are all of equal interval type and can be either open or closed. Let $\{x_i: i = 1, 2, \ldots, k\}$ be a set of equally spaced points with

$$x_{i+1} - x_i = h \quad i = 1, 2, \ldots, k-1 \ . \tag{6.3}$$

Consider the integral

$$I_k = \int_{x_1}^{x_k} f(x) \mathrm{d}x \tag{6.4}$$

and suppose that at each point x_i the value $f_i = f(x_i)$ is available or can be computed. Then a Newton–Cotes closed rule, R_k, is obtained by replacing the integrand in (6.4) with the Lagrange interpolating polynomial of degree $k - 1$

$$p_{k-1}(x) = \sum_{i=1}^{k} \phi_i(x) f_i$$

(see section 5.2). Thus

$$R_k = \int_{x_1}^{x_k} p_{k-1}(x) \mathrm{d}x = \int_{x_1}^{x_k} \sum_{i=1}^{k} \phi_i(x) f_i \mathrm{d}x = \sum_{i=1}^{k} f_i \int_{x_1}^{x_k} \phi_i(x) \mathrm{d}x$$

and this rule is of the form (6.2) with

$$w_i = \int_{x_1}^{x_k} \phi_i(x) \mathrm{d}x \ .$$

The members of the sequence of Newton–Cotes closed rules are obtained by taking $k = 2, 3, \ldots$, in turn.

If $k = 2$ we have

$$R_2 = \int_{x_1}^{x_2} \left(\frac{x - x_2}{x_1 - x_2} f_1 + \frac{x - x_1}{x_2 - x_1} f_2 \right) \mathrm{d}x$$

$$= -\frac{1}{h} f_1 \int_{x_1}^{x_2} (x - x_2) \mathrm{d}x + \frac{1}{h} f_2 \int_{x_1}^{x_2} (x - x_1) \mathrm{d}x$$

$$= \frac{h}{2}\{f_1 + f_2\} \qquad (6.5)$$

using (6.3). This is the *trapezium rule* (discussed briefly in section 1.1) which has a simple geometrical interpretation. Referring to Fig. 6.1, we see that the function $f(x)$ is approximated between the points $x = x_1$ and $x = x_2$ by the straight line linking the coordinates (x_1, f_1) and (x_2, f_2). Integrating this straight line approximation we obtain the area of the shaded trapezium.

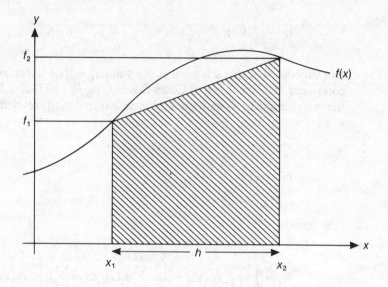

Fig. 6.1

For the case $k = 3$ we have

$$R_3 = \int_{x_1}^{x_3} p_2(x) \, dx$$

where

$$p_2(x) = \frac{(x-x_2)(x-x_3)}{(x_1-x_2)(x_1-x_3)} f_1 + \frac{(x-x_1)(x-x_3)}{(x_2-x_1)(x_2-x_3)} f_2 + \frac{(x-x_1)(x-x_2)}{(x_3-x_1)(x_3-x_2)} f_3 .$$

Carrying out the integration and using (6.3) again we find that

$$R_3 = \frac{h}{3}\{f_1 + 4f_2 + f_3\} \tag{6.6}$$

which is *Simpson's rule*.

EXAMPLE 6.3
For the integral

$$\int_1^2 e^x \, dx = [e^x]_1^2 = e^2 - e = 7.38906 - 2.71828 = 4.67078$$

we have the following approximations.

(i) Trapezium rule ($x_1 = 1$, $x_2 = 2$)

$$R_2 = \frac{1}{2}\{e^1 + e^2\} = 5.05367$$

error $= 4.67078 - 5.05367 = -0.38289$.

(ii) Simpson's rule ($x_1 = 1$, $x_2 = 1.5$, $x_3 = 2$)

$$R_3 = \frac{1}{6}\{e^1 + 4e^{1.5} + e^2\} = 4.67235$$

error $= -0.00157$.

Of the two rules, Simpson's rule is clearly the more accurate (for this integrand at least).

Generating further closed Newton–Cotes rules is a straightforward, if tedious, process and the reader is referred to Ralston and Rabinowitz (1978) for the formulae for R_4, R_5, ..., R_9. It transpires that the weights in these rules are positive only for $k \leq 8$ and for $k = 10$; thus, taking numerical stability considerations into account, the rules R_9 and R_k: $k > 10$ are unlikely to be of interest.

In section 5.2 we made reference to the error in the Lagrange interpolating polynomial. Let

$$f(x) = p_k(x) + e_k(x) \ .$$

Then, integrating both sides we have

$$I_k = \int_{x_1}^{x_k} f(x)\mathrm{d}x = \int_{x_1}^{x_k} p_k(x)\mathrm{d}x + \int_{x_1}^{x_k} e_k(x)\mathrm{d}x = R_k + E_k$$

where

$$E_k = \int_{x_1}^{x_k} e_k(x)\mathrm{d}x \ .$$

Hence, it is possible to obtain an expression for the truncation error $E_k = I_k - R_k$. We shall not pursue the details further but rather state the results

$$E_2 = -\frac{h^3}{12}f''(\xi) \qquad x_1 < \xi < x_2 \tag{6.7}$$

for the trapezium rule and

$$E_3 = -\frac{h^5}{90}f^{(\mathrm{iv})}(\eta) \qquad x_1 < \eta < x_3 \tag{6.8}$$

for Simpson's rule.

The trapezium rule is based on the approximation of the integrand by a straight line and we would therefore expect it to have precision 1. From (6.7) the truncation error associated with this rule is proportional to the second derivative of $f(x)$ and since $f''(x) \equiv 0$ for $f(x) = 1$ and $f(x) = x$ but not for $f(x) = x^2$, our supposition is verified. Equation (6.8) says that Simpson's rule has precision 3 even though the rule is based on the approximation of the integrand by a quadratic. It turns out that the rule R_k has precision k when k is even, but when k is odd we have a 'bonus' in that the rule has precision $k+1$.

In (6.7) and (6.8) we have no way of knowing what the points ξ and η are. However, if, for example, we can assert that

$$|f''(x)| \leq M_2 \qquad x \in \,]x_1, x_2[$$

we can obtain the bound

$$|E_2| \leq \frac{h^3}{12} M_2$$

on the truncation error for R_2. Similarly

$$|E_3| \leq \frac{h^5}{90} M_4$$

where M_4 is a bound on the magnitude of the fourth derivative of $f(x)$.

EXAMPLE 6.4
Consider the integral of Example 6.3. The second and fourth derivatives of e^x are e^x itself, and

$$\max_{x \in [1,2]} e^x = e^2 = 7.38906 \ .$$

Hence

$$|E_2| \leq \frac{1^3}{12} \times 7.38906 = 0.61576$$

$$|E_3| \leq \left(\frac{1}{2}\right)^5 \times 7.38906/90 = 0.00257$$

and we note that the observed errors given in Example 6.3 are within these bounds.

Newton–Cotes open rules may be derived in a similar fashion. Here we form the Lagrange interpolating polynomial which passes through the coordinates $\{(x_i, f_i): i = 2, 3, \ldots, k-1\}$. When $k = 3$ we obtain the *midpoint*, or *rectangle*, rule

$$\bar{R}_3 = 2hf_2 \ . \tag{6.9}$$

Referring to Fig. 6.2, this quantity is the area of the shaded rectangle. The truncation error, $\bar{E}_3$, associated with $\bar{R}_3$ is given by

$$\bar{E}_3 = \frac{h^3}{3} f''(\xi) \qquad \xi \in]x_1, x_3[\tag{6.10}$$

and we conclude that the midpoint rule has precision 1. We shall make use of this rule in Chapter 7 as well as

$$\bar{R}_5 = \frac{4h}{3} \{2f_2 - f_3 + 2f_4\} \tag{6.11}$$

which corresponds to the case $k = 5$. The truncation error associated with

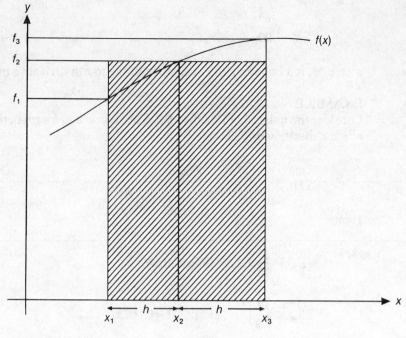

Fig. 6.2

this rule (*Milne's rule*) is

$$\bar{E}_5 = \frac{14}{45} h^5 f^{(iv)}(\eta) \qquad \eta \in \,]x_1, x_5[\, . \tag{6.12}$$

EXAMPLE 6.5
For the integral of Example 6.3 we have:

(i) Midpoint rule

$$\bar{R}_3 = 2 \times \frac{1}{2} \times e^{1.5} = 4.48169$$

error = 0.18909.

From (6.10) we obtain the error bound

$$|\bar{E}_3| \leq (\tfrac{1}{2})^3 e^2/3 = 0.30788 \, .$$

(ii) Milne's rule

$$\bar{R}_5 = \frac{4}{3} \times \frac{1}{4}\{2e^{1.25} - e^{1.5} + 2e^{1.75}\} = 4.66940$$

error = 0.00137 .

Using (6.12) we obtain the error bound

$$|\bar{E}_5| \leq \frac{14}{45}(\tfrac{1}{4})^5 e^2 = 0.00224 .$$

Before going any further it is worthwhile noting that it is sometimes possible to predict the sign of the error. For example, all derivatives of the integrand of Example 6.3 are positive on the interval of integration. Using (6.8) we would therefore expect Simpson's rule to give an overestimate of the true integral and indeed we saw that this was the case.

6.2.2 Composite Newton–Cotes rules

In the previous subsection we saw that, overall, increasing k increases the precision of a closed rule. However, high precision rules are rarely used in practice; instead we subdivide the range of interest and use a low order approximation over each sub-interval. Applying the same approach here, an approximation to (6.1) is obtained as a sum of integrals and a rule derived in this manner is known as a *composite rule*.

To derive the *composite trapezium rule*, suppose $[a, b]$ is dissected into $n - 1$ sub-intervals each of length h (so that $b - a = (n - 1)h$) and let

$$x_i = a + (i - 1)h \qquad i = 1, 2, \ldots, n .$$

Then

$$I = \int_a^b f(x)dx = \int_{x_1}^{x_n} f(x)dx$$

$$= \int_{x_1}^{x_2} f(x)dx + \int_{x_2}^{x_3} f(x)dx + \ldots + \int_{x_{n-1}}^{x_n} f(x)dx . \tag{6.13}$$

We now apply (6.5) to the integral over $[x_1, x_2]$, then over $[x_2, x_3]$ and so on to obtain the rule

$$R = \frac{h}{2}\{f_1 + f_2\} + \frac{h}{2}\{f_2 + f_3\} + \ldots + \frac{h}{2}\{f_{n-1} + f_n\}$$

$$= h \sum_{i=1}^{n}{}'' f_i \tag{6.14}$$

(which is of the form (6.2)) where the double prime on the summation sign indicates that the first and last terms are to be halved. Equation (6.14) defines an n-point closed equal-interval rule.

In the composite trapezium rule (6.14) each integral approximation will involve a truncation error of the form (6.7) and hence the total truncation error incurred by approximating (6.13) with (6.14) is

$$E = -\frac{h^3}{12}\sum_{i=1}^{n-1} f''(\xi_i) \quad \xi_i \in]x_i, x_{i+1}[\ .$$

Now

$$\min_{a \leq x \leq b} f''(x) \leq f''(\xi_i) \leq \max_{a \leq x \leq b} f''(x) \quad i = 1, 2, \ldots, n-1$$

and so

$$(n-1)\min_{a \leq x \leq b} f''(x) \leq \sum_{i=1}^{n-1} f''(\xi_i) \leq (n-1)\max_{a \leq x \leq b} f''(x) \ .$$

By the intermediate value theorem (Theorem 1.5) there exists a point $\eta \in [a, b]$ such that

$$(n-1)f''(\eta) = \sum_{i=1}^{n-1} f''(\xi_i)$$

and hence we have

$$E = -\frac{h^3}{12}(n-1)f''(\eta) = -\frac{b-a}{12}h^2 f''(\eta) \quad \eta \in [a,b] \quad (6.15)$$

which says that the error in the composite trapezium rule is proportional to (amongst other things) h^2. Hence, in the absence of rounding errors, as $h \to 0$ we would expect (6.14) to converge to the true integral. Further, by reducing h by a factor of 2 we would expect the truncation error to reduce by a factor of $2^2 = 4$.

EXAMPLE 6.6
Once again we consider the integral of Example 6.3 and compute composite trapezium rule approximations for $h = 1.0, 0.5$ and 0.25. Table 6.2 summarises the results.

Table 6.2

h	Approximation using (6.14)	Error
1.0	5.053 67	− 0.382 89
0.5	4.767 68	− 0.096 90
0.25	4.695 08	− 0.024 30

It can be seen that reducing h from 1.0 to 0.5 has the effect of reducing the error by a factor of $0.38289/0.09690 \simeq 3.95$. Halving h again results in a further reduction in the error of $0.09690/0.02430 \simeq 3.99$.

If we can assert that

$$|f''(x)| \leq M_2 \quad x \in [a, b]$$

then from (6.15) we have

$$|E| \leq \frac{b-a}{12} h^2 M_2 \tag{6.16}$$

Provided a realistic value for M_2 is available, (6.16) can be used to bound the truncation error for a particular value of h. In addition, (6.16) can be used to choose a value for h which will ensure that the magnitude of the truncation error is less than some prescribed tolerance. Suppose that we wish to obtain a composite trapezium rule approximation to (6.1) for which the magnitude of the error is at most ε. Then $|E| \leq \varepsilon$ if we choose

$$\frac{b-a}{12} h^2 M_2 \leq \varepsilon$$

that is

$$h \leq \sqrt{\frac{12\varepsilon}{(b-a)M_2}}.$$

EXAMPLE 6.7
For the integral of Example 6.3 we can choose $M_2 = 7.38906$. If we require an approximation which is in error by at most 0.005 then make

$$h \leq \sqrt{\frac{12 \times 0.005}{7.38906}} = \sqrt{0.008120} \simeq 0.09 \ .$$

Choose the 13 point composite trapezium rule (so that $h = 1/12 = 0.08333$) to obtain the value 4.67347. The error in this approximation is -0.00269, well within the specified bound.

To derive the *composite Simpson's rule* it is necessary to have an even number of sub-intervals. Let $n = 2m + 1$. Then

$$I = \int_a^b f(x)dx = \int_{x_1}^{x_3} f(x)dx + \int_{x_3}^{x_5} f(x)dx + \ldots + \int_{x_{2m-1}}^{x_{2m+1}} f(x)dx$$

and, using result (6.6) repeatedly for each integral we obtain the rule

$$R = \frac{h}{3}\{f_1 + 4f_2 + f_3\} + \frac{h}{3}\{f_3 + 4f_4 + f_5\} + \ldots$$

$$+ \frac{h}{3}\{f_{2m-1} + 4f_{2m} + f_{2m+1}\}$$

$$= \frac{h}{3}\left\{f_1 + 4\sum_{i=1}^{m} f_{2i} + 2\sum_{i=1}^{m-1} f_{2i+1} + f_{2m+1}\right\} . \tag{6.17}$$

Using (6.8) it can be easily shown that the truncation error associated with (6.17) takes the form

$$E = -\frac{b-a}{180} h^4 f^{(iv)}(\eta) \quad \eta \in [a, b] \tag{6.18}$$

so that if M_4 is a bound on the magnitude of the integrand on the interval of integration,

$$|E| \leq \frac{b-a}{180} h^4 M_4 \tag{6.19}$$

may be used to bound the error. We note that (6.18) says that the truncation error is proportional to h^4. This means that if the distance between two successive points is halved, the error should reduce by a factor of about 16.

EXAMPLE 6.8
For the integral of Example 6.3 we obtain the results shown in Table 6.3.

Table 6.3

h	Approximation using (6.17)	Error
0.5	4.672 349 0	$-0.001\,5747$
0.25	4.670 874 9	$-0.000\,100\,6$

The reduction in the error obtained by halving h is $0.0015747/0.0001006 \simeq 15.7$.

The result (6.19) can be used in principle to choose a value for h which ensures that $|E|$ is less than some prescribed quantity. However, even if we can find $f^{(iv)}(x)$ analytically without too much difficulty, finding a realistic value for M_4 may not be easy. Thus the value of results such as (6.16) and (6.19) is limited since a crude bound for the magnitude of the derivative will lead to an unnecessarily small h and hence to the use of an unnecessarily large number of function evaluations.

It is possible to derive composite rules based on $R_4, R_5, \ldots$. However, the reader will realise that in using R_k the interval of integration will have to be divided into a number of sub-intervals which is an integral multiple of $k-1$ and this may be inconvenient. It is possible also to derive composite rules based on open Newton–Cotes rules but, with the possible exception of the composite midpoint rule, they are not usually of interest.

6.2.3 A program for trapezium rule approximation

In the previous subsection we saw that it is possible to preselect a value for h which will ensure that the truncation error associated with the composite trapezium rule approximation is less than some prescribed tolerance. Another way of approaching the problem is successively to halve the *step length*, h, until two approximations agree to the required accuracy.

Let T_j denote the composite trapezium rule approximation to the definite integral (6.1) obtained using a value for h of $(b-a)/2^j$. Then we have, for example,

$$T_0 = \frac{b-a}{2}\{f(a)+f(b)\} \tag{6.20}$$

$$T_1 = \frac{b-a}{4}\left\{f(a)+2f\left(a+\frac{b-a}{2}\right)+f(b)\right\} \tag{6.21}$$

and so on. From (6.20) and (6.21) we observe that

$$T_1 = \frac{1}{2}T_0 + \frac{b-a}{2}f\left(a+\frac{b-a}{2}\right).$$

This means that, having found T_0 (the two-point approximation), T_1 may be found in terms of T_0 and the newly introduced function value $f(a+(b-a)/2)$. This result can be generalised. Formally we have

$$T_j = \frac{b-a}{2^j}\sum_{i=1}^{2^j+1}{}''f\left(a+(i-1)\frac{b-a}{2^j}\right) \qquad j=0,1,\ldots$$

so that

$$T_{j+1} = \frac{b-a}{2^{j+1}} \sum_{i=1}^{2^{j+1}+1} {}'' f\left(a + (i-1)\frac{b-a}{2^{j+1}}\right)$$

$$= \frac{1}{2}\frac{b-a}{2^{j}} \sum_{i=1}^{2^{j}+1} {}'' f\left(a + (i-1)\frac{b-a}{2^{j}}\right)$$

$$+ \frac{b-a}{2^{j+1}} \sum_{i=1}^{2^{j}} f\left(a + (2i-1)\frac{b-a}{2^{j+1}}\right)$$

$$= \frac{1}{2}T_j + \frac{b-a}{2^{j+1}} \sum_{i=1}^{2^{j}} f\left(a + (2i-1)\frac{b-a}{2^{j+1}}\right) \qquad j = 0, 1, \ldots.$$

Hence when halving the step length we need to

(i) sum the function values at the newly introduced points,
(ii) multiply this sum by the new value of h, and
(iii) add the result to half the previous estimate of the integral.

In this way we avoid any unnecessary recomputation of function values.

The programs of Case Study 6.1 implement the scheme outlined above. As with any iterative process we need to set an upper limit on the number of halvings of the step length. Whilst the code of these programs has been carefully written the programs will not always give the correct answer. For example consider the integral

$$\int_{-\pi}^{\pi} |\sin(x)| \, dx \ .$$

Then T_0 and T_1 are both zero so the programs will return a value of zero even though the true integral is strictly positive. It is far from easy to produce code which will take account of situations such as this. If it is known that the code will fail to work for certain problems the program documentation should indicate such shortcomings.

CASE STUDY 6.1

<u>Algol 68 Program</u>

```
BEGIN
  COMMENT
  a program to compute a composite trapezium rule approximation to a
  definite integral
  COMMENT
  PROC f = ( REAL x ) REAL :
    COMMENT
    the integrand
    COMMENT
    exp ( x ) ;
  REAL a , b , approximation , tol ;
  INT max ;
  BOOL converged , monitorprogress ;
  read ( ( a , b , max , tol , monitorprogress ) ) ;
  print ( ( newline , newline ,
            "lower limit of integration      " , a , newline ,
            "upper limit of integration      " , b , newline ,
            "convergence criterion           " , tol , newline ,
            "maximum number of step halvings " , max , newline ,
            "monitor progress                " , monitorprogress ) ) ;
  PROC trapeziumrule = ( PROC ( REAL ) REAL integrand ,
                         REAL lower , upper , tolerance ,
                         INT maximumdepth , BOOL monitorprogress ,
                         REF BOOL converged , REF REAL newintegral ) VOID :
  BEGIN
    COMMENT
    a procedure for approximating a definite integral using the trapezium rule
    COMMENT
    REAL upperminuslower = upper - lower ;
    REAL steplength := upperminuslower * 0.5 , twicestep , sum , gridpoint ;
    REAL oldintegral := ( integrand ( lower ) + integrand ( upper ) )
                       * steplength ;
    newintegral := steplength * integrand ( lower + steplength )
                   + 0.5 * oldintegral ;
    converged := ABS ( newintegral - oldintegral ) < tolerance ;
    IF monitorprogress
    THEN
      print ( ( newline , newline ,
                "interval length       approximate integral" , newline ,
                newline , upperminuslower , "            " , oldintegral ) )
    FI ;
    INT numberofnewterms := 1 ;
    TO maximumdepth - 1 WHILE NOT converged
    DO
      IF monitorprogress
      THEN
        print ( ( newline , steplength , "            " , newintegral ) )
      FI ;
      sum := 0.0 ;
      twicestep := steplength ;
      steplength *:= 0.5 ;
      gridpoint := lower - steplength ;
      numberofnewterms *:= 2 ;

      COMMENT
      sum the new function evaluations
      COMMENT
      TO numberofnewterms
      DO
        gridpoint +:= twicestep ;
        sum +:= integrand ( gridpoint )
      OD ;
```

```
            oldintegral := newintegral ;
            newintegral := sum * steplength + oldintegral * 0.5 ;
            converged := ABS ( newintegral - oldintegral ) < tolerance
      OD
END ;
trapeziumrule ( f , a , b , tol , max , monitorprogress ,
                                            converged , approximation ) ;
print ( ( newline , newline , newline , newline ,
          "the computed approximation to the given integral is" ,
          newline , newline , approximation , newline , newline ,
          "and convergence to the specified accuracy was" ,
          IF converged THEN " " ELSE " not " FI , "achieved" ,
          newline ) )
END
```

Pascal Program

```
PROGRAM compositetrapeziumrule(input, output);

   {  a program to compute a composite trapezium rule approximation
      to a definite integral  }

   VAR
      a, b, approximation, tol:real;
      max:1..maxint;
      converged, monitorprogress:boolean;
      character:char;

   FUNCTION f(x:real):real;
      {  the integrand  }
   BEGIN
      f:= exp(x)
   END { f } ;

   PROCEDURE trapeziumrule(FUNCTION integrand(x:real):real;
                           lower, upper, tolerance:real;
                           maximumdepth:integer; monitorprogress:boolean;
                           VAR converged:boolean; VAR newintegral:real);
      VAR
         depth, termnumber, numberofnewterms:1..maxint;
         upperminuslower, steplength, oldintegral:real;
         twicestep, sum, gridpoint:real;
   BEGIN
      upperminuslower:= upper - lower;
      steplength:= upperminuslower*0.5;
      oldintegral:= (integrand(lower) + integrand(upper))*steplength;
      newintegral:= steplength*integrand(lower + steplength)
                    + 0.5*oldintegral;
      converged:= abs(newintegral - oldintegral)<tolerance;
      IF monitorprogress
      THEN
         BEGIN
            writeln;
            writeln('interval length       approximate integral');
            writeln;
            writeln(upperminuslower, ' ':11, oldintegral)
         END ;
      numberofnewterms:= 1;
      depth:= 1;
      WHILE NOT converged AND (depth<maximumdepth) DO
      BEGIN
         depth:= depth + 1;
         IF monitorprogress
         THEN
            writeln(steplength, ' ':11, newintegral);
         twicestep:= steplength;
         steplength:= steplength*0.5;
```

```
            numberofnewterms:= numberofnewterms*2;
            gridpoint:= lower - steplength;
            sum:= 0.0;
            {  sum the new function evaluations  }
            FOR termnumber:= 1 TO numberofnewterms DO
              BEGIN
                gridpoint:= gridpoint + twicestep;
                sum:= sum + integrand(gridpoint)
              END ;
            oldintegral:= newintegral;
            newintegral:= sum*steplength + oldintegral*0.5;
            converged:= abs(newintegral - oldintegral)<tolerance
          END
        {  converged OR (depth>=maximumdepth)  }
    END { trapeziumrule } ;

BEGIN
    read(a, b, max, tol);
    REPEAT
        read(character)
    UNTIL character<>' ';
    monitorprogress:= (character='Y') OR (character='y');
    writeln;
    writeln;
    writeln('lower limit of integration      ', a);
    writeln('upper limit of integration      ', b);
    writeln('convergence criterion           ', tol);
    writeln('maximum number of step halvings ', max);
    writeln('monitor progress                ', monitorprogress);
    trapeziumrule(f, a, b, tol, max, monitorprogress,
                                converged, approximation);
    writeln;
    writeln;
    writeln;
    writeln('the computed approximation to the given integral is');
    writeln;
    writeln(approximation);
    writeln;
    write('and convergence to the specified accuracy was');
    IF converged THEN write(' ') ELSE write(' not ');
    writeln('achieved')
END { compositetrapeziumrule } .
```

Sample Data

    1.0   2.0   20 0.00005 Y

Sample Output

lower limit of integration	1.00000e 0
upper limit of integration	2.00000e 0
convergence criterion	5.00000e -5
maximum number of step halvings	20
monitor progress	Y

interval length	approximate integral
1.00000e 0	5.05367e 0
5.00000e -1	4.76768e 0
2.50000e -1	4.69507e 0
1.25000e -1	4.67685e 0
6.25000e -2	4.67229e 0
3.12500e -2	4.67115e 0
1.56250e -2	4.67086e 0
7.81250e -3	4.67079e 0

the computed approximation to the given integral is

4.67076e 0

and convergence to the specified accuracy was achieved

6.3 THE ROMBERG QUADRATURE SCHEME
6.3.1 The Romberg T-table

In subsection 6.2.3 we investigated and programmed a scheme for computing an integral correct to some specified tolerance. In computing T_j using the composite trapezium rule we need to use $2^j + 1$ function evaluations and the method can, therefore, be quite expensive. We now examine a method, *Romberg quadrature*, which improves, or extrapolates on, a sequence of composite trapezium rule approximations. The aim is to produce an algorithm with a reduced number of function calls.

Romberg quadrature depends on the result that the truncation error in the composite trapezium rule estimate (6.14) of (6.1) can be expressed in the form

$$E = A_2 h^2 + A_4 h^4 + A_6 h^6 + \ldots \tag{6.22}$$

for a sufficiently smooth integrand. The quantities $A_2, A_4, A_6, \ldots$ depend on a, b and $f(x)$ but not on h. The derivation of (6.22) is not easy and the reader is referred to Ralston and Rabinowitz (1978) for details.

Suppose we have available a sequence of composite trapezium rule approximations, $T_{0,j}$, obtained using a value for h of $h_j = (b-a)/2^j$. So far all we have done is to add an additional subscript, 0, to the T_js defined in subsection 6.2.3. Then, from (6.22) we have

$$I - T_{0,j} = A_2 h_j^2 + A_4 h_j^4 + A_6 h_j^6 + \ldots \qquad j = 0, 1, \ldots . \tag{6.23}$$

Since this result is true for all j, it follows that

$$I - T_{0,j+1} = A_2 h_{j+1}^2 + A_4 h_{j+1}^4 + A_6 h_{j+1}^6 + \ldots .$$

But $h_{j+1} = h_j/2$ so that

$$I - T_{0,j+1} = A_2 (h_j/2)^2 + A_4 (h_j/2)^4 + A_6 (h_j/2)^6 + \ldots . \tag{6.24}$$

If we now multiply equation (6.24) by 4, subtract (6.23) and divide by 3 we have

$$\frac{4(I - T_{0,j+1}) - (I - T_{0,j})}{3} = \left(4A_2 \frac{h_j^2}{4} + 4A_4 \frac{h_j^4}{16} + 4A_6 \frac{h_j^6}{64} + \ldots \right.$$
$$\left. - A_2 h_j^2 - A_4 h_j^4 - A_6 h_j^6 - \ldots \right)/3$$
$$= B_4 h_j^4 + B_6 h_j^6 + \ldots .$$

The precise form of $B_4, B_6, \ldots$ need not concern us but these quantities are clearly independent of h_j. The left-hand side simplifies to $I - (4T_{0,j+1} - T_{0,j})/3$ and if we define

$$T_{1,j} = (4T_{0,j+1} - T_{0,j})/3 \qquad j = 0, 1, \ldots \tag{6.25}$$

then

$$I - T_{1,j} = B_4 h_j^4 + B_6 h_j^6 + \ldots . \tag{6.26}$$

Equation (6.26) shows that $T_{1,j}$, defined by (6.25) as a linear combination of the composite trapezium rule estimates $T_{0,j}$ and $T_{0,j+1}$, has a leading term in the error of order h_j^4. In fact it is easy to show that $T_{1,j}$ is the composite Simpson's rule approximation with $h = h_{j+1}$. We now repeat the ideas employed in the previous paragraph and eliminate the term involving h_j^4. We have

$$I - T_{1,j+1} = B_4 h_{j+1}^4 + B_6 h_{j+1}^6 + \ldots$$
$$= B_4 (h_j/2)^4 + B_6 (h_j/2)^6 + \ldots \tag{6.27}$$

so that multiplying (6.27) by 16, subtracting (6.26) and dividing by 15 we have

$$I - T_{2,j} = C_6 h_j^6 + \ldots \tag{6.28}$$

where

$$T_{2,j} = (16 T_{1,j+1} - T_{1,j})/15 .$$

The C_ks are independent of h_j and (6.28) shows that $T_{2,j}$ has a leading term in the error of order h_j^6. It can be shown that $T_{2,j}$ is based on the closed Newton–Cotes rule R_5. We can clearly proceed in a similar manner to eliminate the term involving h_j^6, and so on. However, it should be noted that these later estimates do not have any direct interpretation in terms of composite Newton-Cotes rules.

The formal statement of the Romberg scheme is as follows. Given the initial data consisting of the composite trapezium rule estimates $T_{0,j}: j = 0, 1, \ldots, m$, form the quantities

$$T_{i,j} = \frac{1}{4^i - 1}(4^i T_{i-1,j+1} - T_{i-1,j}) .$$

We can imagine these quantities set out in a table, the *Romberg T-table*, as follows, the arrows indicating the interdependence of entries in the table.

$T_{0,0}$
$\quad \searrow$
$T_{0,1} \rightarrow T_{1,0}$
$\quad \searrow \qquad \searrow$
$T_{0,2} \rightarrow T_{1,1} \rightarrow T_{2,0}$

$\ldots \qquad \ldots$
$\quad \searrow \qquad \searrow \qquad \searrow \qquad \qquad \searrow$
$T_{0,m} \rightarrow T_{1,m-1} \rightarrow T_{2,m-2} \cdots \rightarrow T_{m,0}$

EXAMPLE 6.9
If we take the sample output of Case Study 6.1 we have $T_{0,0} = 5.05367$, $T_{0,1} = 4.76768$ and $T_{0,2} = 4.69507$ so that

$$T_{1,0} = (4 \times 4.76768 - 5.05367)/3 = 4.67235$$
$$T_{1,1} = (4 \times 4.69507 - 4.76768)/3 = 4.67087$$
$$T_{2,0} = (16 \times 4.67087 - 4.67235)/15 = 4.67077 .$$

Collecting the results together in the form of a Romberg T-table we have

5.05367

4.76768   4.67235

4.69507   4.67087   4.67077 .

The final figure is very close to the true value at a cost of just five function evaluations.

The entries in the Romberg T-table have some important properties. If $f(x)$ has derivatives of all orders on the interval of integration $[a,b]$, the sequences of entries in successive columns of the table converge to (6.1) with those in the $p + 1$th column usually converging more rapidly than those in the pth column. Further, the sequence of diagonal elements $T_{0,0}$, $T_{1,0}$, $T_{2,0}$, ... also converges to (6.1) and it is usually preferable in practice to monitor the convergence of this sequence. Hence we generate the entries in the table a row at a time continuing until a value of m is reached for which

$$|T_{m,0} - T_{m-1,0}| < \varepsilon$$

or, using a relative accuracy criterion,

$$|T_{m,0} - T_{m-1,0}| < \delta |T_{m,0}| \; .$$

6.3.2 A program for Romberg integration

The programs of Case Study 6.2 use the code of Case Study 6.1 to compute the Romberg T-table. The values produced by the earlier code are, of course, just the entries in the first column of the table. Each time we find a new trapezium rule approximation the entries in the rest of that row must be found. Note that we have managed to save a considerable amount of space by employing a one-dimensional array instead of a two-dimensional one. The code works as follows.

Ignoring unused locations, the array *integrals* is initialised to

$T_{0,0}$

 .

$T_{0,1}$ is then computed and entered into the array

$T_{0,0}$	$T_{0,1}$

 .

$T_{1,0}$ is found and this value overwrites $T_{0,0}$ to give

$T_{1,0}$	$T_{0,1}$

 .

This completes the first iteration. At the start of the second iteration we find $T_{0,2}$ and enter it in the array

$T_{1,0}$	$T_{0,1}$	$T_{0,2}$

 .

From $T_{0,2}$ and $T_{0,1}$ the code finds $T_{1,1}$ and overwrites $T_{0,1}$

$T_{1,0}$	$T_{1,1}$	$T_{0,2}$

 .

and then $T_{2,0}$ is found in terms of $T_{1,0}$ and $T_{1,1}$ and placed in $T_{1,0}$s position, giving

$T_{2,0}$	$T_{1,1}$	$T_{0,2}$

 .

Continuing in this way at the end of each iteration the array contains the next row of the Romberg T-table. All that needs to be done now is to compare the zeroth entry with the value that was in that position at the end of the previous iteration (stored in *oldintegral*).

CASE STUDY 6.2

Algol 68 Program

```
BEGIN
  COMMENT
  a program to compute an approximation to a definite integral using the
  romberg t - table
  COMMENT
  PROC f = ( REAL x ) REAL :
    COMMENT
    the integrand
    COMMENT
    exp ( - x * x ) / x ;
  REAL a , b , approximation , tol ;
  INT itermax , iter ;
  BOOL converged ;
  read ( ( a , b , itermax , tol ) ) ;
  print ( ( newline , newline ,
           "lower limit of integration   " , a , newline ,
           "upper limit of integration   " , b , newline ,
           "convergence criterion        " , tol , newline ,
           "maximum number of iterations " , itermax , newline ) ) ;
  PROC romberg = ( PROC ( REAL ) REAL integrand , REAL lower , upper , tol ,
                   INT itermax , REF BOOL converged , REF INT iter ,
                   REF REAL result ) VOID :
  BEGIN
    COMMENT
    this procedure calculates an approximation to a definite integral
    the romberg t - table is used and convergence to the specified
    tolerance is sought along the leading diagonal
    COMMENT
    REAL h := upper - lower , twoh , step , sum , poweroftwo ,
         poweroftwominus1 , oldintegral ;
    INT newterms ;
    [ 0 : itermax ] REAL integrals ;
    integrals [ 0 ] :=
             h * ( integrand ( lower ) + integrand ( upper ) ) * 0.5 ;
    converged := FALSE ;
    iter := 0 ;
    WHILE NOT converged AND iter < itermax
    DO
      iter +:= 1 ;
      oldintegral := integrals [ 0 ] ;
      sum := 0.0 ;
      twoh := h ;
      h *:= 0.5 ;
      newterms := IF iter = 1 THEN 1 ELSE newterms * 2 FI ;
      step := lower - h ;
      FOR j TO newterms
      DO
        step +:= twoh ;
        sum +:= integrand ( step )
      OD ;
      COMMENT
      compute the next trapezium rule approximation
      COMMENT
      integrals [ iter ] := 0.5 * integrals [ iter - 1 ] + h * sum ;
      COMMENT
      compute the next row of the t - table
      COMMENT
```

```
            poweroftwo := 1 ;
            FOR k TO iter
            DO
               poweroftwo *:= 4 ;
               poweroftwominus1 := poweroftwo - 1 ;
               FOR j FROM iter BY - 1 TO k
               DO
                  integrals [ j - 1 ] := ( poweroftwo * integrals [ j ]
                                           - integrals [ j - 1 ] )
                                         / poweroftwominus1
               OD
            OD ;
            COMMENT
            test for convergence
            COMMENT
            converged := ABS ( oldintegral - integrals [ 0 ] ) < tol
         OD ;
         result := integrals [ 0 ]
   END ;
   romberg ( f , a , b , tol , itermax , converged , iter , approximation ) ;
   print ( ( newline , newline ,
             "the computed approximation to the given integral is" ,
             newline , newline , approximation , newline , newline ,
             "and convergence to the specified accuracy was" ) ) ;
   IF converged
   THEN
      print ( ( " achieved after" , newline , newline , iter + 1 ,
                newline , newline ,
                "rows of the romberg t-table had been computed" , newline ) )
   ELSE
      print ( ( " not achieved" , newline ) )
   FI
END
```

Pascal Program

```
PROGRAM rombergt(input, output);

   {  a program to compute an approximation to a definite integral
      using the romberg t-table  }

   CONST
      maxitermax=50;

   TYPE
      itertype=0..maxitermax;   itermaxtype=1..maxitermax;

   VAR
      iter:itertype;   itermax:itermaxtype;
      a, b, approximation, tol:real;
      converged:boolean;

   FUNCTION f(x:real):real;
      {  the integrand  }
   BEGIN
      f:= exp(-x*x)/x
   END { f } ;

   PROCEDURE romberg(FUNCTION integrand(x:real):real;
                     lower, upper, tol:real;   itermax:itermaxtype;
                     VAR converged:boolean;   VAR iter:itertype;
                     VAR result:real);
      {  this procedure calculates an approximation to a definite integral
         the romberg t-table is used and convergence to the specified
         tolerance is sought along the leading diagonal  }
      VAR
         h, twoh, step, sum, power, powerminus1, oldintegral:real;
         j, newterms:1..maxint;
         k:itermaxtype;
         integrals:ARRAY[ itertype ] OF real;
```

```
        BEGIN
            h:= upper - lower;
            integrals[0]:= h*(integrand(lower) + integrand(upper))*0.5;
            iter:= 0;
            REPEAT
                iter:= iter + 1;
                oldintegral:= integrals[0];
                twoh:= h;
                h:= h*0.5;
                IF iter=1 THEN newterms:= 1 ELSE newterms:= newterms*2;
                step:= lower - h;
                sum:= 0.0;
                FOR j:= 1 TO newterms DO
                BEGIN
                    step:= step + twoh;
                    sum:= sum + integrand(step)
                END ;
                { compute the next trapezium rule approximation }
                integrals[iter]:= 0.5*integrals[iter - 1] + h*sum;
                { compute the next row of the t-table }
                power:= 1;
                FOR k:= 1 TO iter DO
                BEGIN
                    power:= power*4;
                    powerminus1:= power - 1;
                    FOR j:= iter DOWNTO k DO
                        integrals[j - 1]:= (power*integrals[j] - integrals[j - 1])
                                                                /powerminus1
                END ;
                { test for convergence }
                converged:= abs(oldintegral - integrals[0])<tol
            UNTIL converged OR (iter=itermax);
            result:= integrals[0]
        END { romberg } ;

BEGIN
    read(a, b, itermax, tol);
    writeln;
    writeln;
    writeln('lower limit of integration    ', a);
    writeln('upper limit of integration    ', b);
    writeln('convergence criterion         ', tol);
    writeln('maximum number of iterations ', itermax);
    romberg(f, a, b, tol, itermax, converged, iter, approximation);
    writeln;
    writeln;
    writeln('the computed approximation to the given integral is');
    writeln;
    writeln(approximation);
    writeln;
    write('and convergence to the specified accuracy was');
    IF converged
    THEN
        BEGIN
            writeln(' achieved after');
            writeln;
            writeln(iter + 1);
            writeln;
            writeln('rows of the romberg t-table had been computed')
        END
    ELSE
        writeln(' not achieved')
END { rombergt } .
```

Sample Data

    1.0   2.0   20 0.00005

Sample Output

```
lower limit of integration      1.00000e  0
upper limit of integration      2.00000e  0
convergence criterion           5.00000e -5
maximum number of iterations        20
```

the computed approximation to the given integral is

   1.07781e -1

and convergence to the specified accuracy was achieved after

   9

rows of the romberg t-table had been computed

6.4 GAUSS QUADRATURE

6.4.1 Gauss Legendre quadrature

We now give an introduction to the derivation and use of a sequence of quadrature rules which are of great practical importance. We suppose the interval of integration is $[-1, 1]$ so that we are attempting to estimate

$$I = \int_{-1}^{1} f(x) dx \ . \tag{6.29}$$

(If the interval is not $[-1, 1]$ we recall from subsection 4.4.1 that it may be transformed to $[-1, 1]$ using a simple transformation.) Consider the two-point rule

$$R = w_1 f(x_1) + w_2 f(x_2) \ . \tag{6.30}$$

We now ask whether it is possible to choose the weights w_1 and w_2 and the points x_1 and x_2 so that the rule (6.30) is exact when $f(x)$ is $1, x, x^2$ and x^3. If this is so then we must have

$$\text{for } f(x) = 1 \qquad w_1 + w_2 = \int_{-1}^{1} 1 \, dx = 2$$

$$\text{for } f(x) = x \qquad w_1 x_1 + w_2 x_2 = \int_{-1}^{1} x \, dx = 0$$

for $f(x) = x^2$ $\quad w_1x_1^2 + w_2x_2^2 = \int_{-1}^{1} x^2 dx = 2/3$

for $f(x) = x^3$ $\quad w_1x_1^3 + w_2x_2^3 = \int_{-1}^{1} x^3 dx = 0$

which gives a set of four simultaneous equations in the four unknowns w_1, w_2, x_1 and x_2. It is easy to verify that

$$w_1 = w_2 = 1 \; ; \quad x_1 = -x_2 = -1/\sqrt{3} \tag{6.31}$$

is a solution to this system. The values (6.31) define the rule

$$R = f(-1/\sqrt{3}) + f(1/\sqrt{3}) \tag{6.32}$$

which we assert has precision three.

Developing the argument further we now consider the three-point rule

$$R = w_1 f(x_1) + w_2 f(x_2) + w_3 f(x_3)$$

and ask whether it is possible to choose w_1, w_2, w_3, x_1, x_2 and x_3 so that the rule is exact for the integrands $f(x) = 1, x, x^2, x^3, x^4, x^5$. The reader is encouraged to write down the six equations which must be satisfied by the w_is and x_is and verify that

$$w_1 = w_3 = 5/9 \; ; \quad w_2 = 8/9$$

$$x_1 = -x_3 = -\sqrt{3/5} \; ; \quad x_2 = 0$$

is a solution. We can therefore assert that the rule

$$R = \frac{5}{9}f\left(-\sqrt{\frac{3}{5}}\right) + \frac{8}{9}f(0) + \frac{5}{9}f\left(\sqrt{\frac{3}{5}}\right) \tag{6.33}$$

(introduced in Example 6.1) has precision five. The rules (6.32) and (6.33) are known as *Gauss Legendre rules* and we will refer to them as G_2 and G_3 respectively, the suffix indicating the number of integration points involved.

We could attempt to proceed in the same way to derive the rules G_4, G_5, ... (of precision seven, nine, ...) but this would be tedious and a much more satisfactory approach is possible. We shall not develop this fully but we observe that the integration points in G_2 and G_3 are respectively the zeros of the Legendre polynomials $P_2(x)$ and $P_3(x)$ which were introduced in subsection 4.4.1. This observation could be no more than coincidence but in fact it has a deep significance which we now try to indicate.

Suppose that the values $f_i: i = 1, 2, ..., n$ of a function and $f'_i: i = 1$,

2, ..., n of its first derivative are available at the n distinct points x_i: $i = 1$, 2, ..., n. Then, from subsection 5.4.1 equation (5.12), we have that the Hermite interpolating polynomial for this data is

$$p_{2n-1}(x) = \sum_{i=1}^{n} \alpha_i(x) f_i + \sum_{i=1}^{n} \beta_i(x) f'_i . \qquad (6.34)$$

Replacing the integrand in (6.29) with $p_{2n-1}(x)$ we have the rule

$$R = \sum_{i=1}^{n} \left(\int_{-1}^{1} \alpha_i(x) \mathrm{d}x \right) f_i + \sum_{i=1}^{n} \left(\int_{-1}^{1} \beta_i(x) \mathrm{d}x \right) f'_i \qquad (6.35)$$

which has precision $2n - 1$. However, this is not really helpful since it involves the undesirable notion of having available the n values $\{f'_i: i = 1, 2, ..., n\}$. We now ask whether it is possible to choose the points $\{x_i: i = 1, 2, ..., n\}$ in such a way that the integrals

$$\bar{w}_i = \int_{-1}^{1} \beta_i(x) \mathrm{d}x \qquad i = 1, 2, ..., n$$

are each equal to zero, since the rule (6.35) would then involve function values only. In fact it turns out (see Ralston and Rabinowitz (1978) for details) that the required result can be achieved if the points $\{x_i: i = 1, 2, ..., n\}$ are chosen to be the n zeros of the Legendre polynomial $P_n(x)$ and these are all real, distinct and lie in $]-1, 1[$. The rule (6.35) then takes the standard form

$$R = \sum_{i=1}^{n} w_i f_i = G_n$$

with

$$w_i = \int_{-1}^{1} \alpha_i(x) \mathrm{d}x \qquad i = 1, 2, ..., n .$$

The important property to note about the n-point Gauss Legendre rule G_n is that it has precision $2n - 1$ and we are unlikely to be able to do better with the n points and n weights at our disposal. Although maximum precision (related, as it is, to the exact integration of polynomials) is not always an accurate measure of the efficiency of a quadrature rule, we can expect an n-point Gauss Legendre rule to be effective for some value of n when the integrand is 'polynomial-like' on the interval $[-1, 1]$.

EXAMPLE 6.10
We have already derived the two- and three-point Gauss Legendre rules (see equations (6.32) and (6.33)). G_4 is defined by

$$x_1 = -x_4 = -0.861136312 \; ; \quad x_2 = -x_3 = -0.339981044$$
$$w_1 = w_4 = 0.347854845 \; ; \quad w_2 = w_3 = 0.652145155 \; .$$

To evaluate the integral of Example 6.3 we need to make the change of variable $z = 2x - 3$. Thus

$$\int_1^2 e^x dx = \int_{-1}^1 \tfrac{1}{2} e^{(z+3)/2} dz \; .$$

Applying G_2, G_3 and G_4 to this integral we obtain the results shown in Table 6.4.

Table 6.4

Rule	Approximation
G_2	4.669 726 508
G_3	4.670 772 030
G_4	4.670 774 268

Given that, to nine decimal places the true value is 4.670774270 the figure for G_4 is very accurate indeed.

The Gauss points and weights have been extensively tabulated (see Stroud and Secrest, 1966) and will be available for use in a library of mathematical software. Note that it is necessary to store only half of the points and weights. The points are symmetrically distributed about the origin (so that $x_1 = -x_n, x_2 = -x_{n-1}, \ldots$) and if n is odd the middle point is 0. Moreover, $w_1 = w_n$, $w_2 = w_{n-1}, \ldots$. In order to estimate (6.29) to a given accuracy, it is usual to use a sequence of rules G_4, G_8, G_{12}, ... continuing until two (or preferably three) successive estimates agree to within the desired accuracy or the available rules are exhausted in which case the last three estimates would be output. However, except for the point $x = 0$ when n is odd, the Gauss rules have no integration points in common. This means that when we generate a sequence of Gauss rule approximations it is not possible to make use of previously calculated values of the integrand. The situation should be compared with that in the Romberg scheme where, in calculating the elements of the first column of the table, we are able to make use of previously made function calls (see subsection 6.2.3). We do not suggest that this necessarily makes a Gauss scheme inferior to Romberg but the reader should be aware of the possible disadvantage. We shall raise this matter again in section 6.6.

It is possible to obtain an expression for the error $I - G_n$ although we shall not derive it. Since G_n is of precision $2n - 1$, the reader will not be surprised to learn that the error involves $f^{(2n)}(x)$, the $2n$th derivative of the integrand (see Ralston and Rabinowitz (1978) for further details). Finding this derivative analytically and then bounding its magnitude is likely to be extremely difficult and the result is not usually of any practical value.

One further observation is that it is possible to employ a composite scheme along the lines of subsection 6.2.2 in which the interval of integration is divided into a number of sub-intervals and a low precision Gauss Legendre rule is used (after an appropriate transformation) for each sub-interval.

6.4.2 A program for Gauss Legendre integration

Case Study 6.3 contains programs which compute an n-point Gauss Legendre approximation to a definite integral. The value of n is read in together with the points and weights which constitute G_n, the symmetry of the rule being used to keep the storage required as low as possible. Having stored the points and weights it is a simple matter to calculate the Gauss Legendre rule approximation. The code of Case Study 6.3 accepts any finite range of integration and uses a suitable transformation to map the range onto $[-1, 1]$.

Many of the NAG Algol 68 routines are more or less direct translations of the corresponding Fortran code. However the chapter on quadrature, d01, makes extensive use of the Algol 68 *MODE* declaration facility. To give a flavour of this, in the Algol 68 program of Case Study 6.3, the information about the Gauss Legendre rule is stored in the structured mode *GAUSS-RULE* which has three fields, *terms*, *points* and *weights*. Note that the arrays are stored on the heap. The reader is now encouraged to consult the documentation for the d01 chapter of the NAG Algol 68 library where a complete discussion is given of the more comprehensive facilities available. A further discussion will be given in section 6.10.

CASE STUDY 6.3

Algol 68 Program

```
BEGIN
  COMMENT
  a program to compute an approximation to a definite integral using
  gauss legendre quadrature
  COMMENT
  PROC f = ( REAL x ) REAL :
    COMMENT
    this procedure defines the integrand
    COMMENT
    exp ( x ) ;
```

278 NUMERICAL INTEGRATION [Ch. 6]

```
      REAL a , b ;
      PROC map = ( REAL x ) REAL :
        COMMENT
        this procedure defines the transformation of the range of integration
        onto [ - 1 , 1 ]
        COMMENT
        ( x * ( b - a ) + a + b ) * 0.5 ;
      COMMENT
      define a mode for storing the gauss points and weights
      COMMENT
      MODE GAUSSRULE = STRUCT ( INT terms , REF [ ] REAL points , weights ) ;
      GAUSSRULE gaussrule ;
      read ( terms OF gaussrule ) ;
      INT termsdiv2 = ( terms OF gaussrule ) % 2 ;
      BOOL even = termsdiv2 * 2 = terms OF gaussrule ;
      INT npoints = IF even THEN termsdiv2 ELSE termsdiv2 + 1 FI ;
      points OF gaussrule := HEAP [ 1 : npoints ] REAL ;
      weights OF gaussrule := HEAP [ 1 : npoints ] REAL ;
      read ( ( points OF gaussrule , weights OF gaussrule , a , b ) ) ;
      print ( ( newline , newline ,
                "the points and weights for the" , terms OF gaussrule ,
                "   point gauss rule have been read in" , newline ,
                "and they are" , newline , newline , "points" , newline ,
                points OF gaussrule , newline , newline , "weights" , newline ,
                weights OF gaussrule , newline , newline ) ) ;
      print ( ( newline , newline ,
                "lower limit of integration      " , a , newline ,
                "upper limit of integration      " , b , newline ) ) ;
      REAL mapderivative = ( b - a ) * 0.5 ;
      REAL gridpoint1 , gridpoint2 , point , integral ;
      COMMENT
      compute the gauss legendre rule approximation
      COMMENT
      integral := IF even THEN 0.0
                  ELSE
                    ( weights OF gaussrule ) [ 1 ]
                    * f ( map ( ( points OF gaussrule ) [ 1 ] ) )
                  FI ;
      FOR i FROM IF even THEN 1 ELSE 2 FI TO npoints
      DO
        point := ( points OF gaussrule ) [ i ] ;
        gridpoint1 := map ( - point ) ;
        gridpoint2 := map ( point ) ;
        integral +:= ( weights OF gaussrule ) [ i ]
                     * ( f ( gridpoint1 ) + f ( gridpoint2 ) )
      OD ;
      integral *:= mapderivative ;
      print ( ( newline , newline ,
                "the approximation to the given integral is" , newline ,
                newline , integral , newline ) )
END
```

Pascal Program

```
PROGRAM gausslegendre(input, output);

   CONST
      maxnpoints=50;

   VAR
      terms:0..maxint;
      points, weights:ARRAY[ 1..maxnpoints ] OF real;
      termsdiv2, npoints, i:0..maxint;
      startingpoint:1..2;
      termsisodd:boolean;
      a, b, mapderivative, integral, point, gridpt1, gridpt2:real;
```

```
FUNCTION f(x:real):real;
    { this function defines the integrand }
BEGIN
    f:= exp(x)
END { f } ;

FUNCTION map(x:real):real;
    { this procedure defines the transformation of the range of
      integration onto [-1,1] }
BEGIN
    map:= (x*(b - a) + a + b)*0.5
END { map } ;

BEGIN
    read(terms);
    termsdiv2:= terms DIV 2;
    termsisodd:= odd(terms);
    IF termsisodd
    THEN
        npoints:= termsdiv2 + 1
    ELSE
        npoints:= termsdiv2;
    FOR i:= 1 TO npoints DO
        read(points[i]);
    FOR i:= 1 TO npoints DO
        read(weights[i]);
    read(a, b);
    writeln;
    writeln;
    writeln('the points and weights for the', terms,
            '   point gauss rule have been read in');
    writeln('and they are');
    writeln;
    writeln('points');
    FOR i:= 1 TO npoints DO
        write(points[i]);
    writeln;
    writeln;
    writeln('weights');
    FOR i:= 1 TO npoints DO
        write(weights[i]);
    writeln;
    writeln;
    writeln;
    writeln;
    writeln('lower limit of integration         ', a);
    writeln('upper limit of integration         ', b);
    mapderivative:= (b - a)*0.5;
    { compute the gauss legendre rule approximation }
    IF termsisodd
    THEN
        BEGIN
            integral:= weights[1]*f(map(points[i]));
            startingpoint:= 2
        END
    ELSE
        BEGIN
            integral:= 0.0;
            startingpoint:= 1
        END ;
    FOR i:= startingpoint TO npoints DO
    BEGIN
        point:= points[i];
        gridpt1:= map(-point);
        gridpt2:= map( point);
        integral:= integral + weights[i]*(f(gridpt1) + f(gridpt2))
    END ;
```

```
        integral:= integral*mapderivative;
        writeln;
        writeln;
        writeln('the approximation to the given integral is');
        writeln;
        writeln(integral)
END { gausslegendre } .
```

Sample Data
```
    4
    0.339981    0.861136
    0.652145    0.347855
    1.0   2.0
```

Sample Output
```
the points and weights for the     4   point gauss rule have been read in
and they are

points
  3.39981e -1   8.61136e -1

weights
  6.52145e -1   3.47855e -1


lower limit of integration         1.00000e  0
upper limit of integration         2.00000e  0

the approximation to the given integral is

  4.67077e  0
```

6.5 OTHER GAUSS QUADRATURE RULES

In this section we make a brief reference to other rules which are based on the Gauss approach. The first of these is used to obtain approximations to the integral

$$I = \int_{-1}^{1} \frac{f(x)}{\sqrt{(1-x^2)}} dx .$$

If we replace $f(x)$ with the Hermite interpolating polynomial (6.34) we obtain the rule

$$R = \sum_{i=1}^{n} w_i f_i + \sum_{i=1}^{n} \bar{w}_i f'_i$$

with

$$w_i = \int_{-1}^{1} \frac{\alpha_i(x)}{\sqrt{(1-x^2)}} dx$$

$$\bar{w}_i = \int_{-1}^{1} \frac{\beta_i(x)}{\sqrt{(1-x^2)}} dx$$

$i = 1, 2, \ldots, n$

and it can be shown that $\bar{w}_i = 0: i = 1, 2, \ldots, n$ if the points $\{x_i: i = 1, 2, \ldots, n\}$ are chosen to be the zeros of the nth Chebyshev polynomial $T_n(x)$ introduced in subsection 4.5.3. Choosing the points in this way we obtain a *Gauss Chebyshev rule* (of precision $2n - 1$) which takes the standard form (that is, a linear combination of function values).

For the integral

$$I = \int_0^{\infty} e^{-x} f(x) dx$$

we proceed in the same way and choose as integration points the roots of the nth Laguerre polynomial $L_n(x)$ (see section 4.6) to obtain a *Gauss Laguerre rule* of precision $2n - 1$. The same approach applied to the infinite integral

$$I = \int_{-\infty}^{\infty} e^{-x^2} f(x) dx$$

leads to a rule of the standard form if the integration points are chosen to be the roots of the nth Hermite polynomial (again, see section 4.6). This rule, a *Gauss Hermite rule*, will have precision $2n - 1$.

Further details of the derivation of Gauss Chebyshev, Gauss Laguerre and Gauss Hermite rules may be found in Ralston and Rabinowitz (1978) where it is shown that the choice of integration points is based on the orthogonality properties of the Chebyshev, Laguerre and Hermite polynomials respectively. The points and weights for these rules have been extensively tabulated (see Stroud and Secrest, 1966) and again the reader is recommended to examine the facilities which are available in the NAG library.

Although we shall not develop any mathematical details, we must also make reference to the so-called Gauss Rational rules which are designed to deal with integrals of the type

$$\int_a^\infty \frac{f(x)}{(x+b)^n} dx \quad a,b > 0 .$$

The reader is referred to the discussion in Stroud and Secrest (1966).

6.6 PATTERSON QUADRATURE RULES

In section 6.4 we referred to the computational inefficiency associated with the use of a sequence of Gauss Legendre rules caused by the fact that, apart from $x = 0$ when n is odd, the rules have no integration points in common. At the expense of some loss of precision, the *Patterson rules* are designed to overcome this deficiency and we give an outline of how the sequence of rules is derived.

The starting point is the Gauss Legendre rule, G_3, defined by equation (6.33), which we write as

$$G_3 = \sum_{i=1}^{3} w_i f(x_i) .$$

The idea is now to derive a seven point rule in which three of the points are the x_1, x_2 and x_3 of G_3. This rule will have the form

$$R = \sum_{i=1}^{3} w'_i f(x_i) + \sum_{i=4}^{7} w'_i f(x_i) \qquad (6.36)$$

where the four additional points x_4, x_5, x_6 and x_7 and the seven weights $\{w'_i: i = 1, 2, \ldots, 7\}$ are to be chosen so that the rule has as high a precision as possible. (Note that we no longer impose any ordering on the points $\{x_i: i = 1, 2, \ldots, 7\}$.) Recalling the intuitive approach given at the beginning of subsection 6.4.1 for the derivation of the Gauss Legendre rules, we observe that there are eleven parameters at our disposal and we can hope that it is possible to choose them so that (6.36) will integrate $f(x) = 1, x, x^2, \ldots, x^{10}$ exactly, that is, so that the rule has precision ten. This can in fact be done and it turns out that the rule actually has precision eleven.

The next stage is to obtain a fifteen point rule of the form

$$R = \sum_{i=1}^{7} w''_i f(x_i) + \sum_{i=8}^{15} w''_i f(x_i)$$

in which the seven points $x_1, x_2, \ldots, x_7$ are those appearing in (6.36) and in which the eight new points $x_8, x_9, \ldots, x_{15}$ and the fifteen new weights $\{w''_i: i =$

1, 2, ..., 15} are chosen so that the precision is as great as possible. Once again we get a bonus in that the rule so derived has precision 23, rather than the expected 22. In this way a sequence of rules, known as Patterson rules, can be derived, each member making use of the function evaluations used to compute its predecessor. Rules corresponding to $n = 7, 15, 63, 127$ and 255 points are known. For a full discussion of these rules, the reader is referred to Davis and Rabinowitz (1975) and to the original paper by Patterson (1968).

These rules are of considerable practical importance and can be expected to appear in a library of mathematical software. Although for a given n the precision of a Patterson rule is less than that of the corresponding Gauss Legendre rule, the efficiency arising from the use of previously calculated function values may well outweigh this disadvantage for certain types of integrand. As with the Gauss rules discussed in the previous two sections it is not possible to derive an expression for the truncation error which is generally of practical use.

6.7 CLENSHAW–CURTIS QUADRATURE

It is not possible to give here a complete account of this method but the reader should be aware of its existence and be prepared to read further by consulting Davis and Rabinowitz (1975) and the original papers by Clenshaw and Curtis (1960), by Smith (1965) and O'Hara and Smith (1968). We assume that the integral is of the form

$$I = \int_{-1}^{1} f(x)\,dx \tag{6.37}$$

so that if the interval of integration is other than $[-1,1]$ it will need to be transformed. The basis of the method is to suppose that $f(x)$ is expanded as a linear combination of Chebyshev polynomials. Let

$$p_n(x) = \sum_{i=0}^{n}{''} a_i T_i(x) \tag{6.38}$$

be such an approximation, where the double prime indicates that the first and last terms are to be halved. (We have already seen, in subsection 4.7.3, that it is possible to obtain straightforward expressions for the expansion coefficients $\{a_i: i = 0, 1, ..., n\}$.) The quadrature rule is obtained by replacing the integrand in (6.37) with the approximation (6.38) to obtain the rule

$$R = \int_{-1}^{1} p_n(x)\mathrm{d}x = \sum_{i=0}^{n}{''} w_i f\left(\cos\frac{i\pi}{n}\right) \qquad (6.39)$$

which is an $n + 1$-point rule. We do not here give the precise form for the weights $\{w_i : i = 0, 1, \ldots, n\}$ but their computation is not difficult and can be achieved efficiently through recurrence relations.

An extremely important and attractive feature of these rules is that it is possible to obtain a useful bound for the truncation error in terms of simple functions of n and values of the integrand which have already been calculated. In this way a simple strategy for the estimation of (6.37) can be implemented. Setting $n = 4, 8, 16, 32 \ldots$ in (6.39) successive approximations are obtained until the error estimate indicates that the most recent estimate of the integral is satisfactory (or until a specified maximum value of n has been reached). Note that, as with Romberg integration, it is possible to make use of previously made function calls in calculating the next approximation to (6.37).

6.8 ADAPTIVE QUADRATURE

6.8.1 Adaptive Simpson's rule

In our discussion of quadrature so far, we have said very little about the integrand $f(x)$ apart from assuming it to have the necessary analytical properties for the methods we have described to be valid. Suppose that we are able to sketch the integrand for $a \leqslant x \leqslant b$ and the result is as shown in Fig. 6.3. Roughly speaking, we can describe the behaviour of the integrand as being 'smooth' on the sub-intervals $[a, a_1]$ and $[a_2, a_3]$, as being rapidly varying and including a peak on $[a_1, a_2]$ and again as being rapidly varying (possibly due to an infinity (singularity) of the integrand just outside the interval of integration) on $[a_3, b]$. In trying to estimate the integral of $f(x)$ over $[a, b]$ by, for example, the composite Simpson's rule (6.17), we would have to use a small step length, h, in order to take account of the regions where $f(x)$ is varying rapidly. However, for the regions where $f(x)$ is slowly varying, it is likely that a larger value of h could be tolerated and so, if the small value of h is used throughout, there will be unnecessary evaluations of the integrand. This suggests that we ought to treat the four sub-intervals $[a, a_1], [a_1, a_2], [a_2, a_3]$ and $[a_3, b]$ separately, obtaining an approximation of sufficient accuracy for the integral over each of them.

In practice, it is unlikely that we shall be able to sketch the integrand and we now ask and try to answer the question whether it is possible to set up a scheme which will *adapt* itself to the behaviour of the integrand on the interval of integration. This is a subject which has been extensively investigated in recent years and all we can do is to give some flavour of the idea by describing a scheme which is based on the use of Simpson's rule.

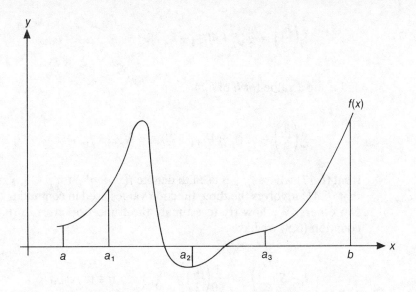

Fig. 6.3

We therefore attempt to calculate an approximation R_S to (6.1) which is such that

$$|I - R_S| < \varepsilon$$

and in such a way that the number of evaluations of the integrand is kept as low as possible.

We suppose initially that the interval of integration $[a, b]$ is subdivided into a number of sub-intervals, $n - 1$, say, which are not necessarily of equal length. Suppose $[x_j, x_{j+1}]$ is one such sub-interval and let

$$h_j = x_{j+1} - x_j \ .$$

Further, define

$$I_j = \int_{x_j}^{x_{j+1}} f(x) \, dx \ .$$

Then the Simpson's rule estimate of I_j obtained using a value for h of $h_j/2$ is, from (6.6),

$$S_j\left(\frac{h_j}{2}\right) = \frac{h_j}{6}\{f_j + 4f_{j+\frac{1}{2}} + f_{j+1}\} \tag{6.40}$$

and using a value for h of $h_j/4$

$$S_j\left(\frac{h_j}{4}\right) = \frac{h_j}{12}\{f_j + 4f_{j+\frac{1}{4}} + 2f_{j+\frac{1}{2}} + 4f_{j+\frac{3}{4}} + f_{j+1}\} \tag{6.41}$$

from (6.17) where f_{j+p} is used to denote $f(x_j + ph_j)$ for $p = \frac{1}{4}, \frac{1}{2}$ and $\frac{3}{4}$. Note that (6.41) involves the three function values used in computing (6.40) plus two others. We now try to estimate the truncation error in (6.41). From equation (6.8) we have

$$I_j - S_j\left(\frac{h_j}{2}\right) = -\frac{1}{90}\left(\frac{h_j}{2}\right)^5 f^{(iv)}(\eta) \quad \eta \in \,]x_j, x_{j+1}[\,. \tag{6.42}$$

Similarly,

$$I_j - S_j\left(\frac{h_j}{4}\right) = -\frac{1}{90}\left(\frac{h_j}{4}\right)^5 f^{(iv)}(\eta_1) -$$
$$- \frac{1}{90}\left(\frac{h_j}{4}\right)^5 f^{(iv)}(\eta_2) \quad \begin{matrix} \eta_1 \in \,]x_j, x_{j+\frac{1}{2}}[\\ \eta_2 \in \,]x_{j+\frac{1}{2}}, x_{j+1}[\end{matrix} . \tag{6.43}$$

If we now assume that $f^{(iv)}(x)$ is approximately constant on $[x_j, x_{j+1}]$ and let M be an estimate of its value we have, on subtracting (6.43) from (6.42)

$$S_j\left(\frac{h_j}{4}\right) - S_j\left(\frac{h_j}{2}\right) = -\frac{M}{90}\left(\frac{h_j}{4}\right)^5 (32 - 2)$$

or

$$-\frac{M}{90}\left(\frac{h_j}{4}\right)^5 = \frac{1}{30}\left\{S_j\left(\frac{h_j}{4}\right) - S_j\left(\frac{h_j}{2}\right)\right\} . \tag{6.44}$$

If (6.44) is now substituted into (6.43) with $f^{(iv)}(\eta_1)$ and $f^{(iv)}(\eta_2)$ both replaced by M we have

$$I_j - S_j\left(\frac{h_j}{4}\right) = \frac{1}{15}\left\{S_j\left(\frac{h_j}{4}\right) - S_j\left(\frac{h_j}{2}\right)\right\} \tag{6.45}$$

as an estimate of the truncation error in (6.41).

Now set $x_1 = a$, $x_n = b$ and let $x_2, x_3, \ldots, x_{n-1}$ be the interior points which make the required dissection. Then if

$$\left|I_j - S_j\left(\frac{h_j}{4}\right)\right| \leq \frac{h_j}{b-a}\varepsilon \qquad j = 1, 2, \ldots, n-1$$

(where ε is the accuracy tolerance), or rather, using (6.45),

$$\frac{1}{15}\left|S_j\left(\frac{h_j}{4}\right) - S_j\left(\frac{h_j}{2}\right)\right| \leq \frac{h_j}{b-a}\varepsilon \tag{6.46}$$

the truncation error for the rule

$$R_S = \sum_{j=1}^{n-1} S_j\left(\frac{h_j}{4}\right)$$

has the bound

$$|I - R_S| = \left|\sum_{j=1}^{n-1}\left(I_j - S_j\left(\frac{h_j}{4}\right)\right)\right|$$

$$\leq \sum_{j=1}^{n-1}\left|I_j - S_j\left(\frac{h_j}{4}\right)\right|$$

$$\leq \sum_{j=1}^{n-1} \frac{h_j \varepsilon}{b-a} = \frac{\varepsilon}{b-a}\sum_{j=1}^{n-1} h_j = \varepsilon$$

since the sum of the interval lengths, h_j, is the entire range.

The way in which we can proceed is as follows. We consider first the extreme right-hand sub-interval $[x_{n-1}, x_n]$ ($= [x_{n-1}, b]$). The estimates (6.40) and (6.41) are formed (with $j = n-1$) and a test made to see if condition (6.46) is satisfied. If it is, the next interval, $[x_{n-2}, x_{n-1}]$, is considered and the process repeated. We continue in this way until the extreme left-hand sub-interval has been similarly considered. If we encounter a sub-interval $[x_k, x_{k+1}]$, say, for which (6.46) is not satisfied, we introduce the additional point $x_{k+\frac{1}{2}}$ and form the two Simpson's rule estimates (6.40) and (6.41) for the sub-interval $[x_{k+\frac{1}{2}}, x_{k+1}]$ with $h_j = h_k/2$, $x_j = x_{k+\frac{1}{2}}$ and $x_{j+1} = x_k$. The criterion (6.46) is examined and if it is satisfied we proceed to the interval $[x_k, x_{k+\frac{1}{2}}]$; if it is not the interval $[x_{k+\frac{1}{2}}, x_{k+1}]$ must be further sub-divided.

At this stage it might be helpful to give a numerical example.

EXAMPLE 6.11
We consider the integral

$$I = \int_1^9 \frac{1}{x} dx \ . \tag{6.47}$$

The integrand is sketched in Fig. 6.4. We observe that the curve is relatively steep near $x = 1$ but as x increases the gradient decreases in absolute value. We would therefore expect an adaptive scheme to concentrate the grid points towards the start of the range of integration. Suppose that we start with no interior grid points. Then we compute the following two estimates of (6.47)

$$S_1 = \frac{8}{6}\left\{\frac{1}{1} + 4\frac{1}{5} + \frac{1}{9}\right\} = 2.54815 \tag{6.48}$$

$$S_2 = \frac{8}{12}\left\{\frac{1}{1} + 4\frac{1}{3} + 2\frac{1}{5} + 4\frac{1}{7} + \frac{1}{9}\right\} = 2.27725 \ . \tag{6.49}$$

(6.48) is Simpson's rule with $h = 4$ whilst (6.49) is the composite Simpson's rule with $h = 2$. We now compute the quantity

$$\frac{1}{15}|S_2 - S_1| = 0.01806$$

which is an estimate of the magnitude of the truncation error in (6.49). Suppose that we are looking for an accuracy of 0.001. Then we find that

$$0.01806 > 8 \times 0.001/8 = 0.001$$

and conclude that we must split $[1, 9]$ into $[1, 5]$ and $[5, 9]$.

To estimate the integral over $[5, 9]$ we compute

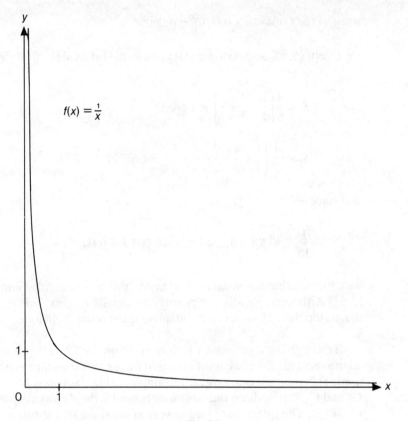

Fig. 6.4

$$S_1 = \frac{4}{6}\left\{\frac{1}{5} + \frac{4}{7} + \frac{1}{9}\right\} = 0.58836$$

$$S_2 = \frac{4}{12}\left\{\frac{1}{5} + \frac{4}{6} + \frac{2}{7} + \frac{4}{8} + \frac{1}{9}\right\} = 0.58783 \ . \tag{6.50}$$

and then form

$$\frac{1}{15}|S_2 - S_1| = 0.00004$$

which is an estimate of the truncation error in (6.50). Since

$$0.00004 < 4 \times 0.001/8 = 0.0005$$

we accept (6.50) and turn our attention to the interval [1, 5]. Here, we have

$$S_1 = \frac{4}{6}\left\{\frac{1}{1} + \frac{4}{3} + \frac{1}{5}\right\} = 1.68889$$

$$S_2 = \frac{4}{12}\left\{\frac{1}{1} + \frac{4}{2} + \frac{2}{3} + \frac{4}{4} + \frac{1}{5}\right\} = 1.62222$$

and since

$$\frac{1}{15}|S_2 - S_1| = 0.0044 \not< 4 \times 0.001/8 = 0.0005$$

we conclude that the interval [1, 5] must itself be sub-divided into [1, 3] and [3, 5]. Although we will not pursue the details further, this example has illustrated the behaviour of an adaptive quadrature technique.

Before giving a program which is based on the foregoing description of adaptive quadrature, we must note that rules other than Simpson's could be used. Schemes are available, for example, using the Gauss Legendre rules G_2 and G_3, but perhaps the most widely used is the *doubly adaptive* scheme of Oliver. The quadrature rules involved are those of Clenshaw and Curtis introduced in section 6.7 where we noted that it is possible to obtain an estimate of the truncation error. In the Oliver scheme, interval subdivision along the lines we have described is part of the strategy but, when a local error requirement is not satisfied, the use of a higher order quadrature formula is also investigated as an alternative to subdivision. We cannot here pursue the details and the interested reader is referred to the original paper by Oliver (1972) and to Davis and Rabinowitz (1975).

6.8.2 A program for adaptive quadrature

The programs of Case Study 6.4 implement the adaptive quadrature technique based on Simpson's rule outlined in the previous subsection. The main procedure of this code is *adaptivesimpson* which, given an initial interval subdivision supplied by the user, attempts to integrate the function f by applying the adaptive algorithm to each sub-interval. For each sub-interval there is a call to *integraloversubinterval*. This procedure is somewhat different to those we have looked at so far in that the procedure body contains calls to itself; that is, this is an example of a *recursive* procedure. If the current sub-interval is found to be too large, then it is halved and integration over each half of the sub-interval is achieved by a recursive call. The value held in *minimumsubinterval* is used as a check to ensure that an excessively small sub-interval is not used. The use of recursive procedures often leads to code which is easy to follow. However, there are problems involved and sometimes it is best to code a recursive process as an iterative one. A particularly relevant discussion on the merits, or otherwise, of recursive procedures is given by Atkinson and Harley (1983) who consider recursive and iterative algorithms for adaptive quadrature based on the trapezium rule.

Both *adaptivesimpson* and *integraloversubinterval* call the function *simpson* which, given three function values and an interval length, evaluate the corresponding Simpson's rule approximation to an integral.

CASE STUDY 6.4

Algol 68 Program

```
BEGIN
  COMMENT
  a program which implements the adaptive quadrature technique
  based on simpsons rule
  COMMENT
  PROC f = ( REAL x ) REAL :
    COMMENT
    this procedure defines the integrand
    COMMENT
    1.0 / x ;
  PROC simpson = ( REF [] REAL functionvalues , REAL interval ) REAL :
    COMMENT
    given three function values and an interval length this procedure
    evaluates the simpsons rule approximation to an integral
    COMMENT
    interval * ( functionvalues [ 1 ] + 4.0 * functionvalues [ 2 ]
            + functionvalues [ 3 ] ) / 6.0 ;
  PROC adaptivesimpson = ( PROC ( REAL ) REAL f , REF [] REAL subdivision ,
                           REAL minimumsubinterval , epsilon ,
                           REF BOOL errorflag ,
                           REF REAL totalintegral ) VOID :
  BEGIN
    COMMENT
    this procedure attempts to evaluate an integral correct to a specified
    tolerance using an adaptive technique based on simpsons rule
    COMMENT
    REAL left , subinterval , halfsubinterval , partintegral ;
    REAL right := subdivision [ 1 ] ;
    [ 1 : 3 ] REAL functionvalues ;
    functionvalues [ 3 ] := f ( right ) ;
    INT u = UPB subdivision ;
    REAL epsilonoverbminusa = epsilon / ( subdivision [ u ]
                                         - subdivision [ 1 ] ) ;
    COMMENT
    integrate over each subinterval of the original sub-division
    COMMENT
    totalintegral := 0.0 ;
    FOR i FROM 2 TO u
    DO
      left := right ; right := subdivision [ i ] ;
      subinterval := right - left ; halfsubinterval := subinterval * 0.5 ;
      functionvalues [ 1 ] := functionvalues [ 3 ] ;
```

```
            functionvalues [ 2 ] := f ( left + halfsubinterval ) ;
            functionvalues [ 3 ] := f ( right ) ;
            integraloversubinterval ( f , functionvalues , subinterval ,
                                      halfsubinterval , left ,
                                      minimumsubinterval , epsilonoverbminusa ,
                                      simpson ( functionvalues , subinterval ) ,
                                      errorflag , partintegral ) ;
        totalintegral +:= partintegral
    OD
END ;

PROC integraloversubinterval = ( PROC ( REAL ) REAL f ,
                                 REF [] REAL functionvalues ,
                                 REAL subinterval , halfsubinterval , left ,
                                 minimumsubinterval , epsilonoverbminusa ,
                                 simpsonusinghover2 ,
                                 REF BOOL errorflag ,
                                 REF REAL partintegral ) VOID :
BEGIN
  COMMENT
    this procedure attempts to evaluate the integral over a subinterval
    correct to the specified tolerance
  COMMENT
  REAL quartersubinterval = subinterval * 0.25 ;
  REAL quarterpoint = left + quartersubinterval ,
       threequarterpoint = quarterpoint + halfsubinterval ;
  REAL fquarterpoint = f ( quarterpoint ) ,
       fthreequarterpoint = f ( threequarterpoint ) ;
  [ 1 : 3 ] REAL functionvaluesleft := ( functionvalues [ 1 ] ,
                                         fquarterpoint ,
                                         functionvalues [ 2 ] ) ,
            functionvaluesright := ( functionvalues [ 2 ] , fthreequarterpoint ,
                                     functionvalues [ 3 ] ) ;
  COMMENT
    halve the subinterval and integrate using simpsons rule twice
  COMMENT
  REAL simpsonleft = simpson ( functionvaluesleft , halfsubinterval ) ,
       simpsonright = simpson ( functionvaluesright , halfsubinterval ) ;
  REAL simpsonusinghover4 = simpsonleft + simpsonright ;
  COMMENT
    estimate the truncation error associated with the current estimate of
    the integral
  COMMENT
  REAL truncationerror = ( simpsonusinghover4 - simpsonusinghover2 )
                         / 15.0 ;
  COMMENT
    test whether the truncation error is small enough
  COMMENT
  IF ABS truncationerror < epsilonoverbminusa * subinterval
  THEN
    COMMENT
      the current estimate of the integral is accurate enough
    COMMENT
    partintegral := simpsonusinghover4
  ELIF halfsubinterval < minimumsubinterval
  THEN
    COMMENT
      an interval length smaller than the minimum permitted is required
      to achieve the desired accuracy
      accept the current estimate of the integral and warn the user that
      something has gone wrong
    COMMENT
    errorflag := TRUE ;
    partintegral := simpsonusinghover4
  ELSE
    COMMENT
      subdivide the current subinterval and integrate over each half in turn
    COMMENT
    REAL lefthandintegral , righthandintegral ;
    REAL subint = halfsubinterval ; REAL halfsubint = subint * 0.5 ;
```

```
            integraloversubinterval ( f , functionvaluesleft , subint ,
                            halfsubint , left ,
                            minimumsubinterval , epsilonoverbminusa ,
                            simpsonleft ,
                            errorflag , lefthandintegral ) ;
            integraloversubinterval ( f , functionvaluesright , subint ,
                            halfsubint , left + subint ,
                            minimumsubinterval , epsilonoverbminusa ,
                            simpsonright ,
                            errorflag , righthandintegral ) ;
            partintegral := lefthandintegral + righthandintegral
      FI
    END ;
END ;
BOOL errorflag := FALSE ;
INT numberofsubdivisionpoints ; read ( numberofsubdivisionpoints ) ;
[ 1 : numberofsubdivisionpoints ] REAL subdivision ;
REAL minimumsubinterval , epsilon ;
read ( ( subdivision , minimumsubinterval , epsilon ) ) ;
REAL requiredintegral ;
adaptivesimpson ( f , subdivision , minimumsubinterval ,
            epsilon , errorflag , requiredintegral ) ;
print ( ( newline , newline ,
         "the suggested initial subdivision is" , newline , newline ,
         subdivision , newline , newline ,
         "minimum subinterval size      " , minimumsubinterval , newline ,
         "accuracy criterion            " , epsilon , newline , newline ,
         "approximate integral using adaptive simpson quadrature" , newline ,
         newline , requiredintegral , newline , newline ) ) ;
IF errorflag
THEN
   print ( ( "in order to achieve the specified accuracy a subinterval " ,
            newline , "smaller than the minimum allowed is required" ,
            newline ) )
   FI
END
```


Pascal Program

```
PROGRAM exampleofadaptivequadrature(input, output);

   {  a program which implements the adaptive quadrature technique
      based on simpsons rule  }

   CONST
      maxnumberofsubdivisionpoints=50;

   TYPE
      pointsindextype=1..maxnumberofsubdivisionpoints;
      pointsarraytype=ARRAY[ pointsindextype ] OF real;
      fvaluestype=ARRAY[ 1..3 ] OF real;

   VAR
      errorflag:boolean;
      i, numberofsubdivisionpoints:pointsindextype;
      subdivision:pointsarraytype;
      minimumsubinterval, epsilon, requiredintegral:real;


   FUNCTION f(x:real):real;
      {  this function defines the integrand  }
   BEGIN
      f:= 1.0/x
   END { f } ;

   FUNCTION simpson(VAR fvalues:fvaluestype; interval:real):real;
         {  given three function values and an interval length
            this procedure evaluates the simpsons rule approximation
            to an integral  }
```

```
BEGIN
    simpson:= interval*(fvalues[1] + 4.0*fvalues[2] + fvalues[3])/6.0
END { simpson } ;
PROCEDURE integraloversubinterval(FUNCTION f(x:real):real;
                                  VAR fvalues:fvaluestype;
                                  subinterval, halfsubinterval, left,
                                  minimumsubinterval, epsilonoverbminusa,
                                  simphby2:real;
                                  VAR errorflag:boolean;
                                  VAR partintegral:real);
    { this procedure attempts to evaluate the integral over a
      subinterval correct to the specified tolerance }
    VAR
        quartersubinterval, quarterpoint, threequarterpoint, fquarterpoint,
        fthreequarterpoint, simpleft, simpright, simphby4,
        truncationerror, newsubinterval, newhalfsubinterval,
        lefthandintegral, righthandintegral:real;
        fvaluesleft, fvaluesright:fvaluestype;
BEGIN
    quartersubinterval:= subinterval*0.25;
    quarterpoint:= left + quartersubinterval;
    threequarterpoint:= quarterpoint + halfsubinterval;
    fquarterpoint:= f(quarterpoint);
    fthreequarterpoint:= f(threequarterpoint);
    fvaluesleft[1]:= fvalues[1];
    fvaluesleft[2]:= fquarterpoint;
    fvaluesleft[3]:= fvalues[2];
    fvaluesright[1]:= fvalues[2];
    fvaluesright[2]:= fthreequarterpoint;
    fvaluesright[3]:= fvalues[3];
    { halve the subinterval and integrate using simpsons rule twice }
    simpleft:= simpson(fvaluesleft, halfsubinterval);
    simpright:= simpson(fvaluesright, halfsubinterval);
    simphby4:= simpleft + simpright;
    { estimate the truncation error associated with the current
      estimate of the integral }
    truncationerror:= (simphby4 - simphby2)/15.0;
    { test whether the truncation error is small enough }
    IF abs(truncationerror)<epsilonoverbminusa*subinterval
    THEN
        { the current estimate of the interval is accurate enough }
        partintegral:= simphby4
    ELSE
        IF halfsubinterval<minimumsubinterval
        THEN
            BEGIN
                { an interval length smaller than the minimum permitted
                  is required to achieve the desired accuracy
                  accept the current estimate of the integral and warn
                  the user that something has gone wrong }
                errorflag:= true;
                partintegral:= simphby4
            END
        ELSE
            BEGIN
                { subdivide the current subinterval and integrate over
                  each half in turn }
                newsubinterval:= halfsubinterval;
                newhalfsubinterval:= newsubinterval*0.5;
                integraloversubinterval(f, fvaluesleft,
                        newsubinterval, newhalfsubinterval,
                        left, minimumsubinterval,
                        epsilonoverbminusa, simpleft,
                        errorflag, lefthandintegral);
                integraloversubinterval(f, fvaluesright,
                        newsubinterval, newhalfsubinterval,
                        left + newsubinterval, minimumsubinterval,
                        epsilonoverbminusa, simpright,
                        errorflag, righthandintegral);
```

```
                    partintegral:= lefthandintegral + righthandintegral
            END
END { integraloversubinterval } ;

PROCEDURE adaptivesimpson(FUNCTION f(x:real):real;
                          u:pointsindextype;
                          VAR subdivision:pointsarraytype;
                          minimumsubinterval, epsilon:real;
                          VAR errorflag:boolean;
                          VAR totalintegral:real);
    { this procedure attempts to evaluate an integral correct to a
      specified tolerance using an adaptive technique based on
      simpsons rule }
    VAR
        left, right, subinterval, halfsubinterval, partintegral:real;
        epsilonoverbminusa:real;
        fvalues:fvaluestype;
        i:pointsindextype;
BEGIN
    right:= subdivision[1];
    totalintegral:= 0.0;
    fvalues[3]:= f(right);
    epsilonoverbminusa:= epsilon/(subdivision[u] - subdivision[1]);
    { integrate over each subinterval of the original sub-division }
    FOR i:= 2 TO u DO
    BEGIN
        left:= right;  right:= subdivision[i];
        subinterval:= right - left;  halfsubinterval:= subinterval*0.5;
        fvalues[1]:= fvalues[3];
        fvalues[2]:= f(left + halfsubinterval);
        fvalues[3]:= f(right);
        integraloversubinterval(f, fvalues, subinterval,
                                halfsubinterval, left,
                                minimumsubinterval, epsilonoverbminusa,
                                simpson(fvalues, subinterval),
                                errorflag, partintegral);
        totalintegral:= totalintegral + partintegral
    END
END { adaptivesimpson } ;

BEGIN
    read(numberofsubdivisionpoints);
    FOR i:= 1 TO numberofsubdivisionpoints DO
        read(subdivision[i]);
    read(minimumsubinterval, epsilon);
    adaptivesimpson(f, numberofsubdivisionpoints, subdivision,
        minimumsubinterval, epsilon, errorflag, requiredintegral);
    writeln;
    writeln;
    writeln('the suggested initial subdivision is');
    writeln;
    FOR i:= 1 TO numberofsubdivisionpoints DO
        write(subdivision[i]);
    writeln;
    writeln;
    writeln('minimum subinterval size     ', minimumsubinterval);
    writeln('accuracy criterion           ', epsilon);
    writeln;
    writeln('approximate integral using adaptive simpson quadrature');
    writeln;
    writeln(requiredintegral);
    writeln;
    IF errorflag
    THEN
        BEGIN
            writeln('in order to achieve the specified accuracy a subinterval');
            writeln('smaller than the minimum allowed is required')
        END
END { exampleofadaptivequadrature } .
```

Sample Data

```
3
1.0    5.0    9.0
0.0005
0.0005
```

Sample Output

```
the suggested initial subdivision is

    1.00000e  0   5.00000e  0   9.00000e  0

minimum subinterval size        5.00000e -4
accuracy criterion              5.00000e -4

approximate integral using adaptive simpson quadrature

    2.19730e  0
```

6.9 THE COMPUTATIONAL ERROR

In the introductory section we pointed out that, because of the finite word length of the computer (resulting, in general, in a rounding error in the representation of data and in the result of arithmetic operations) and because of the error introduced in the evaluation of functions, we actually compute a quantity R^* which is an approximation to R.

We shall not attempt to give a complete account of how to estimate $|R^* - R|$ but an elementary investigation will be sufficient to show that the computational error is usually not a significant quantity. We will suppose that double length facilities (see section 1.9) are available on the computer used so that we are justified in assuming that error in the eventual single length sum of products is due essentially to the errors in the representation of w_i and x_i and in the computation of $f(x_i)$ to a finite precision. Thus, if we actually use $w_i + \theta_i$ and compute $f(x_i) + \rho_i$, the quantity R^* is given by

$$R^* = \sum_{i=1}^{n} (w_i + \theta_i)(f(x_i) + \rho_i)$$

which, on neglecting second order terms (that is, terms of the form $\theta_i \rho_i$), gives

$$R^* \simeq \sum_{i=1}^{n} w_i f(x_i) + \sum_{i=1}^{n} \rho_i w_i + \sum_{i=1}^{n} \theta_i f(x_i)$$

so that

$$R^* - R \simeq \sum_{i=1}^{n} \rho_i w_i + \sum_{i=1}^{n} \theta_i f(x_i) \ . \qquad (6.51)$$

If we can now assert that

$$\rho = \max_{1 \leq i \leq n} |\rho_i| \ ; \quad \theta = \max_{1 \leq i \leq n} |\theta_i| \ ; \quad M = \max_{1 \leq i \leq n} |f(x_i)|$$

then, on taking moduli in (6.51), applying the triangle inequality to the right-hand side and simplifying, we have

$$|R^* - R| \leq \rho \sum_{i=1}^{n} |w_i| + n\theta M \ . \qquad (6.52)$$

Now, if the rule is exact when $f(x) \equiv 1$ (and this is true for all the rules we have discussed), we have

$$\sum_{i=1}^{n} w_i = b - a$$

and if, further, the weights are all positive, we can assert from (6.52) that

$$|R^* - R| \leq \rho(b - a) + n\theta M \ . \qquad (6.53)$$

The bound given by the right-hand side of (6.53) is likely to be a considerable overestimate but it is worthwhile to look at its structure. The quantity ρ is likely to be no more than a few machine round-off units so that first term, $\rho(b - a)$, will be small. The quantity θ in the second term will correspond to the machine round-off unit. Since this second term depends linearly on n the effect of rounding error may be significant only if a rule with a large number of integration points is used. What we can certainly assert from (6.53) is that it would be dangerous to attempt to estimate I through the rule R if an accuracy which is close to the limit of machine accuracy is demanded.

The reader should note that in passing from (6.52) to (6.53), the assumption was made that all the weights are positive. If this is not the case then there is no guarantee that the sum of the absolute values of the weights will remain bounded as $n \to \infty$. This, together with our earlier remarks concerning possible problems due to differencing, means that rules in which the weights are positive are to be preferred.

6.10 THE QUADRATURE CHAPTER OF THE NAG ALGOL 68 LIBRARY

In subsection 6.4.2 we gave an introduction to the workings of the NAG Algol 68 library Gauss quadrature routines. Recall that the points and weights for a particular rule are, in Case Study 6.3, stored in an item of mode *GAUSSRULE*. It is natural to gather together rules associated with a particular set of orthogonal polynomials (such as Legendre or Hermite) and so we introduce the mode

GAUSSRULESET = STRUCT(REF[ ]GAUSSRULE rules,
  PROC([ ]REAL,REAL,REAL,REF[ ]REAL)VOID map) .

Here *map* defines the transformation that needs to be applied to the points and weights. Thus, for example, if we wish to evaluate a definite integral on the finite interval $[a, b]$ using Gauss Legendre quadrature the points would need to be mapped and the weights modified to take account of the derivative. The parameters of *map* are, in turn

(i) a real array which specifies the range of integration (that is, $[a, b]$ for integration on a finite interval),
(ii) a real number for specifying an x-coordinate,
(iii) a real number specifying a weight, and
(iv) a real array returning the new point and weight.

Obviously, only a limited number of rules can be stored. When a user specifies that a k-point rule, say, is required, the rule which uses at least k points (subject to a maximum) should be employed.

The actual modes used in the NAG Algol 68 library are somewhat more involved than the ones described here. Nevertheless it is quite straightforward for the user to augment a set of rules, or, indeed, to introduce a new set. The reader is now encouraged to become fully conversant with the d01 chapter documentation and to make use of the facilities available.

EXERCISES

6.1 Verify that the rule $\frac{2}{3}\left\{f\left(-\frac{1}{\sqrt{2}}\right) + f(0) + f\left(\frac{1}{\sqrt{2}}\right)\right\}$

for estimating $\int_{-1}^{1} f(x)dx$ has precision 3.

6.2 Euler's rule for estimating $\int_a^b f(x)dx$ takes the form $R = (b-a) f(a)$.

Use this rule to estimate the integral of e^x over $[1,2]$. Given that the truncation error for this rule is $E = \frac{1}{2} f'(\xi)$, where $\xi \in]a, b[$, compute a bound for the error in your estimate and show that the actual error is within this bound. Obtain a further estimate by using Euler's rule over the two sub-intervals $[1, 1.5]$ and $[1.5, 2]$ and monitor the reduction in the error.

6.3 Calculate a value for h which will ensure that the truncation error in the composite trapezium rule approximation to $I = \int_1^2 \frac{1}{x} dx$ is at most 0.01 in absolute value. Hence compute an estimate of I which is correct to this accuracy and show that the actual error is within the specified bound.

6.4 Show that in the Romberg T-table,

$$T_{2,0} = \frac{b-a}{90}\left\{7f(a) + 32f\left(a + \frac{b-a}{4}\right) + 12f\left(a + \frac{b-a}{2}\right) \right.$$

$$\left. + 32f\left(a + 3\frac{b-a}{4}\right) + 7f(b)\right\}.$$

6.5 Compute the two-point Gauss Legendre approximation to $\int_{-1}^{1} e^x dx$. Find the error in this approximation.

6.6 Show that the three-point Gauss Chebyshev rule integrates

$$\int_{-1}^{1} \frac{x^4}{\sqrt{1-x^2}} dx$$

exactly. (The rule is defined by

$$\int_{-1}^{1} \frac{f(x)}{\sqrt{1-x^2}} dx \simeq R_3 = w_1 f(x_1) + w_2 f(x_2) + w_3 f(x_3)$$

where $w_1 = w_2 = w_3 = \pi/3$ and

$$x_i = \cos\left(\frac{2i-1}{3}\frac{\pi}{2}\right).$$

Use the change of variable $x = \cos(\theta)$ and the identity $\cos^2(\theta) = \frac{1}{2}(1 + \cos(2\theta))$ to find the integral analytically.)

6.7 Write a program based on Simpson's rule which successively halves h until two estimates agree to some user specified tolerance. Make sure that no function evaluations are repeated by storing in separate sums the function values at the end points, at the even interior points and at the odd interior points. Test your program using $\int_{-1}^{1} e^x dx$.

6.8 Write a program, similar to that of Case Study 6.3, which computes an n-point Gauss Laguerre approximation to $\int_{a}^{\infty} e^{-x} f(x) dx$. (You will need to map the interval $[a, \infty[$ onto $[0, \infty[$ using the transformation $z = x - a$.) Use the program to find a four-point approximation to $\int_{0}^{\infty} e^{-x} \sin(x) dx$. (The points and weights are

$x_1 = 0.322548$ $w_1 = 0.603154$
$x_2 = 1.74576$ $w_2 = 0.357419$
$x_3 = 4.53662$ $w_3 = 0.0388879$
$x_4 = 9.39507$ $w_4 = 0.000539295$

to six significant figures.)

6.9 Use the formula for the truncation error in the trapezium rule (equation (6.7)) to derive an adaptive algorithm based on this rule. Program your algorithm and use the code to obtain an estimate of $\int_{1}^{2} \frac{1}{x} dx$ which is in error by at most 0.005 in absolute value.

7
Ordinary Differential Equations

7.1 INTRODUCTION

There are many situations of interest in the physical and life sciences and in other subjects which can be modelled through an ordinary differential equation (o.d.e.) or a system of simultaneous o.d.e.'s. The simplest model of this sort involves the *first order initial value problem* (*i.v.p.*). Here we seek a solution, $y(x)$, to the first order differential equation

$$y' = f(x, y) \qquad x \geq a \tag{7.1}$$

where $f(x, y)$ is a known function, subject to the *initial condition*

$$y(a) = \alpha \tag{7.2}$$

where $y(x)$ is required to have the specified value α at the point $x = a$. The conditions which the function $f(x, y)$ must satisfy in order to ensure the existence of a unique, continuous and differentiable function $y(x)$ satisfying (7.1) and (7.2) in some interval $a \leq x \leq b$ are the subject of a well-known theorem and the reader is referred to Henrici (1962).

In many cases encountered in practice, it will not be possible to obtain an analytical solution to an initial value problem and we have to think in terms of employing approximate techniques. Even if an analytical solution can be obtained, it may not be easy to evaluate $y(x)$ for a given value of x and an approach through an approximate method may be preferable.

EXAMPLE 7.1
(i) Given the initial condition $y(0) = \alpha$ the equation $y' = \lambda y$ has the solution $y(x) = \alpha e^{\lambda x}$.
(ii) For the equation $y' = 1 + 2xy$ with the initial condition $y(0) = 0$ the

solution may be written as

$$y(x) = e^{x^2} \int_0^x e^{-t^2}\, dt.$$

In order to evaluate $y(x)$ for a particular value of x we would have to use a quadrature method to evaluate the integral.

(iii) It is not possible to obtain an analytical solution to the equation $y' = x^2 + y^4$ with initial condition $y(0) = 1$.

In practice we are usually interested in the solution $y(x)$ on some finite interval $[a, b]$. In an approximate method the idea is to try to set up a scheme which will enable an approximation to $y(x)$ to be obtained on the point set $\{x_j : j = 1, 2, \ldots, N\}$, where $x_N = b$. It is natural to introduce the additional point $x_0 = a$ so that, from (7.2), we already know $y(x_0)$. We shall use Y_j to denote an approximation to $y(x_j)$ obtained by some approximation scheme, assuming that all the computations involved are performed exactly. The approximate methods we consider are known as *step-by-step* methods. Each Y_j is obtained from information known at one or more of the preceding points $\{x_i : i = 0, 1, \ldots, j - 1\}$. Having found Y_j, we proceed to the next point x_{j+1} and generate Y_{j+1}, the process being continued until Y_N is found. Owing to the inevitable presence of rounding errors, it will not be possible to find Y_j exactly and what we actually calculate is a quantity Y_j^*. We would like to be able to say something about the *truncation error* $y(x_j) - Y_j$ and also the *computational error* $Y_j - Y_j^*$. These terms were introduced, in the context of quadrature, in Chapter 6 and again we shall find that it is very difficult to obtain useful bounds on the magnitude of these two quantities. In any case, new phenomena are encountered which can make a computed approximation completely valuelesss.

In later discussions we shall from time to time make use of *test problems*, whose analytical solutions are known, to illustrate the performance of a particular method. In the context of the first order i.v.p. we shall use

$$y' = \lambda y \quad x \geq a$$
$$y(a) = \alpha \qquad\qquad\qquad\qquad (7.3)$$

where λ is a real constant. The reader can easily verify that the analytical solution to (7.3) is

$$y(x) = \alpha\, e^{\lambda(x-a)}.$$

7.2 EULER'S METHOD

7.2.1 Derivation of Euler's method

We describe this method (and those of sections 7.3 and 7.4) in the context of obtaining an approximation Y_j to $y(x_j)$ at a set of equally spaced points $\{x_j : j = 1, 2, \ldots, N\}$ with

$$x_{j+1} - x_j = h \qquad j = 1, 2, \ldots, N-1.$$

The quantity h is often referred to as the *step length*. We discuss the method, not because it is widely used in practice, but rather because it is possible to introduce some important ideas without too much mathematical difficulty.

We suppose that a value for h has been chosen and we first consider obtaining an approximation Y_1 to $y(x_1)$, where $x_1 = x_0 + h = a + h$. Using Taylor's theorem (Theorem 1.1) we have

$$y(x_1) = y(x_0) + hy'(x_0) + \frac{h^2}{2} y''(\xi_0) \qquad \xi_0 \in {]}x_0, x_1{[} \, . \tag{7.4}$$

Using the initial value (7.2) and the differential equation (7.1) we may rewrite (7.4) as

$$y(x_1) = \alpha + hf(x_0, \alpha) + \frac{h^2}{2} y''(\xi_0) \, . \tag{7.5}$$

The approximation Y_1 is now obtained by neglecting the term $h^2 y''(\xi_0)/2$ in (7.5) giving

$$Y_1 = \alpha + hf(x_0, \alpha) \, . \tag{7.6}$$

Since we know α and h, the right-hand side of (7.6) may be found by evaluating $f(x, y)$ for the arguments x_0 and α. Subtracting (7.6) from (7.5) we have

$$y(x_1) - Y_1 = \frac{h^2}{2} y''(\xi_0)$$

which gives an expression for the truncation error in Y_1.

Having found Y_1 we now attempt to obtain an expression for Y_2. Using Taylor's theorem again we have

$$y(x_2) = y(x_1) + hy'(x_1) + \frac{h^2}{2} y''(\xi_1) \qquad \xi_1 \in {]}x_1, x_2{[}$$

$$= y(x_1) + hf(x_1, y(x_1)) + \frac{h^2}{2} y''(\xi_1) \ . \tag{7.7}$$

The approximation Y_2 is obtained by neglecting the term $h^2 y''(\xi_1)/2$ and replacing $y(x_1)$ by Y_1 in (7.7) giving

$$Y_2 = Y_1 + hf(x_1, Y_1) \ . \tag{7.8}$$

Having already determined Y_1, the right-hand side of (7.8) may be found by evaluating $f(x, y)$ for $x = x_1$ and $y = Y_1$. Subtracting (7.8) from (7.7) we have

$$y(x_2) - Y_2 = \frac{h^2}{2} y''(\xi_1) + y(x_1) - Y_1 + h(f(x_1, y(x_1)) - f(x_1, Y_1))$$

which says that the truncation error for Y_2 is $(h^2/2)y''(\xi_1)$ *plus* the errors introduced by the use of Y_1 in place of $y(x_1)$ in (7.8). Using a similar argument we have

$$Y_3 = Y_2 + hf(x_2, Y_2)$$

and the truncation error associated with Y_3 will be $(h^2/2)y''(\xi_2)$ (where $\xi_2 \in \,]x_2, x_3[$) plus errors introduced by using Y_2 in place of $y(x_2)$. Extending the process further we can formally state *Euler's method* as

$$\begin{aligned} Y_0 &= \alpha \\ Y_j &= Y_{j-1} + hf(x_{j-1}, Y_{j-1}) \qquad j = 1, 2, \ldots, N. \end{aligned} \tag{7.9}$$

Equation (7.9) is used in a step-by-step fashion until the approximation Y_N is generated.

EXAMPLE 7.2
Consider the initial value problem

$$\begin{aligned} y' &= y^2 \qquad x \in [0, 1] \\ y(0) &= 0.25 \end{aligned}$$

which has the solution

$$y(x) = 1/(4 - x).$$

Choose $h = 0.2$. Then, using Euler's method, we have

$$\begin{aligned} Y_0 &= 0.25 \\ Y_1 &= Y_0 + hY_0^2 = 0.25 + 0.2 \times 0.25^2 = 0.2625 \end{aligned}$$

$$Y_2 = Y_1 + hY_1^2 = 0.276281$$
$$Y_3 = Y_2 + hY_2^2 = 0.291548$$
$$Y_4 = Y_3 + hY_3^2 = 0.308548$$
$$Y_5 = Y_4 + hY_4^2 = 0.327588.$$

Since $y(1) = 1/3 = 0.333333$ the error in Y_5 is 0.005745.

Euler's method is said to be an *explicit, one-step* scheme. By explicit we mean that each Y_j may be obtained directly and by one-step we mean that Y_j is found using information available at the immediately preceding point x_{j-1} only. Having introduced these adjectives, the reader will not be surprised to learn that in due course we shall consider *implicit* methods (in which an equation for Y_j will need to be solved) and *k-step* methods (which define Y_j in terms of information at $x_{j-1}, x_{j-2}, \ldots, x_{j-k}$).

7.2.2 A program implementing Euler's method

Constructing a program which implements Euler's method is a simple matter (see Case Study 7.1). All that we need to do is to define a function to evaluate $f(x, y)$ of equation (7.1) and then to employ a loop which evaluates the Y_js using (7.9). In the programs of Case Study 7.1 we have used the code to solve problem (7.3) with $\lambda = 1$. The values of a and α are read from data into the variables *a* and *initialcondition*.

CASE STUDY 7.1

Algol 68 Program

```
BEGIN
  COMMENT
  a program to solve a first order initial value problem using eulers method
  COMMENT
  INT numberofyvalues ;
  REAL steplength , a , b , initialcondition , xvalue , yvalue ;
  PROC f = ( REAL x , y ) REAL :
    y ;
  read ( ( a , b , initialcondition , numberofyvalues ) ) ;
  print ( ( newline , newline ,
            "lower limit of range               " , a , newline ,
            "upper limit of range               " , b , newline ,
            "number of approximate values required  " ,
            numberofyvalues , newline ,
            "initial condition                  " ,
            initialcondition , newline , newline , newline ,
            "    x-value              approximate solution" , newline ,
            newline ) ) ;
  steplength := ( b - a ) / numberofyvalues ;
  xvalue := a ; yvalue := initialcondition ;
```

```
   TO numberofyvalues
   DO
      yvalue +:= steplength * f ( xvalue , yvalue ) ;
      xvalue +:= steplength ;
      print ( ( "   " , xvalue , "              " , yvalue , newline ) )
   OD
END
```

Pascal Program

```
PROGRAM eulersmethod(input, output);

   { a program to solve a first order initial value problem
     using eulers method }

   VAR
      i, numberofyvalues:1..maxint;
      steplength, a, b, initialcondition, xvalue, yvalue:real;

   FUNCTION f(x, y:real):real;
   BEGIN
      f:= y
   END { f } ;

BEGIN
   read(a, b, initialcondition, numberofyvalues);
   writeln;
   writeln;
   writeln('lower limit of range                  ', a);
   writeln('upper limit of range                  ', b);
   writeln('number of approximate values required  ', numberofyvalues);
   writeln('initial condition                     ', initialcondition);
   writeln;
   writeln;
   writeln(' ':7, 'x-value', ' ':12, 'approximate solution');
   writeln;
   steplength:= (b - a)/numberofyvalues;
   xvalue:= a;  yvalue:= initialcondition;
   FOR i:= 1 TO numberofyvalues DO
   BEGIN
      yvalue:= yvalue + steplength*f(xvalue, yvalue);
      xvalue:= xvalue + steplength;
      writeln('    ', xvalue, ' ':12, yvalue)
   END
END { eulersmethod } .
```

Sample Data

    0.0   1.0   1.0   8

Sample Output

```
lower limit of range                   0.00000e  0
upper limit of range                   1.00000e  0
number of approximate values required       8
initial condition                      1.00000e  0

       x-value           approximate solution

      1.25000e -1           1.12500e  0
      2.50000e -1           1.26562e  0
      3.75000e -1           1.42383e  0
      5.00000e -1           1.60181e  0
      6.25000e -1           1.80203e  0
      7.50000e -1           2.02729e  0
      8.75000e -1           2.28070e  0
      1.00000e  0           2.56578e  0
```

7.2.3 Use of the test equation

In section 7.1 we introduced the test equation (7.3). We choose here to set $a = 0$ so that (7.3) becomes

$$y' = \lambda y$$
$$y(0) = \alpha \qquad (7.10)$$

for which the analytical solution is

$$y(x) = \alpha e^{\lambda x}. \qquad (7.11)$$

Setting $f(x, y) = \lambda y$ in (7.9) we find that Euler's method applied to (7.10) gives

$$Y_j = Y_{j-1} + h\lambda Y_{j-1} = (1 + h\lambda)Y_{j-1} \qquad j = 1, 2, \ldots . \qquad (7.12)$$

We can use (7.12) again to express Y_{j-1} as a multiple of Y_{j-2} to give

$$Y_j = (1 + h\lambda)[(1 + h\lambda)Y_{j-2}] = (1 + h\lambda)^2 Y_{j-2} \qquad j = 2, 3, \ldots .$$

Continuing in this manner we obtain the result

$$Y_j = (1 + h\lambda)^j Y_0 = (1 + h\lambda)^j \alpha \qquad j = 1, 2, \ldots .$$

Now, from (7.11)

$$y(x_j) = \alpha e^{\lambda x_j} = \alpha e^{\lambda j h} \qquad j = 0, 1, 2, \ldots .$$

so the method being used here approximates $e^{\lambda j h}$ by $(1 + h\lambda)^j$. If λ is positive, both $e^{\lambda j h}$ and $(1 + h\lambda)^j$ are positive and increase with j and the approximate solution is exhibiting the right sort of qualitative behaviour. If, however, λ is negative, equal to $-\mu^2$ say, $e^{-\mu^2 j h}$ is being approximated by $(1 - h\mu^2)^j$. Now $e^{-\mu^2 j h}$ is positive but decreases with j but $(1 - h\mu^2)^j$ will exhibit this behaviour only if $0 < 1 - h\mu^2 < 1$. It is easy to see that $1 - h\mu^2 < 1$ for any choice of $h\ (>0)$ and μ^2. However, $0 < 1 - h\mu^2$ demands that the step length used must satisfy $h < 1/\mu^2$. Thus, if the approximate solution is to exhibit the right sort of qualitative behaviour, a restriction on the choice of h is imposed.

EXAMPLE 7.3

We saw from the sample output of Case Study 7.1 that for problem (7.10) the choice $\lambda = 1$, $h = 0.125$, $\alpha = 1$ results in a sequence of approximations Y_j to $y(x_j)$ being computed which reflect the known form of the analytical solution $y(x) = e^x$. Suppose we choose $\alpha = 1$ again but this time let $\lambda = -4.5$. For the approximate solution to exhibit the correct behaviour we must have $h <$ $1/4.5 \simeq 0.222222$. Selecting a value for h of 0.2 gives the sequence $Y_1 = 1 - 4.5 \times 0.2 = 0.1$, $Y_2 = (1 - 4.5 \times 0.2)^2 = 0.01$, ... , or, in general, $Y_j = 10^{-j}$. However, if we let $h = 0.25$ we obtain the sequence $Y_1 = 1 - 4.5 \times 0.25 = -0.125$, $Y_2 = (1 - 4.5 \times 0.25)^2 = (-0.125)^2$, ... , or, in general, $Y_j = (-0.125)^j$. Hence although the analytical solution is known to be everywhere positive, the Y_js alternate in sign.

The above analysis suggests that it may be necessary to impose a restriction on the size of the step length when seeking to approximate the solution of a differential equation by Euler's method when that solution is a decreasing function of x. Since Euler's method is rarely used in practice, this is perhaps not an important observation. Nevertheless, it is the first of a number of warnings we shall have that the choice of h can be a matter of importance. The reader is now encouraged to modify the code of Case Study 7.1 so that it can be used to solve the slightly more general problem (7.3). Several values for the available parameters (α, h, λ, a) should be chosen with a view to demonstrating the phenomenon which has just been outlined.

7.2.4 Estimating the truncation error
In the case of Euler's method it is possible, without too much difficulty, to obtain a bound for the magnitude of the truncation error $y(x_j) - Y_j$ and we give an outline of how this can be done. The result is not of great practical significance but it is of theoretical importance.

Recall that Euler's method may be written as

$$Y_j = Y_{j-1} + h f(x_{j-1}, Y_{j-1}) \qquad j = 1, 2, \ldots . \tag{7.13}$$

Using Taylor's theorem (Theorem 1.1) we have

$$y(x_j) = y(x_{j-1}) + (x_j - x_{j-1}) y'(x_{j-1}) + \frac{(x_j - x_{j-1})^2}{2} y''(\lambda_{j-1})$$

$$\lambda_{j-1} \in]x_{j-1}, x_j[$$

$$= y(x_{j-1}) + h f(x_{j-1}, y(x_{j-1})) + \frac{h^2}{2} y''(\xi_{j-1}). \tag{7.14}$$

Subtracting (7.13) from (7.14) and defining $e_j = y(x_j) - Y_j$ we have

$$e_j = e_{j-1} + h[f(x_{j-1}, y(x_{j-1})) - f(x_{j-1}, Y_{j-1})] + \frac{h^2}{2} y''(\xi_{j-1}). \tag{7.15}$$

Using the mean value theorem (Theorem 1.4) on the second argument of $f(x, y)$, (7.15) becomes

$$e_j = e_{j-1} + h(y(x_{j-1}) - Y_{j-1})\,(\partial f/\partial y)\,(x_{j-1}, \eta_{j-1}) + \frac{h^2}{2} y''(\xi_{j-1})$$

$$= (1 + h(\partial f/\partial y)\,(x_{j-1}, \eta_{j-1}))\,e_{j-1} + \frac{h^2}{2} y''(\xi_{j-1}) \qquad (7.16)$$

where by $(\partial f/\partial y)\,(x_{j-1}, \eta_{j-1})$ we mean the first partial derivative of $f(x, y)$ with respect to its second argument evaluated at $x = x_{j-1}$, $y = \eta_{j-1}$, and η_{j-1} lies between $y(x_{j-1})$ and Y_{j-1}.

In order to make further progress we have to make two assumptions, namely

$$|y''(x)| \leq M \qquad x \in [a, b] \qquad (7.17)$$
$$|(\partial f/\partial y)\,(x, y)| \leq L$$

where $[a, b]$ is the interval of interest. The numbers M and L are, respectively, upper bounds for the magnitude of the second derivative of the analytical solution and the magnitude of the y-partial derivative of $f(x, y)$. If we now take moduli in (7.16), apply the triangle inequality and make use of the bounds (7.17), we obtain

$$|e_j| \leq (1 + hL)|e_{j-1}| + \frac{1}{2} h^2 M \qquad j = 1, 2, \ldots . \qquad (7.18)$$

Now, (7.18) is valid for all $j \geq 1$ so that we may write

$$|e_{j-1}| \leq (1 + hL)|e_{j-2}| + \frac{1}{2} h^2 M \qquad j = 2, 3, \ldots$$

and substituting this result into (7.18) we have

$$|e_j| \leq (1 + hL)\left[(1 + hL)\,|e_{j-2}| + \frac{1}{2} h^2 M\right] + \frac{1}{2} h^2 M$$

$$= (1 + hL)^2 |e_{j-2}| + \frac{1}{2} h^2 M\,[1 + (1 + hL)].$$

Continuing the process we obtain the result

$$|e_j| \leq (1 + hL)^j |e_0| + \frac{1}{2} h^2 M\,[1 + (1 + hL) + \ldots + (1 + hL)^{j-1}]. \qquad (7.19)$$

But $e_0 = y(x_0) - Y_0 = 0$ and the terms within the square brackets in (7.19) constitute a geometrical progression with sum $((1 + hL)^j - 1)/hL$. Further, by using the Taylor series expansion of e^x about the point $x = 0$

$$e^{hL} = 1 + hL + \frac{(hL)^2}{2} + \ldots$$

so we conclude that $e^{hL} > 1 + hL$ ($hL > 0$) and hence that $e^{jhL} > (1 + hL)^j$. Returning to (7.19) we can now write

$$|e_j| < hM(e^{jhL} - 1)/2L$$
$$= hM \, (e^{L(x_j - x_0)} - 1)/2L \, . \tag{7.20}$$

As a result for actually estimating $|y(x_j) - Y_j|$ (7.20) is not particularly useful. It is only a bound and it is in terms of the numbers L and M which will not usually be readily available. However, it is of considerable theoretical significance. It shows that if x_j is kept fixed the truncated error tends to zero as γh, for some constant γ, as $h \to 0$.

EXAMPLE 7.4
Consider again the problem of Example 7.2 but now choose $h = 0.1$. We obtain the following results

$Y_0 = 0.25$
$Y_1 = 0.25625$
$Y_2 = 0.262816$
$Y_3 = 0.269723$
$Y_4 = 0.276998$
$Y_5 = 0.284671$
$Y_6 = 0.292775$
$Y_7 = 0.301346$
$Y_8 = 0.310427$
$Y_9 = 0.320063$
$Y_{10} = 0.330307$.

The final value is in error by 0.003026 so that halving the step length has resulted in a reduction of the error in the approximation of $y(1.0)$ by a factor of about two, emphasising the fact that the truncation error for Euler's method is proportional to h. The reader should now make further comparisons between the figures given here and the results of Example 7.2.

The result (7.20) enables us to introduce the concept of *convergence*.

Formally, we have that for a fixed point $x_j = x_0 + jh$

$$\lim_{h \to 0} Y_j = y(x_j)$$

that is, as we let h tend to zero and let j increase in such a way that $x_0 + jh$ remains fixed, the approximation Y_j tends to the analytical solution $y(x_j)$. This definition is adequate for a one-step method but requires some extension for a k-step method (when $k > 1$).

The reader is now encouraged to make use of the suitably modified version of the code of Case Study 7.1 to monitor the convergence behaviour of Euler's method when used to obtain an approximate solution to the general test problem (7.3).

7.3 THE USE OF TAYLOR SERIES

Provided $y(x)$ is sufficiently differentiable we can write down formally the Taylor series for $y(x)$ expanded about the point $x = x_0$ as

$$y(x) = y(x_0) + (x - x_0)y'(x_0) \\ + \frac{(x - x_0)^2}{2} y''(x_0) + \ldots + \frac{(x - x_0)^{p-1}}{(p-1)!} y^{(p-1)}(x_0) + R_p$$

(7.21)

where

$$R_p = \frac{(x - x_0)^p}{p!} y^{(p)}(\xi) \qquad \xi \in \,]x_0, x[$$

is the remainder. The first stage of Euler's method was to take $p = 2$, choose $x = x_1$, ignore R_2 and generate Y_1 using

$$Y_1 = Y_0 + hf(x_0, \alpha) \, .$$

We might now ask whether it is possible to consider values of p greater than two and so derive a method which will, presumably, give a more accurate estimate of $y(x_1)$. Such a method would require the evaluation of $y''(x_0)$, $y'''(x_0), \ldots$. For the second derivative of $y(x)$ we observe that

$$y''(x) = \frac{d}{dx} y'(x) = \frac{d}{dx} f(x, y)$$

making use of the fact that $y(x)$ satisfies the differential equation. It may be possible to form $df(x, y)/dx$ directly. Alternatively

$$\frac{d}{dx} f(x, y) = \frac{\partial}{\partial x} f(x, y) + y'(x) \frac{\partial}{\partial y} f(x, y)$$

and so it is possible to evaluate $y''(x_0)$ by forming the x- and y-partial derivatives of $f(x, y)$ and evaluating them, and $y'(x) = f(x, y)$, at $x = x_0$, $y = \alpha$. Proceeding further we can obtain an expression for $y'''(x)$ involving x, $y(x)$, $y'(x)$ and $y''(x)$ and, using the values x_0, $y(x_0)$, $y'(x_0)$ and the value $y''(x_0)$ just found, therefore determine $y'''(x_0)$. Next we obtain an expression for $y^{(iv)}(x)$, evaluate $y^{(iv)}(x_0)$, and so on. In principle therefore, we could choose some value for p, evaluate the required number of derivatives at x_0 and obtain an approximation Y_1 to $y(x_1)$ by setting $x = x_1$ in (7.21) and ignoring R_p.

EXAMPLE 7.5
Consider the problem of Example 7.2. Then

$$y'(0) = 0.25^2 = 0.0625$$

$$y''(x) = \frac{d}{dx}(y') = \frac{d}{dx}(y^2) = 2yy' \Rightarrow y''(0) = 2 \times 0.25 \times 0.625 = 0.03125$$

$$y'''(x) = \frac{d}{dx}(y'') = \frac{d}{dx}(2yy') = 2y'y' + 2yy''$$

$$\Rightarrow y'''(0) = 2 \times 0.0625^2 + 2 \times 0.25 \times 0.03125 = 0.0234375 .$$

Hence, if we choose $h = 0.2$

$$Y_1 = 0.25 + 0.2 \times 0.0625 + \frac{0.2^2}{2} \times 0.03125 + \frac{0.2^3}{6} \times 0.0234375 = 0.263156$$

gives an approximation to $y(0.2)$.

The process may be extended in an obvious manner to give an approximation Y_2 to $y(x_2)$. In (7.21) x_0 is replaced by x_1, $y(x_1)$ is approximated by Y_1, $y'(x_1)$ is approximated as $f(x_1, Y_1)$, the further derivatives are evaluated from the analytical forms already found using the approximate information available at x_1 and R_p is again ignored. Alternatively, we could put $x = x_2$ in (7.21) and use the values of $y'(x_0)$, $y''(x_0)$, ... already found. We shall not pursue the merits of these two alternatives for obtaining an approximation to $y(x_2)$ but both may be of use in particular circumstances.

It will be clear that the problem in using the Taylor series method lies in evaluating analytically the successive derivatives of $y(x)$ when the function $f(x, y)$ is complicated. In exceptional circumstances, it may be possible to derive a recurrence relation involving successive derivatives. Further, if a symbolic differentiation scheme is available on the computer to be used, the method may be of value. However, at the present time, Taylor's series does not provide a satisfactory general method for obtaining an approximate solution to a first order i.v.p.. However, we shall see in section 7.5 that it can be used to start other methods off.

7.4 TRAPEZOIDAL METHOD
7.4.1 Deriving the trapezoidal method
Many methods for the solution of a first order i.v.p. may be derived by making use of a quadrature rule. The reader will not be surprised to learn that we intend here to employ the trapezium rule which was discussed in subsection 6.2.1. We start by noting that

$$\int_{x_{j-1}}^{x_j} y'(x)\, dx = y(x_j) - y(x_{j-1}) \qquad j = 1, 2, \ldots . \tag{7.22}$$

Using (6.5) and (6.7) we can also express the integral (7.22) as

$$\int_{x_{j-1}}^{x_j} y'(x)\, dx = \frac{h}{2}\{y'(x_{j-1}) + y'(x_j)\} - \frac{h^3}{12} y'''(\xi_{j-1}) \qquad \xi_{j-1} \in \,]x_{j-1}, x_j[$$

$$= \frac{h}{2} \{f(x_{j-1}, y(x_{j-1})) + f(x_j, y(x_j))\} - \frac{h^3}{12} y'''(\xi_{j-1}) \tag{7.23}$$

the right-hand side being the trapezium rule estimate of the integral plus the associated truncation error. If we let $j=1$ and ignore the term $-h^3 y'''(\xi_0)/12$, an approximation Y_1 to $y(x_1)$ may be obtained by equating the right-hand sides of (7.22) and (7.23) to give

$$Y_1 = Y_0 + \frac{h}{2} \{f(x_0, Y_0) + f(x_1, Y_1)\}. \tag{7.24}$$

(7.24) is an equation which has to be solved for Y_1 and unless $f(x, y)$ has a particularly simple form (for example, if it is linear in y) it will be necessary to use an iterative scheme. We shall return to this point shortly and for now assume that Y_1 has been found. We proceed to the next step by setting $j=2$ in (7.22) and (7.23), ignoring the term $-h^3 y'''(\xi_1)/12$ in (7.23) and using Y_1 in place of $y(x_1)$. This yields the scheme

$$Y_2 = Y_1 + \frac{h}{2} \{f(x_1, Y_1) + f(x_2, Y_2)\}$$

which is an equation which needs to be solved for Y_2, an approximation to

$y(x_2)$. The general step can now be stated as

$$Y_j = Y_{j-1} + \frac{h}{2}\{f(x_{j-1}, Y_{j-1}) + f(x_j, Y_j)\} \qquad j = 1, 2, \ldots \qquad (7.25)$$

which is obtained from (7.22) and (7.23) by ignoring the term $-h^3 y'''(\xi_{j-1})/12$ and using Y_{j-1} in place of $y(x_{j-1})$. We refer to (7.25) as the *trapezoidal method* and we note that it is one-step (only information at the immediately preceding step is required) and implicit (an equation has to be solved for Y_j).

We now consider the problem of solving (7.25) for Y_j. In general an iterative process will have to be employed which will require the provision of an initial approximation $Y_j^{[0]}$ to Y_j. The value Y_{j-1} (the approximation at the previous step) is an obvious suggestion for $Y_j^{[0]}$. Another possibility is $Y_{j-1} + h f(x_{j-1}, Y_{j-1})$. This estimate is obtained by using Euler's rule (see Exercise 6.2) to approximate the left-hand side of (7.22). (Effectively, this is one step of Euler's method). However, there are better choices for the initial approximation and we return to this later. Given a value $Y_j^{[0]}$, equation (7.25) is used in the iterative form

$$Y_j^{[n]} = Y_{j-1} + \frac{h}{2}\{f(x_{j-1}, Y_{j-1}) + f(x_j, Y_j^{[n-1]})\} \qquad n = 1, 2, \ldots$$
$$(7.26)$$

and, under suitable circumstances, the sequence $\{Y_j^{[n]} : n = 0, 1, \ldots\}$ will converge (to Y_j). In order to determine precisely what these circumstances are we appeal to the analysis of iterative methods given in Chapter 2. In order to lighten the notation, we drop the subscript j on $Y_j^{[n]}$ and $Y_j^{[n-1]}$ and write (7.26) as

$$Y^{[n]} = \phi(Y^{[n-1]}) \qquad (7.27)$$

where

$$\phi(Y) = \frac{h}{2} f(x_j, Y) + Y_{j-1} + \frac{h}{2} f(x_{j-1}, Y_{j-1}) \qquad (7.28)$$

the last two terms on the right-hand side of (7.28) being independent of Y. Apart from the fact that the iteration number, n, appears as a superscript instead of a subscript, (7.27) is in the standard form of functional iteration introduced in subsection 2.3.1. We can therefore make use of the convergence theorem, Theorem 2.1, given in subsection 2.3.2 to assert that the iterative process (7.27) will converge provided that $|d\phi/dY| < 1$ on a sufficiently large interval containing the root and the initial approximation. Now

$$\frac{d\phi}{dY} = \frac{h}{2} \frac{d}{dY} f(x_j, Y) = \frac{h}{2} \frac{\partial}{\partial y} f(x, y) \bigg|_{x=x_j, y=Y}$$

that is, $d\phi/dY$ is $h/2$ times the y-partial derivative of $f(x, y)$ evaluated at $x = x_j$

Sec. 7.4] TRAPEZOIDAL METHOD

and $y = Y$. Hence, for convergence, we require

$$\frac{h}{2}\left|\frac{\partial f}{\partial y}\right| < 1$$

so that the step length h must be chosen to satisfy

$$h < 2/|\partial f/\partial y| . \qquad (7.29)$$

We shall not pursue the consequences of (7.29) further except to note that the restriction imposed on h is not usually of great importance since an even smaller value for h is often required for the trapezoidal method to provide an acceptable accuracy in the overall approximate solution.

In practice, then, the trapezoidal method can be used in a step-by-step manner, the progression from x_{j-1} to x_j being provided by (7.25) which is solved for Y_j through the iterative scheme (7.26) where a suitable convergence criterion might be to continue until a value for n is reached such that $|Y_j^{[n]} - Y_j^{[n-1]}| < \varepsilon$. $Y_j^{[n]}$ is then accepted as the estimate of Y_j (and hence of $y(x_j)$) and we proceed to the next point.

EXAMPLE 7.6

Consider the problem of Example 7.2. We can make use of the known form of the analytical solution to deduce that $\max_{x \in [0,1]} |y(x)| = \frac{1}{3}$. Since $\partial f/\partial y = 2y$ the upper bound on h given by (7.29) is 3. Hence, if we solve this problem using the trapezoidal method, a value for h of 0.2 should certainly guarantee convergence of the iterative process (7.26). Using Euler's rule to provide a value for $Y_j^{[0]}$ and iterating until two successive approximations to Y_j agree to five decimal places we have the following results.

$Y_0 = 0.25$

$Y_1^{[0]} = Y_0 + hY_0^2 = 0.25 + 0.2 \times 0.25^2 = 0.2625$

$Y_1^{[1]} = Y_0 + \frac{h}{2}[Y_0^2 + Y_1^{[0]^2}] = 0.25 + 0.1[0.25^2 + 0.2625^2] = 0.263141$

$Y_1^{[2]} = Y_0 + \frac{h}{2}[Y_0^2 + Y_1^{[1]^2}] = 0.263174$

$Y_1^{[3]} = Y_0 + \frac{h}{2}[Y_0^2 + Y_1^{[2]^2}] = 0.263176$

$Y_1^{[4]} = Y_0 + \frac{h}{2}[Y_0^2 + Y_1^{[3]^2}] = 0.263176 = Y_1 .$

Continuing in this manner we obtain the results $Y_2 = 0.27782$, $Y_3 = 0.29419$, $Y_4 = 0.31262$ and $Y_5 = 0.33352$, the error in the last value being just -0.00019.

If y occurs only linearly in $f(x, y)$, it will not be necessary to iterate since (7.25) can be solved immediately for Y_j. In particular the reader should verify, following the approach given in subsection 7.2.3, that the test problem (7.3) gives

$$Y_j = \left(\frac{1+\lambda h/2}{1-\lambda h/2}\right)^j \alpha.$$

Hence $e^{\lambda jh}$ is being approximated by $((1+\lambda h/2)/(1-\lambda h/2))^j$ and for $\lambda > 0$ this approximation will be positive (and increase with j) provided $1 - \lambda h/2 > 0$ that is, $h < 2/\lambda$. The reader should now investigate the case $\lambda < 0$ and obtain the appropriate result.

As a further observation on the trapezoidal method we state that it can be shown to be convergent in the sense that this term was defined in subsection 7.2.4.

7.4.2 A program implementing the trapezoidal method

The programs of Case Study 7.2 implement the trapezoidal method as described in the previous subsection. It is used here to solve the problem of Example 7.2. The procedure which calculates the Y_js consists of a simple loop to step along the x_js within which is embedded code evaluating $y_j^{[0]}$ using Euler's rule and a further loop implementing the iterative process (7.26). As with any iterative scheme, a limit on the total number of iterations used at each step must be imposed. Should convergence fail to be achieved at any of the x_js the user is informed of this via the parameter *successfulconvergence*. The procedure *trapezoidal* returns all the Y_js (and their associated x_js) via the parameter list.

Running the program again but asking for only ten approximate values to be computed, we obtain 0.333378 as an estimate of $y(1.0)$, the approximation being in error by -0.000045. Comparing this with the figure obtained in Example 7.6 we see that a halving of the step length reduces the error by a factor of about four, suggesting that the truncation error for the trapezoidal method is proportional to h^2 (and this is actually the case).

7.5 AN INTRODUCTION TO PREDICTOR-CORRECTOR METHODS

7.5.1 A simple predictor-corrector pair

We start by maintaining our original assumption, given at the beginning of section 7.2, that the x_js are equally spaced although we shall shortly want to consider relaxing this restriction. Employing the approach used in subsection 7.4.1 it is possible to derive what might be called the *midpoint method* by

CASE STUDY 7.2

Algol 68 Program

```
BEGIN
  COMMENT
  a program to solve a first order initial value problem using the
  trapezoidal method
  COMMENT
  PROC trapezoidal = ( PROC ( REAL , REAL ) REAL f ,
                      REAL a , b , initialcondition , convergencecriterion ,
                      REF [] REAL xvalues , yvalues ,
                      INT itermax ,
                      REF BOOL successfulconvergence ) VOID :
  BEGIN
    COMMENT
    this procedure computes an approximation to a first order initial
    value problem at a set of points using the trapezoidal method
    eulers method is used to start the iteration process which defines
    the approximate solution at a point
    COMMENT
    INT numberofyvalues = UPB yvalues ;
    REAL steplength = ( b - a ) / numberofyvalues ;
    REAL halfsteplength = steplength * 0.5 ;
    REAL xvalue := a , yvalue := initialcondition ,
      yatpreviouspoint , oldyvalue , partialsum ;
    BOOL converged ;
    successfulconvergence := TRUE ;
    COMMENT
    step along the points
    COMMENT
    FOR j TO numberofyvalues
    DO
      yatpreviouspoint := yvalue ;
      partialsum := halfsteplength * f ( xvalue , yvalue ) ;
      COMMENT
      compute a starting value using eulers method
      COMMENT
      yvalue := yatpreviouspoint + 2.0 * partialsum ;
      partialsum +:= yatpreviouspoint ;
      xvalue +:= steplength ;
      converged := FALSE ;
      COMMENT
      use functional iteration to find a solution to the equation
      defining the approximate solution at the current point
      COMMENT
      TO itermax WHILE NOT converged
      DO
        oldyvalue := yvalue ;
        yvalue := partialsum + halfsteplength * f ( xvalue , oldyvalue ) ;
        converged := ABS ( yvalue - oldyvalue ) < convergencecriterion
      OD ;
      IF NOT converged
      THEN
        successfulconvergence := FALSE
      FI ;
      xvalues [ j ] := xvalue ; yvalues [ j ] := yvalue
    OD
  END ;
```

318 ORDINARY DIFFERENTIAL EQUATIONS [Ch. 7]

```
    PROC f = ( REAL x , y ) REAL :
      y * y ;
    REAL a , b , initialcondition , epsilon ;
    INT itermax , numberofvaluesrequired ;
    read ( ( a , b , initialcondition , epsilon , itermax ,
              numberofvaluesrequired ) ) ;
    print ( ( newline , newline ,
              "lower limit of range                    " , a , newline ,
              "upper limit of range                    " , b , newline ,
              "initial condition                       " ,
              initialcondition , newline ,
              "iteration convergence criterion         " , epsilon , newline ,
              "maximum number of iterations permitted  " , itermax , newline ,
              "number of approximate y-values required " ,
              numberofvaluesrequired , newline , newline ) ) ;
    [ 1 : numberofvaluesrequired ] REAL xvalues , yvalues ;
    BOOL successfulconvergence ;
    trapezoidal ( f , a , b , initialcondition , epsilon , xvalues , yvalues ,
                  itermax , successfulconvergence ) ;
    print ( ( "         x-value              approximate solution" , newline ,
              newline ) ) ;
    FOR j TO numberofvaluesrequired
    DO
       print ( ( "     " , xvalues [ j ] , "                " , yvalues [ j ] ,
                 newline ) )
    OD ;
    IF NOT successfulconvergence
    THEN
       print ( ( newline , newline ,
                 "functional iteration failed to converge to within" ,
                 newline , "the specified accuracy at at least one point" ,
                 newline ) )
    FI
END
```

Pascal Program

```
PROGRAM exampleusingthetrapezoidalmethod(input, output);

   { a program to solve a first order initial value problem
     using the trapezoidal method }

   CONST
      maxnumberofvaluesrequired=50;

   TYPE
      indextype=1..maxnumberofvaluesrequired;
      valuestype=ARRAY[ indextype ] OF real;
      itermaxtype=1..maxint;

   VAR
      a, b, initialcondition, epsilon:real;
      j, numberofvaluesrequired:indextype;
      itermax:itermaxtype;
      successfulconvergence:boolean;
      xvalues, yvalues:valuestype;

   FUNCTION f(x, y:real):real;
   BEGIN
      f:= y*y
   END { f } ;

   PROCEDURE trapezoidal(FUNCTION f(x, y:real):real;
                         a, b, initialcondition, convergencecriterion:real;
                         numberofvalues:indextype;
                         VAR xvalues, yvalues:valuestype;
                         itermax:itermaxtype;
                         VAR successfulconvergence:boolean);
      { this procedure computes an approximation to a first order initial
        value problem at a set of points using the trapezoidal method
```

Sec. 7.5]   AN INTRODUCTION TO PREDICTOR-CORRECTOR METHODS   319

```
              eulers method is used to start the iteration process which defines
              the approximate solution at a point  }
         VAR
            thisoneconverged:boolean;
            steplength, halfsteplength, xvalue, yvalue, yatpreviouspoint,
               oldyvalue, partialsum:real;
            iter:0..maxint;
            j:indextype;
      BEGIN
         steplength:= (b - a)/numberofyvalues;
         halfsteplength:= steplength*0.5;
         xvalue:= a;   yvalue:= initialcondition;
         successfulconvergence:= true;
         {  step along the points  }
         FOR j:= 1 TO numberofyvalues DO
            BEGIN
              yatpreviouspoint:= yvalue;
              partialsum:= halfsteplength*f(xvalue, yvalue);
              {  compute a starting value using eulers method  }
              yvalue:= yatpreviouspoint + 2.0*partialsum;
              partialsum:= partialsum + yatpreviouspoint;
              xvalue:= xvalue + steplength;
              {  use functional iteration to find a solution to the equation
                 defining the approximate solution at the current point  }
              iter:= 0;
              REPEAT
                 iter:= iter + 1;
                 oldyvalue:= yvalue;
                 yvalue:= partialsum + halfsteplength*f(xvalue, oldyvalue);
                 thisoneconverged:= abs(yvalue - oldyvalue)<convergencecriterion
              UNTIL thisoneconverged OR (iter=itermax);
              IF NOT thisoneconverged
                 THEN
                    successfulconvergence:= false;
              xvalues[j]:= xvalue;   yvalues[j]:= yvalue
            END
      END { trapezoidal } ;

BEGIN
   read(a, b, initialcondition, epsilon, itermax, numberofyvaluesrequired);
   writeln;
   writeln;
   writeln('lower limit of range                   ', a);
   writeln('upper limit of range                   ', b);
   writeln('initial condition                      ', initialcondition);
   writeln('iteration convergence criterion        ', epsilon);
   writeln('maximum number of iterations permitted ', itermax);
   writeln('number of approximate y-values required ',
            numberofyvaluesrequired);
   writeln;
   trapezoidal(f, a, b, initialcondition, epsilon, numberofyvaluesrequired,
      xvalues, yvalues, itermax, successfulconvergence);
   writeln(' ':7, 'x-value', ' ':14, 'approximate solution');
   writeln;
   FOR j:= 1 TO numberofyvaluesrequired DO
      writeln('     ', xvalues[j], ' ':14, yvalues[j]);
   IF NOT successfulconvergence
      THEN
         BEGIN
            writeln;
            writeln;
            writeln('functional iteration failed to converge to within');
            writeln('the specified accuracy at at least one point')
         END
END { exampleusingthetrapezoidalmethod } .
```

Sample Data

```
   0.0    1.0
   0.25
   0.000005
   15
   16
```

Sample Output

```
lower limit of range                          0.00000e  0
upper limit of range                          1.00000e  0
initial condition                             2.50000e -1
iteration convergence criterion               5.00000e -6
maximum number of iterations permitted             15
number of approximate y-values required            16
```

x-value	approximate solution
6.25000e -2	2.53969e -1
1.25000e -1	2.58065e -1
1.87500e -1	2.62297e -1
2.50000e -1	2.66669e -1
3.12500e -1	2.71189e -1
3.75000e -1	2.75865e -1
4.37500e -1	2.80706e -1
5.00000e -1	2.85719e -1
5.62500e -1	2.90915e -1
6.25000e -1	2.96303e -1
6.87500e -1	3.01895e -1
7.50000e -1	3.07702e -1
8.12500e -1	3.13736e -1
8.75000e -1	3.20012e -1
9.37500e -1	3.26545e -1
1.00000e 0	3.33349e -1

making use of the midpoint quadrature rule introduced in subsection 6.2.1. From equations (6.9) and (6.10) we have

$$\int_{x_{j-2}}^{x_j} y'\,dx = 2hy'(x_{j-1}) + \frac{h^3}{3} y'''(\xi_{j-2}) \quad \xi_{j-2} \in \,]x_{j-2}, x_j[\quad j=2, 3, \ldots$$

$$= 2hf(x_{j-1}, y(x_{j-1})) + \frac{h^3}{3} y'''(\bar{\xi}_{j-2}). \qquad (7.30)$$

But

$$\int_{x_{j-2}}^{x_j} y'\,dx = y(x_j) - y(x_{j-2}) \qquad j=2, 3, \ldots \qquad (7.31)$$

so by combining (7.30) and (7.31) we derive the scheme

$$Y_j = Y_{j-2} + 2hf(x_{j-1}, Y_{j-1}) \qquad j=2, 3, \ldots \qquad (7.32)$$

for finding an approximation Y_j to $y(x_j)$. It is clear that (7.32) defines an explicit (Y_j can be found immediately) two-step (information at x_{j-1} and x_{j-2} is used) method. There is, however, an immediate difficulty in implementing (7.32) for, if we write down the first step (corresponding to $j = 2$) we have

$$Y_2 = Y_0 + 2hf(x_1, Y_1) = \alpha + 2hf(x_1, Y_1)$$

and we do not yet have a value for Y_1. An estimate of $y(x_1)$ has to be provided by some other means and, in principle, we could use the Taylor series approach described in section 7.3. Having obtained Y_1, (7.32) can be employed in the usual step-by-step manner. For reasons we shall discuss later the midpoint method is not satisfactory and we now consider an alternative way of making use of (7.32).

We observed in subsection 7.4.1 when discussing the trapezoidal method that iteration is usually necessary to estimate Y_j and consequently an initial approximation $Y_j^{[0]}$ must be available. We have already considered the use of Y_{j-1} and the Euler's rule approximation to start off the iterative process (7.26). Clearly we could also make use of (7.32) to provide the initial approximation

$$Y_j^{[0]} = Y_{j-2} + 2hf(x_{j-1}, Y_{j-1}) \tag{7.33}$$

where Y_{j-2} and Y_{j-1} are trapezoidal estimates. The equations (7.33) and (7.26) are known as a *predictor-corrector* pair. The predictor is an explicit method and its function is to provide the first estimate of the approximation at the new step, Y_j. The corrector is an implicit method and is used iteratively to provide the final estimate of Y_j. For the method currently under discussion there is still the problem of providing the estimate Y_1 and we refer to this as a *starting value*.

EXAMPLE 7.7
We return to the problem of Example 7.2 and again choose $h = 0.2$. The value obtained in Example 7.5 can be used to provide the starting value Y_1. The predictor-corrector pair (7.33) and (7.26) is used to estimate Y_2 as follows

$$Y_2^{[0]} = Y_0 + 0.4 Y_1^2 = 0.25 + 0.4 \times 0.263156^2 = 0.277700$$
$$Y_2^{[1]} = Y_1 + 0.1 \, [Y_1^2 + Y_2^{[0]2}]$$
$$\quad = 0.263156^2 + 0.1 \, [0.263156^2 + 0.277700^2] = 0.277793$$
$$Y_2^{[2]} = Y_1 + 0.1 \, [Y_1^2 + Y_2^{[1]2}] = 0.277798$$
$$Y_2^{[3]} = Y_1 + 0.1 \, [Y_1^2 + Y_2^{[2]2}] = 0.277798$$

where, as in Example 7.6, we have iterated until, when rounded, two successive estimates agree to five decimal places. Continuing in this manner we obtain the values $Y_3 = 0.29417$, $Y_4 = 0.31259$, $Y_5 = 0.33349$.

7.5.2 The local truncation error

The reader may be wondering why (7.33) is used to provide $Y_j^{[0]}$ in view of the disadvantage of having to obtain the starting value Y_1. The first reason is that the estimate (7.33) might be expected to be more accurate than the alternative proposed in subsection 7.4.1 with the consequent advantage that the number of iterations required for the convergence of the corrector may be less. There is an important second advantage, however: the use of the midpoint method and the trapezoidal method together enables an estimate to be made of what is known as the *local truncation error* (*l.t.e.*) (as opposed to the (*global*) *truncation error* considered in earlier sections). If we make the assumption that all previously calculated values are exact, equation (7.33) may be written as

$$Y_j^{[0]} = y(x_{j-2}) + 2hf(x_{j-1}, y(x_{j-1})). \tag{7.34}$$

But, from (7.30) and (7.31)

$$y(x_j) = y(x_{j-2}) + 2hf(x_{j-1}, y(x_{j-1})) + \frac{h^3}{3} y'''(\bar{\xi}_{j-2}). \tag{7.35}$$

Subtracting (7.34) from (7.35) gives

$$y(x_j) - Y_j^{[0]} = \frac{h^3}{3} y'''(\bar{\xi}_{j-2}). \tag{7.36}$$

which we define as the l.t.e. for the midpoint method. Using the same approach, from (7.23) it can be seen that the l.t.e. for the trapezoidal method can be written as

$$y(x_j) - Y_j = -\frac{h^3}{12} y'''(\xi_{j-1}). \tag{7.37}$$

The important thing to notice here is that both l.t.e.'s are of the same form — they both involve h^3 and the third derivative of $y(x)$. If we now make the assumption that $y'''(x)$ is not varying greatly over the interval in question (so that $y'''(\bar{\xi}_{j-2})$ and $y'''(\xi_{j-1})$ are approximately equal), subtracting (7.37) from (7.36) gives

$$Y_j - Y_j^{[0]} \simeq \frac{5h^3}{12} y'''(\xi_{j-1}). \tag{7.38}$$

and hence

$$y(x_j) - Y_j \simeq -\frac{1}{5}(Y_j - Y_j^{[0]})$$

which gives an estimate T_j, say, of the l.t.e. in Y_j in terms of the initial value $Y_j^{[0]}$ and the finally corrected value Y_j, assuming we iterate until convergence is achieved. The importance of this measure is that it can be used to control the step length h as the integration proceeds. If it is too large in the context of the accuracy being demanded in the approximate solution, a smaller value of h is called for; if it is small, a larger value of h can perhaps be tolerated. Decreasing and increasing the value of h is usually carried out by halving and doubling the interval and this adds considerably to the complexity of a computer program.

There is a further way in which an estimate of the l.t.e. can be of use. Under the usual assumptions we have from (7.38)

$$Y_{j-1} - Y_{j-1}^{[0]} \simeq \frac{5h^3}{12} y'''(\xi_{j-2}).$$

Combining this with (7.36) and again assuming the third derivative of $y(x)$ is fairly constant we find

$$y(x_j) - Y_j^{[0]} \simeq \frac{4}{5} (Y_{j-1} - Y_{j-1}^{[0]}) \tag{7.39}$$

which gives an estimate of the l.t.e. in the predicted value, $Y_j^{[0]}$, in terms of the predicted and corrected (to convergence) values at the previous point. Equation (7.39) suggests that we should form the *modified initial estimate*

$$\tilde{Y}_j = Y_j^{[0]} + \frac{4}{5} (Y_{j-1} - Y_{j-1}^{[0]}) \tag{7.40}$$

and then use the corrector (7.26) for $n = 2, 3, \ldots$ with $Y_j^{[1]} = \tilde{Y}_j$. The scheme (7.40) is known as *Milne's device* and can be used at $x_3, x_4, \ldots$.

EXAMPLE 7.8
Once again we return to the problem of Example 7.2 and make use of the values $Y_1 = 0.263156$ found in Example 7.5 and $Y_2 = 0.277798$ found in Example 7.7. (It is not possible to employ Milne's device when estimating Y_2 since we do not have a value $Y_1^{[0]}$ or a Y_1 obtained from the trapezoidal method.) We now form

$$Y_3^{[0]} = Y_1 + 0.4 Y_2^2 = 0.263156 + 0.4 \times 0.277798^2 = 0.294025$$
$$Y_3^{[1]} = \tilde{Y}_3 = Y_3^{[0]} + \frac{4}{5} (Y_2 - Y_2^{[0]})$$
$$= 0.294025 + 0.8(0.277798 - 0.277700) = 0.294103$$

$$Y_3^{[2]} = Y_2 + 0.1 \left[Y_2^2 + Y_3^{[1]^2} \right] = 0.277798 + 0.1[0.277798^2 + 0.294103^2]$$
$$= 0.294165$$

$$Y_3^{[3]} = Y_2 + 0.1 \left[Y_2^2 + Y_3^{[2]^2} \right] = 0.294168$$

and deduce $Y_3 = 0.29417$.

The reader is encouraged to use the process to calculate approximations to $y(0.8)$ and $y(1.0)$.

7.6 STABILITY

In order to introduce this topic we return to the suggested use of the midpoint method as a scheme for solving a first order i.v.p. We consider the test problem (7.3) with $\lambda = -2$, $a = 0$ and $\alpha = 1$ so that we have

$$y' = -2y \qquad x \geq 0 \qquad (7.41)$$

$$y(0) = 1$$

the solution to this problem being $y(x) = e^{-2x}$. Employing the scheme (7.32) with a value for h of 0.1 and using the starting value $Y_1 = e^{-0.2}$ (the exact solution at $x = 0.1$) we obtain the results listed in Table 7.1. So far, there

Table 7.1

j	x_j	$y(x_j)$	Y_j	$y(x_j) - Y_j$
2	0.2	0.670320	0.672508	−0.002188
3	0.3	0.548812	0.549728	−0.000916
4	0.4	0.449329	0.452617	−0.003288
5	0.5	0.367879	0.368681	−0.000802
6	0.6	0.301194	0.305144	−0.003950
7	0.7	0.246597	0.246623	−0.000026
8	0.8	0.201897	0.206495	−0.004598
9	0.9	0.165299	0.164025	+0.001274
10	1.0	0.135335	0.140885	−0.005550

appears to be no real trouble in that the Y_js are reasonable estimates of the $y(x_j)$s, although we observe that the relative error in the approximations appears, overall, to be increasing. Curiously Y_9 is an under-estimate of $y(x_9)$ whereas all other computed approximations are over-estimates. In fact if we continue the process we find that a pattern has now been established, namely $Y_j > y(x_j)$ for j even and $Y_j < y(x_j)$ for j odd. We pick up the process again at $j = 21$ (see Table 7.2). Clearly the errors in the Y_js have become completely

Table 7.2

j	x_j	$y(x_j)$	Y_j	$y(x_j) - Y_j$
21	2.1	0.0149958	−0.0187296	0.0337254
22	2.2	0.0122773	0.0542680	−0.0419907
23	2.3	0.0100518	−0.0404369	0.0504887
24	2.4	0.00822975	0.0704428	−0.0622131
25	2.5	0.00673795	−0.0686140	0.0753520
26	2.6	0.00551656	0.0978884	−0.0923718
27	2.7	0.00451658	−0.107769	0.112286
28	2.8	0.00369786	0.140996	−0.137298
29	2.9	0.00302755	−0.164168	0.167196
30	3.0	0.00247875	0.206663	−0.204184

unacceptable and we must investigate further.

The midpoint method (7.32) applied to (7.41) leads to the recurrence relation

$$Y_j = Y_{j-2} - 4hY_{j-1}. \tag{7.42}$$

We now seek a solution to (7.42) of the form

$$Y_j = c\mu^j. \tag{7.43}$$

Substituting (7.43) into (7.42) gives

$$c\mu^j = c\mu^{j-2} - 4hc\mu^{j-1}$$

which we choose to write as

$$c\mu^{j-2}(\mu^2 + 4h\mu - 1) = 0.$$

Hence a solution to (7.42) of the form (7.43) exists if μ is one of the roots of the quadratic $\mu^2 + 4h\mu - 1 = 0$, that is, if $\mu = \mu_1 = -2h + \sqrt{(4h^2 + 1)}$ or $\mu = \mu_2 = -2h - \sqrt{(4h^2 + 1)}$. We deduce that the general solution of (7.42) is of the form

$$Y_j = c_1\mu_1^j + c_2\mu_2^j \quad j = 0, 1, 2, \ldots. \tag{7.44}$$

Setting $j = 0$ and $j = 1$ in (7.44) gives

$$\begin{aligned} Y_0 &= c_1 + c_2 \\ Y_1 &= c_1\mu_1 + c_2\mu_2 \end{aligned} \tag{7.45}$$

where, of course, Y_0 is the initial condition and Y_1 is the starting value for the midpoint method. It is easy to see that

$$c_1 = (Y_1 - \mu_2 Y_0)/(\mu_1 - \mu_2)$$
$$c_2 = (\mu_1 Y_0 - Y_1)/(\mu_1 - \mu_2)$$

is the solution to the system (7.45) and hence the general solution to (7.42) has now been found in terms of Y_0, Y_1 and h.

We now consider the consequences of the above analysis. Consider first the root $\mu_1 = -2h + \sqrt{(4h^2 + 1)}$. Then, for all $h > 0$, $0 < \mu_1 < 1$ and so the first term on the right-hand side of (7.44) decreases as j increases, that is, this part of the solution to (7.42) mimics the behaviour of the solution to our original differential equation (7.41). However, $\mu_2 < -1$ for all $h > 0$ and hence the second term on the right-hand side of (7.44) increases with j and is alternately positive and negative. This second part of the solution to (7.42) does not behave like the known solution of (7.41) and its growth means that the term $c_1 \mu_1^j$ in (7.44) is eventually swamped. The form of the results of Table 7.1 and Table 7.2 is now explained. The reader should investigate the case $\lambda = 2$, $a = 0$ and $\alpha = 1$ for the test problem (7.3) and provide numerical evidence supporting the conclusions drawn.

Because of the possible poor behaviour of the midpoint method (which is independent of the step length, h) it is said to be *unstable*. Stability is an extremely important concept and the reader should be aware of its existence. We cannot here pursue the subject further but direct the reader to the books by Ortega and Poole (1981) and, for a more comprehesive discussion, Lambert (1973).

7.7 FURTHER PREDICTOR-CORRECTOR METHODS

The reader will not be surprised to learn that the predictor-corrector midpoint-trapezoidal pair introduced in subsection 7.5.1 is but one member of a whole family of methods available to us. We introduce the general *linear multistep method* (*l.m.m.*)

$$\sum_{i=0}^{k} \alpha_i Y_{j-i} - h \sum_{i=0}^{k} \beta_i f(x_{j-i}, Y_{j-i}) = 0 \qquad (7.46)$$

(so that, for example, the trapezoidal method (7.25) corresponds to the choice $k = 1$, $\alpha_0 = 1$, $\alpha_1 = -1$, $\beta_0 = \beta_1 = \frac{1}{2}$). Equation (7.46) defines a k-step method which will be explicit if $\beta_0 = 0$ and implicit otherwise. In order to use such a method we need to have some means of providing the $k - 1$ starting values $Y_1, Y_2, \ldots, Y_{k-1}$. A predictor-corrector pair will consist of an explicit l.m.m. to provide an initial estimate $Y_j^{[0]}$ and an implicit l.m.m. which can be used to improve $Y_j^{[0]}$ using an iterative process. If we match the predictor and corrector so that they are both of the same *order* (that is, the l.t.e.'s are both proportional to the same power of h and the same derivative of $y(x)$) then, as with the midpoint-trapezoidal pair, it will be possible to derive an estimate of the l.t.e in the corrector and hence derive a step-control scheme. In addition it will be possible to obtain an estimate of the l.t.e. in the predictor and hence to form a modified initial estimate (equivalent to (7.40)).

In subsection 6.2.1 we introduced Milne's rule (equation (6.11)) which suggests the explicit four-step method (*Milne's method*)

$$Y_j = Y_{j-4} + \frac{4h}{3} \{2f(x_{j-3}, Y_{j-3}) - f(x_{j-2}, Y_{j-2}) + 2f(x_{j-1}, Y_{j-1})\} \quad (7.47)$$

which is in the form (7.46) with $k=4$, $\alpha_0 = 1$, $\alpha_1 = \alpha_2 = \alpha_3 = 0$, $\alpha_4 = -1$, $\beta_0 = \beta_4 = 0$, $\beta_1 = \beta_3 = \frac{8}{3}$ and $\beta_2 = -\frac{4}{3}$. The l.t.e. for (7.47) takes the form

$$\frac{14}{45} h^5 y^{(v)} (\eta_{j-4})$$

and we could use this method as a predictor for the implicit two-step method (*Simpson's method*)

$$Y_j = Y_{j-2} + \frac{h}{3} \{f(x_{j-2}, Y_{j-2}) + 4f(x_{j-1}, Y_{j-1}) + f(x_j, Y_j)\} \quad (7.48)$$

which the reader will recognise as being derived from Simpson's rule (equation (6.6)). Once again we have a rule of the form (7.46), with $\alpha_0 = -\alpha_2 = 1$, $\alpha_1 = 0$ and $\beta_0 = \frac{1}{4}$ $\beta_1 = \beta_2 = \frac{1}{3}$. The associated l.t.e. is $-\frac{1}{90} h^5 y^{(v)} (\xi_{j-2})$ which means that if (7.47) is used to provide an initial value $Y_j^{[0]}$ and (7.48) is used iteratively until convergence is achieved, the quantity $-(Y_j - Y_j^{[0]})/29$ may be used as an estimate of the l.t.e. in (7.48). In addition, $\frac{28}{29} (Y_{j-1} - Y_{j-1}^{[0]})$ may be used as an estimate of the l.t.e. in (7.47) which suggests that

$$\tilde{Y}_j = Y_j^{[0]} + \frac{28}{29} (Y_{j-1} - Y_{j-1}^{[0]}) \quad (7.49)$$

should be used as a modified initial estimate. Before proceeding the reader should verify the stated formulae for the l.t.e. estimates and derive the scheme (7.49) for himself. The three-step implicit method

$$Y_j = \frac{1}{8}(-Y_{j-3} + 9Y_{j-1}) + \frac{3h}{8}\{-f(x_{j-2}, Y_{j-2}) + 2f(x_{j-1}, Y_{j-1})$$
$$+ f(x_j, Y_j)\} \qquad (7.50)$$

(*Hamming's method*) is often preferred to (7.48) because of its superior stability characteristics. Given that the l.t.e. for (7.50) is of the form $-\frac{1}{40} h^5 y^{(v)}(\xi_{j-2})$, if (7.50) is employed as a corrector using (7.47) to provide an initial estimate, the quantity $-\frac{9}{121}(Y_j - Y_j^{[0]})$ provides an estimate of this l.t.e.. Further

$$\tilde{Y}_j = Y_j^{[0]} + \frac{112}{121}\left(Y_{j-1} - Y_{j-1}^{[0]}\right)$$

may be used as a modified initial estimate.

There are available to us a number of l.m.m.'s which are classified as being Adams–Bashforth (explicit) and Adams–Moulton (implicit). We can therefore form a predictor-corrector method based on a compatible Adams-–Bashforth–Adams–Moulton pair. An example of such a method is provided by the four-step explicit method

$$Y_j = Y_{j-1} + \frac{h}{24}\{-9f(x_{j-4}, Y_{j-4}) + 37f(x_{j-3}, Y_{j-3})$$
$$- 59f(x_{j-2}, Y_{j-2}) + 55f(x_{j-1}, Y_{j-1})\} \qquad (7.51)$$

and the three-step implicit method

$$Y_j = Y_{j-1} + \frac{h}{24}\{f(x_{j-3}, Y_{j-3}) - 5f(x_{j-2}, Y_{j-2}) + 19f(x_{j-1}, Y_{j-1})$$
$$+ 9f(x_j, Y_j)\}. \qquad (7.52)$$

Given that the l.t.e.'s for (7.51) and (7.52) are of the form $\frac{251}{720} h^5 y^{(v)}(\eta_{j-4})$ and $-\frac{19}{720} h^5 y^{(v)}(\xi_{j-3})$ respectively, the reader is encouraged to derive estimates of these l.t.e.'s together with a scheme providing a modified initial estimate.

7.8 RUNGE–KUTTA METHODS

We now give an indication of the derivation and use of a new class of methods for the first order i.v.p.. Of the approaches we have discussed so far, only the predictor-corrector methods are generally useful. Euler's method requires a very small step length in order to achieve any reasonable accuracy and the extension of this method based on Taylor series has the serious disadvantage of requiring the explicit algebraic form of a number of successive derivatives of $y(x)$. The aim of the *Runge–Kutta* approach is to obtain greater accuracy without having to obtain these higher derivatives and (in their simplest form) to retain the considerable advantage of Euler's method in that the provision of starting values is avoided. This aim is achieved by evaluating $f(x, y)$ in (7.1) at appropriate points in the sub-interval $[x_{j-1}, x_j]$.

The simplest of the Runge–Kutta methods for stepping from x_{j-1} to x_j takes the form

$$Y_j = Y_{j-1} + w_1 k_1 + w_2 k_2 \tag{7.53}$$

where

$$k_1 = hf(x_{j-1}, Y_{j-1}); \quad k_2 = hf(x_{j-1} + \theta h, Y_{j-1} + \phi k_1)$$

and w_1, w_2, θ and ϕ are constants at our disposal. These constants are to be determined so that (7.53) agrees with as many terms as possible in the Taylor series expansion (7.21) with x_0 replaced by x_{j-1} and x by x_j. In order to proceed further we must expand the term $f(x_{j-1} + \theta h, Y_{j-1} + \phi k_1)$ which appears in the definition of k_2 about $x = x_{j-1}$ and $y = Y_{j-1}$ using Taylor's theorem. We do not here pursue the details (which may be found in Ortega and Poole, 1981) but state the result that when $f(x_{j-1} + \theta h, Y_{j-1} + \phi k_1)$ is expanded and (7.53) is compared with (7.21) it is found that the parameters w_1, w_2, θ and ϕ should be chosen to satisfy

$$\begin{aligned} w_1 + w_2 &= 1 \\ w_2 \theta &= \tfrac{1}{2} \\ w_2 \phi &= \tfrac{1}{2} \ . \end{aligned} \tag{7.54}$$

The system (7.54) gives three equations in four unknowns which means that one of the unknowns may be chosen arbitrarily and the other three then found. The choices $\theta = \dfrac{1}{2}, \dfrac{2}{3}$ and 1 give rise to the following Runge–Kutta methods.

(i) $\left(\theta = \dfrac{1}{2}\right)$ $Y_j = Y_{j-1} + hf\left(x_{j-1} + \dfrac{h}{2}, Y_{j-1} + \dfrac{h}{2} f(x_{j-1}, Y_{j-1})\right)$

(ii) $\left(\theta = \frac{2}{3}\right)$ $Y_j = Y_{j-1} + \frac{h}{4}\left\{f(x_{j-1}, Y_{j-1})\right.$

$$\left. + 3f\left(x_{j-1} + \frac{2h}{3}, Y_{j-1} + \frac{2h}{3} f(x_{j-1}, Y_{j-1})\right)\right\}$$

(iii) $(\theta = 1)$ $Y_j = Y_{j-1} + \frac{h}{2} \{f(x_{j-1}, Y_{j-1})$

$$+ f(x_{j-1} + h, Y_{j-1} + hf(x_{j-1}, Y_{j-1}))\}.$$

We have already met method (iii) albeit in a somewhat different guise. It corresponds to the use of Euler's rule to provide an initial value for the trapezoidal method (7.25) with only one iteration being performed.

EXAMPLE 7.9
We apply method (i) to the problem of Example 7.2. Choosing $h = 0.2$ we have

$Y_0 = 0.25$
$Y_1 = Y_0 + 0.2(Y_0 + 0.1Y_0^2)^2 = 0.25 + 0.2(0.25 + 0.1 \times 0.25^2)^2$
$\quad = 0.263133$
$Y_2 = Y_1 + 0.2(Y_1 + 0.1Y_1^2)^2 = 0.277719$
$Y_3 = Y_2 + 0.2(Y_2 + 0.1Y_2^2)^2 = 0.294013$
$Y_4 = Y_3 + 0.2(Y_3 + 0.1Y_3^2)^2 = 0.312333$
$Y_5 = Y_4 + 0.2(Y_4 + 0.1Y_4^2)^2 = 0.333081.$

The reader may care to formulate method (i) in terms of Euler's method (7.9) and the midpoint method (7.32).

The general Runge–Kutta method takes the form

$$Y_j = Y_{j-1} + \sum_{i=1}^{p} w_i k_i \tag{7.55}$$

where

$$k_i = hf\left(x_{j-1} + \theta_i h, Y_{j-1} + \sum_{l=1}^{i-1} \phi_{il} k_l\right) \quad i = 1, 2, \ldots, p.$$

As with (7.53), the available parameters $\{\theta_i : i = 1, 2, \ldots, p\}$, $\{w_i : i = 1, 2, \ldots, p\}$ and $\{\phi_{il} : l = 1, 2, \ldots, i-1; i = 1, 2, \ldots, p\}$ are chosen so that when all terms have been expanded about $x = x_{j-1}$ and $y = Y_{j-1}$ the result agrees with as many terms as possible in the Taylor series expansion (7.21). One of the most widely used Runge–Kutta methods is the scheme

$$Y_j = Y_{j-1} + \frac{1}{6}\{k_1 + 2k_2 + 2k_3 + k_4\} \qquad (7.56)$$

where

$$k_1 = hf(x_{j-1}, Y_{j-1})$$
$$k_2 = hf(x_{j-1} + h/2, Y_{j-1} + (1/2)k_1)$$
$$k_3 = hf(x_{j-1} + h/2, Y_{j-1} + (1/2)k_2)$$
$$k_4 = hf(x_{j-1} + h, Y_{j-1} + k_3) .$$

By choosing p large enough in (7.55) it is possible to derive one-step (and hence, self-starting) explicit Runge–Kutta methods corresponding to any desired number of terms in the expansion (7.21). Hence, for any k-step l.m.m. it is possible to obtain the starting values $Y_1, Y_2, \ldots, Y_{k-1}$ using a Runge–Kutta method with a l.t.e. proportional to the same power of h and the same derivative of $y(x)$ as the l.m.m. itself.

7.9 SECOND ORDER EQUATIONS

7.9.1 Introduction

We conclude this chapter with a brief introduction to some of the ways in which an approximation may be obtained to $y(x)$, the solution of the second order equation

$$y'' = g(x, y, y') \qquad (7.57)$$

(where $g(x, y, y')$ is known), with two given conditions. These conditions may both be given at the same point and take the form

$$y(a) = \alpha; \qquad y'(a) = \beta \qquad (7.58)$$

that is, the values of $y(x)$ and of its first derivative are specified at the point $x = a$. We refer to (7.57) and (7.58) as a *second order initial value problem* in which (7.58) forms a set of *initial conditions*. As with a first order i.v.p. we are usually interested in the solution to (7.57) and (7.58) on some interval $a \le x \le b$. The two conditions may, alternatively, take the form

$$y(a) = \alpha; \qquad y(b) = \beta \qquad (7.59)$$

that is, the value of $y(x)$ is specified at two points, $x = a$ and $x = b$. We refer to (7.57) and (7.59) as a *second order boundary value problem (b.v.p.)* and the conditions (7.59) as the *boundary conditions*. Here we are interested in the solution on $]a, b[$. (Although we shall not discuss the cases, the boundary conditions can also take a form in which $y'(x)$ or a linear combination of $y(x)$ and $y'(x)$ is specified at $x = a$ and $x = b$.) The question of the existence of a solution is beyond the scope of the present discussion and the interested reader is referred to Keller (1968).

As with the first order i.v.p., we are concerned with trying to set up schemes which will yield an approximation to $y(x)$ on a set of discrete points (which we shall assume to be equally spaced). In the case of a b.v.p. these points are contained in $]a, b[$.

At this point, we observe that it is possible to express a second order i.v.p. in a somewhat different form. Let us rename $y(x)$ as $y_1(x)$ and denote $y'(x)$ by $y_2(x)$. Then $y''(x) = y_2'(x)$ so that (7.57) may be written as the pair of *simultaneous first order equations*

$$y_1' = y_2$$
$$y_2' = g(x, y_1, y_2) \tag{7.60}$$

and the initial conditions (7.58) become

$$y_1(a) = \alpha; \quad y_2(a) = \beta . \tag{7.61}$$

If we now introduce a vector **y** with components y_1 and y_2, a vector **G** with components y_2 and $g(x, y_1, y_2)$ and a vector **s** with components α and β, we can write (7.60) as

$$\mathbf{y}' = \mathbf{G}(x, \mathbf{y}) \tag{7.62}$$

(where by $\mathbf{y}'$ we mean the vector with components y_1' and y_2') and (7.61) as

$$\mathbf{y}(a) = \mathbf{s} . \tag{7.63}$$

We refer to (7.62) and (7.63) as the vector form of (7.60) and (7.61) and note that with vectors replacing scalars, the statements are equivalent to (7.1) and (7.2).

The equations (7.60) are a special case of a pair of simultaneous first order equations which can be more generally written as

$$y_1' = g_1(x, y_1, y_2)$$
$$y_2' = g_2(x, y_1, y_2) \tag{7.64}$$

with the conditions (7.61). These can clearly be written in the form (7.62) with **G** having components $g_1(x, y_1, y_2)$ and $g_2(x, y_1, y_2)$. The extension to more than two simultaneous equations can be made easily.

In our discussion of the first order i.v.p. we made an appeal from time to time to the test problem (7.3), which we recall was $y' = \lambda y$ with the initial condition $y(a) = \alpha$. In the context of (7.64) we choose a test problem in which the functions $g_1(x, y_1, y_2)$ and $g_2(x, y_1, y_2)$ have no explicit dependence on x and are each linear in y_1 and y_2. That is, our test problem is of the form (7.64) with

$$g_1(x, y_1, y_2) = a_{11}y_1 + a_{12}y_2$$
$$g_2(x, y_1, y_2) = a_{21}y_1 + a_{22}y_2$$

where a_{11}, a_{12}, a_{21} and a_{22} are some chosen numbers. The vector **G** can therefore be written as

$$\mathbf{G} = A\mathbf{y}$$

where A is the 2×2 matrix with components $A_{ij} = a_{ij} : i, j = 1, 2$ so that our test problem becomes

$$\mathbf{y}' = A\mathbf{y} \tag{7.65}$$

subject to the initial conditions (7.63).

7.9.2 A pair of simultaneous first order equations

As we have just seen, a second order i.v.p. is a special case of a pair of simultaneous first order equations. All the methods that we have discussed in the earlier parts of this chapter for the solution of (7.1), (7.2) may be readily extended to (7.62), (7.63). Hence, Euler's method (7.9) may be written as

$$\mathbf{Y}_j = \mathbf{Y}_{j-1} + h\mathbf{G}(x_{j-1}, \mathbf{Y}_{j-1}) \tag{7.66}$$

where $\mathbf{Y}_j^T = (Y_{1,j}, Y_{2,j})$ is a vector of approximations to $(y_1(x_j), y_2(x_j))$, or, in component form, as

$$\begin{aligned} Y_{1,j} &= Y_{1,j-1} + hg_1(x_{j-1}, Y_{1,j-1}, Y_{2,j-1}) \\ Y_{2,j} &= Y_{2,j-1} + hg_2(x_{j-1}, Y_{1,j-1}, Y_{2,j-1}) \end{aligned} \quad j = 1, 2, \ldots$$

where $\mathbf{Y}_0^T = (\alpha, \beta)$.

EXAMPLE 7.10
Consider the problem

$$\begin{aligned} y'' &= y \quad x \geq 0 \\ y(0) &= 1; \quad y'(0) = -1 \end{aligned} \tag{7.67}$$

which has the solution $y(x) = e^{-x}$. Using the material of the previous subsection we rewrite (7.67) as

$$y'_1 = y_2 \qquad x \geq 0 \tag{7.68}$$
$$y'_2 = y_1$$
$$y_1(0) = 1; \quad y_2(0) = -1 \, .$$

Applying Euler's method (7.66) to (7.68) we obtain the following results with $h = 0.2$:

$$Y_{1,0} = 1; \quad Y_{2,0} = -1$$
$$Y_{1,1} = Y_{1,0} + hY_{2,0} = 1 + 0.2 \times (-1) = 0.8$$
$$Y_{2,1} = Y_{2,0} + hY_{1,0} = -1 + 0.2(1) = -0.8 \, .$$

Proceeding further we find $Y_{1,2} = 0.64 = -Y_{2,2}$, $Y_{1,3} = 0.512 = -Y_{2,3}$, $Y_{1,4} = 0.4096 = -Y_{2,4}$ and $Y_{1,5} = 0.32768 = -Y_{2,5}$. The values $Y_{1,1}$, $Y_{1,2}$, $Y_{1,3}$, $Y_{1,4}$ and $Y_{1,5}$ represent approximations to $y_1(x)$, and hence to the solution, $y(x)$, of (7.67) at the points $x = 0.2, 0.4, 0.6, 0.8$ and 1.0. On the other hand, the values $Y_{2,1}, Y_{2,2}, Y_{2,3}, Y_{2,4}$ and $Y_{2,5}$ are approximations to $y_2(x)$, and hence $y'(x)$, at the same points.

For the system (7.62) the trapezoidal method (7.25) becomes

$$\mathbf{Y}_j = \mathbf{Y}_{j-1} + \frac{h}{2} \{\mathbf{G}(x_{j-1}, \mathbf{Y}_{j-1}) + \mathbf{G}(x_j, \mathbf{Y}_j)\}$$

or, again in component form,

$$Y_{1,j} = Y_{1,j-1} + \frac{h}{2} \{g_1(x_{j-1}, Y_{1,j-1}, Y_{2,j-1}) + g_1(x_j, Y_{1,j}, Y_{2,j})\}$$
$$Y_{2,j} = Y_{2,j-1} + \frac{h}{2} \{g_2(x_{j-1}, Y_{1,j-1}, Y_{2,j-1}) + g_2(x_j, Y_{1,j}, Y_{2,j})\} \, .$$

This is a pair of simultaneous non-linear equations in $Y_{1,j}$ and $Y_{2,j}$ and iteration will involve the ideas of section 2.9. The midpoint method can be added to give the extension of the predictor-corrector method of section 7.5.

As a further example we can state the extension of the Runge–Kutta method (7.56) to the system (7.62) as

$$\mathbf{Y}_j = \mathbf{Y}_{j-1} + \frac{1}{6} \{\mathbf{k}_1 + 2\mathbf{k}_2 + 2\mathbf{k}_3 + \mathbf{k}_4\} \tag{7.69}$$

where

$$\mathbf{k}_1 = h\mathbf{G}(x_{j-1}, \mathbf{Y}_{j-1})$$
$$\mathbf{k}_2 = h\mathbf{G}\left(x_{j-1} + \frac{h}{2}, \mathbf{Y}_{j-1} + \frac{1}{2}\mathbf{k}_1\right)$$
$$\mathbf{k}_3 = h\mathbf{G}\left(x_{j-1} + \frac{h}{2}, \mathbf{Y}_{j-1} + \frac{1}{2}\mathbf{k}_2\right)$$
$$\mathbf{k}_4 = h\mathbf{G}(x_{j-1} + h, \mathbf{Y}_{j-1} + \mathbf{k}_3).$$

The reader is now encouraged to express (7.69) in component form.

7.9.3 A program for a system of first order ordinary differential equations

In Case Study 7.3 we give programs for the solution of a system of first order ordinary differential equations. The Runge–Kutta scheme (7.69) is used to advance from one step to the next. Note that the procedure *rungekutta* is able to handle any number of simultaneous equations. We have deliberately coded the Algol 68 version of this procedure in such a way that it is almost exactly the same as the procedure we would have written for a single equation. In order to accomplish this we have used the following devices:

(i) The vector-valued function **G** is represented by a *PROC* which returns a result of mode [ ] *REAL*.
(ii) Operators have been introduced to add two vectors and to multiply a vector by a scalar.
(iii) A parameter of mode *REF* [ ] [ ] *REAL* has been used in order to return the computed solution values (*yvalues*). Effectively this is a two-dimensional array although formally it is a one-dimensional array of one-dimensional arrays.

It is not possible to achieve this effect in the Pascal program because of the inability to do (i) and (ii) in Pascal.

We have used the procedure *rungekutta* to compute a solution to the second order initial value problem

$$y''(x) + 1001y'(x) + 1000y(x) = 0 \tag{7.70}$$

which has a solution of the form

$$y(x) = Ae^{-x} + Be^{-1000x}. \tag{7.71}$$

The initial conditions

$$y(0) = 1; \quad y'(0) = -1 \tag{7.72}$$

means that A and B take the values 1 and 0 respectively. Using (7.60) we

CASE STUDY 7.3

Algol 68 Program

```
BEGIN
  COMMENT
  a program to solve a system of first order initial value problems using a
  runge-kutta method
  COMMENT
  INT systemorder = 2 ;
  PROC g = ( REAL x , [] REAL yvalues ) [] REAL :
      ( yvalues [ 2 ] , - 1001.0 * yvalues [ 2 ] - 1000.0 * yvalues [ 1 ] ) ;
  OP + = ( [] REAL vector1 , vector2 ) [] REAL :
  BEGIN
    COMMENT
    an operator to add two one-dimensional arrays
    COMMENT
    INT l = LWB vector1 , u = UPB vector2 ;
    [ l : u ] REAL result ;
    FOR i FROM l TO u
    DO
       result [ i ] := vector1 [ i ] + vector2 [ i ]
    OD ;
    result
  END ;
  OP * = ( REAL multiplier , [] REAL vector ) [] REAL :
  BEGIN
    COMMENT
    an operator to multiply a one-dimensional array by a scalar
    COMMENT
    INT l = LWB vector , u = UPB vector ;
    [ l : u ] REAL result ;
    FOR i FROM l TO u
    DO
       result [ i ] := multiplier * vector [ i ]
    OD ;
    result
  END ;
  PROC rungekutta = ( PROC ( REAL , [] REAL ) [] REAL g ,
                      REAL lowerlimit , upperlimit ,
                      REF [] REAL initialcondition , xvalues ,
                      REF [ ] [ ] REAL yvalues ) VOID :
  BEGIN
    COMMENT
    this procedure computes approximations to a system of first order
    initial value problems at a set of points using a runge-kutta method
    COMMENT
    INT numberofsteps = UPB xvalues ;
    REAL steplength = ( upperlimit - lowerlimit ) / numberofsteps ;
    REAL halfsteplength = steplength * 0.5 ;
    REAL steplengthover6 = steplength / 6.0 ;
    xvalues [ 0 ] := lowerlimit ;
    yvalues [ 0 ] := initialcondition ;
    REAL halfwaypoint ;
    INT systemorder = UPB initialcondition ;
    [ 1 : systemorder ] REAL k1 , k2 , k3 , k4 ;
    COMMENT
    step along the points
    COMMENT
    FOR i TO numberofsteps
    DO
       xvalues [ i ] := xvalues [ i - 1 ] + steplength ;
       halfwaypoint := xvalues [ i - 1 ] + halfsteplength ;
       k1 := g ( xvalues [ i - 1 ] , yvalues [ i - 1 ] ) ;
       k2 := g ( halfwaypoint , yvalues [ i - 1 ] + halfsteplength * k1 ) ;
       k3 := g ( halfwaypoint , yvalues [ i - 1 ] + halfsteplength * k2 ) ;
       k4 := g ( xvalues [ i ] , yvalues [ i - 1 ] + steplength * k3 ) ;
```

```
            COMMENT
            find the next set of solution values
            COMMENT
            yvalues [ i ] := yvalues [ i - 1 ]
                            + steplengthover6 * ( k1 + 2.0 * k2 + 2.0 * k3 + k4 )
   OD
END ;
REAL lower , upper ;
INT numberofsteps ;
[ 1 : systemorder ] REAL initialcondition ;
read ( ( lower , upper , numberofsteps , initialcondition ) ) ;
print ( ( newline , newline ,
          "lower limit of range              " , lower , newline ,
          "upper limit of range              " , upper , newline ,
          "number of steps                   " , numberofsteps , newline ,
          newline , newline , "initial conditions" , newline ,
          initialcondition , newline , newline ) ) ;
[ 0 : numberofsteps ] REAL xvalues ;
[ 0 : numberofsteps ] [ 1 : systemorder ] REAL yvalues ;
rungekutta ( g , lower , upper , initialcondition , xvalues , yvalues ) ;
print ( ( newline , newline ,
          "    x-value              solution values" , newline ) ) ;
FOR i FROM 0 TO numberofsteps
DO
   print ( ( newline , xvalues [ i ] , "        " , yvalues [ i ] ) )
OD ;
print ( newline )
END
```

Pascal Program

```
PROGRAM exampleofrungekutta(input, output);

    { a program to solve a system of first order initial value problems
      using a runge-kutta method  }

    CONST
       systemorder=2;
       maxnumberofsteps=100;

    TYPE
       ordertype=1..systemorder;
       systemtype=ARRAY[ ordertype ] OF real;
       steptype=0..maxnumberofsteps;
       xvaluestype=ARRAY[ steptype ] OF real;
       yvaluestype=ARRAY[ steptype ] OF systemtype;

    VAR
       lower, upper:real;
       numberofsteps:steptype;
       initialcondition:systemtype;
       xvalues:xvaluestype;
       yvalues:yvaluestype;
       i:steptype;
       j:ordertype;

    PROCEDURE g(x:real; VAR y:systemtype; VAR k:systemtype);
    BEGIN
       k[1]:= y[2];
       k[2]:= -1001.0*y[2] - 1000.0*y[1]
    END { g } ;

    PROCEDURE rungekutta(PROCEDURE g(x:real; VAR y:systemtype; VAR k:systemtype);
                         lowerlimit, upperlimit:real;
                         VAR initialcondition:systemtype;
                         numberofsteps:steptype;
                         VAR xvalues:xvaluestype;
                         VAR yvalues:yvaluestype);
```

```
        { this procedure computes approximations to a system of
          first order initial value problems at a set of points
          using a runge-kutta method }
        VAR
            step, halfstep, halfwaypoint, stepby6:real;
            y, k1, k2, k3, k4:systemtype;
            i:steptype;
            j:ordertype;

        BEGIN
            step:= (upperlimit - lowerlimit)/numberofsteps;
            halfstep:= step*0.5;
            stepby6:= step/6.0;
            xvalues[0]:= lowerlimit;
            yvalues[0]:= initialcondition;
            { step along the points }
            FOR i:= 1 TO numberofsteps DO
            BEGIN
                xvalues[i]:= xvalues[i - 1] + step;
                halfwaypoint:= xvalues[i - 1] + halfstep;
                FOR j:= 1 TO systemorder DO
                    y[j]:= yvalues[i - 1, j];
                g(xvalues[i - 1], y, k1);
                FOR j:= 1 TO systemorder DO
                    y[j]:= yvalues[i - 1, j] + halfstep*k1[j];
                g(halfwaypoint, y, k2);
                FOR j:= 1 TO systemorder DO
                    y[j]:= yvalues[i - 1, j] + halfstep*k2[j];
                g(halfwaypoint, y, k3);
                FOR j:= 1 TO systemorder DO
                    y[j]:= yvalues[i - 1, j] + step*k3[j];
                g(xvalues[i], y, k4);
                { find the next set of solution values }
                FOR j:= 1 TO systemorder DO
                    yvalues[i, j]:= yvalues[i - 1, j] +
                              stepby6*(k1[j] + 2.0*k2[j] + 2.0*k3[j] + k4[j])
            END
        END { rungekutta } ;

BEGIN
    read(lower, upper, numberofsteps);
    FOR j:= 1 TO systemorder DO
        read(initialcondition[j]);
    writeln;
    writeln;
    writeln('lower limit of range            ', lower);
    writeln('upper limit of range            ', upper);
    writeln('number of steps                 ', numberofsteps);
    writeln;
    writeln;
    writeln('initial conditions');
    FOR j:= 1 TO systemorder DO
        write(initialcondition[j]);
    writeln;
    writeln;

    rungekutta(g, lower, upper, initialcondition,
               numberofsteps, xvalues, yvalues);

    writeln;
    writeln;
    writeln('    x-value', ' ':15, 'solution values');
    writeln;
    FOR i:= 0 TO numberofsteps DO
    BEGIN
        write(xvalues[i], ' ':8);
        FOR j:= 1 TO systemorder DO
            write(yvalues[i, j]);
        writeln
    END
END { exampleofrungekutta } .
```

Sec. 7.9] SECOND ORDER EQUATIONS 339

Sample Data

```
0.0    0.05
25
1.0   -1.0
```

Sample Output

```
lower limit of range                    0.00000e  0
upper limit of range                    5.00000e -2
number of steps                              25


initial conditions
  1.00000e  0 -1.00000e  0


       x-value                        solution values

     0.00000e  0              1.00000e  0 -1.00000e  0
     2.00000e -3              9.98002e -1 -9.98002e -1
     4.00000e -3              9.96008e -1 -9.96008e -1
     6.00000e -3              9.94018e -1 -9.94018e -1
     8.00000e -3              9.92032e -1 -9.92032e -1
     9.99999e -3              9.90049e -1 -9.90049e -1
     1.20000e -2              9.88071e -1 -9.88071e -1
     1.40000e -2              9.86097e -1 -9.86097e -1
     1.60000e -2              9.84127e -1 -9.84127e -1
     1.80000e -2              9.82160e -1 -9.82160e -1
     2.00000e -2              9.80198e -1 -9.80198e -1
     2.20000e -2              9.78240e -1 -9.78240e -1
     2.40000e -2              9.76285e -1 -9.76285e -1
     2.60000e -2              9.74334e -1 -9.74334e -1
     2.80000e -2              9.72387e -1 -9.72387e -1
     3.00000e -2              9.70445e -1 -9.70445e -1
     3.20000e -2              9.68506e -1 -9.68506e -1
     3.40000e -2              9.66570e -1 -9.66571e -1
     3.59999e -2              9.64639e -1 -9.64639e -1
     3.79999e -2              9.62712e -1 -9.62712e -1
     3.99999e -2              9.60788e -1 -9.60788e -1
     4.19999e -2              9.58868e -1 -9.58868e -1
     4.39999e -2              9.56953e -1 -9.56953e -1
     4.59999e -2              9.55040e -1 -9.55040e -1
     4.79999e -2              9.53132e -1 -9.53133e -1
     4.99999e -2              9.51228e -1 -9.51228e -1
```

express (7.70) in the form

$$y'_1 = y_2$$
$$y'_2 = -1000y_1 - 1001y_2 \qquad (7.73)$$

and the initial conditions (7.72) become

$$y_1(0) = 1; \quad y_2(0) = -1 \qquad (7.74)$$

Solving (7.73) subject to the conditions (7.74), we obtain a sequence of approximations which behave like the true solution $y(x) = e^{-x}$ (see the first column of the solution values in the sample output). For example, the program produces an approximation to $y(0.05)$ of 0.951228. This is very close to the value of $e^{-0.05}$. However, we have used a very small step length. Changing the data so that a step length of 0.005 (instead of 0.002) is used results in an approximation to $y(0.05)$ of -14.2362 being returned. We must now try to discover why this behaviour occurs.

7.9.4 Stiffness

The general solution (7.71) of (7.70) has two components, Ae^{-x} and Be^{-1000x}. Clearly the latter term decays extremely rapidly and soon its contribution to the solution becomes negligible. When computing an approximate solution using the transformed version (7.73) we would expect to need to use a small step length near to the origin in order to obtain reasonable results. However, once the effect of the component Be^{-1000x} becomes insignificant, we would hope to be able to use a much larger interval between successive grid points. Unfortunately this is not so; stability considerations mean that the maximum permitted step length for the entire range of interest is governed by the most rapidly changing component of the solution. Here, it can be shown (see Gear, 1971) that we must have $h < 0.0028$. Systems which are severely restricted in this way are said to be *stiff*.

A full discussion of stiff problems (and of methods for their solution) is well beyond the scope of this book but we make a few observations with respect to the test problem (7.65). The stiffness of a problem of this form is governed by the *eigenvalues* of the matrix A. Consider again problem (7.70) which can be written in the form (7.65) with

$$A = \begin{pmatrix} 0 & 1 \\ -1000 & -1001 \end{pmatrix}. \tag{7.75}$$

Now

$$\begin{pmatrix} 0 & 1 \\ -1000 & -1001 \end{pmatrix} \begin{pmatrix} 1 \\ -1 \end{pmatrix} = \begin{pmatrix} -1 \\ 1 \end{pmatrix} = -\begin{pmatrix} 1 \\ -1 \end{pmatrix}$$

and

$$\begin{pmatrix} 0 & 1 \\ -1000 & -1001 \end{pmatrix} \begin{pmatrix} -\dfrac{1}{1000} \\ 1 \end{pmatrix} = \begin{pmatrix} 1 \\ -1000 \end{pmatrix} = -1000 \begin{pmatrix} -\dfrac{1}{1000} \\ 1 \end{pmatrix}$$

that is, if A is of the form (7.75), then

$$A\mathbf{x} = \lambda \mathbf{x}$$

if $\lambda = -1$ and $\mathbf{x}^T = (1, -1)$ or $\lambda = -1000$ and $\mathbf{x}^T = (-1/1000, 1)$. The values -1 and -1000 are said to be the eigenvalues of the matrix (7.75) and $(1, -1)^T$ and $(-1/1000, 1)^T$ are known as the associated *eigenvectors*. If, in the test problem (7.65), the eigenvalues of the matrix A are negative and widely separated (and we would say that the values -1 and -1000 satisfy this condition) then the problem is said to be stiff. The ratio of the moduli of the maximum and minimum absolute eigenvalues of A, the *stiffness ratio*, is used as a measure of the stiffness of a problem. For (7.70) the stiffness ratio is 1000 and this is considered to be fairly large. However, problems possessing stiffness ratios far greater than this are not uncommon and clearly this is an extremely important phenomenon. The reader may care to consult Lambert (1973) for further details.

7.9.5 The second order boundary value problem
We conclude this section by setting down some elementary ideas about the second order b.v.p.. Referring to (7.57), suppose $g(x, y, y')$ is independent of y' and is linear in y. The problem may therefore be expressed in the form

$$y'' = q(x)y + r(x) \tag{7.76}$$

subject to the boundary conditions (7.59), where $q(x)$ and $r(x)$ are given functions which we suppose can be evaluated for any desired values of x. Assume that $[a, b]$ has been subdivided using a set of equally spaced points $\{x_j : j = 0, 1, \ldots, N\}$, distance h apart. Then, from (7.76) we have

$$y''(x_j) = q(x_j) y(x_j) + r(x_j) . \tag{7.77}$$

We now make use of finite difference techniques to approximate $y''(x_j)$ (see section 5.3). From Taylor's theorem (Theorem 1.1) we have

$$y(x_{j+1}) = y(x_j) + hy'(x_j) + \frac{h^2}{2} y''(x_j) + \frac{h^3}{6} y'''(x_j)$$
$$+ \frac{h^4}{24} y^{(iv)} (\xi_1) \qquad \xi_1 \in]x_j, x_{j+1}[$$

$$y(x_{j-1}) = y(x_j) - hy'(x_j) + \frac{h^2}{2} y''(x_j) - \frac{h^3}{6} y'''(x_j)$$
$$+ \frac{h^4}{24} y^{(iv)} (\xi_2) \qquad \xi_2 \in]x_{j-1}, x_j[$$

so that, ignoring the terms involving h^4, we have

$$\frac{y(x_{j+1}) - 2y(x_j) + y(x_{j-1})}{h^2} \simeq y''(x_j).$$

Substituting this result into (7.77), multiplying throughout by h^2 and rearranging, we have

$$Y_{j+1} - (2 + h^2 q_j) Y_j + Y_{j-1} = h^2 r_j \qquad j = 1, 2, \ldots, N-1 \qquad (7.78)$$

where, as usual, Y_j denotes an approximation to $y(x_j)$ and q_j and r_j represent the values $q(x_j)$ and $r(x_j)$ respectively. The system (7.78) gives $N-1$ equations in the $N-1$ unknowns $Y_1, Y_2, \ldots, Y_{N-1}$. In detail we have

$$Y_2 - (2 + h^2 q_1) Y_1 + \alpha = h^2 r_1$$
$$Y_3 - (2 + h^2 q_2) Y_2 + Y_1 = h^2 r_2$$
$$Y_4 - (2 + h^2 q_3) Y_3 + Y_2 = h^2 r_3$$

.

.

.

$$Y_{N-1} - (2 + h^2 q_{N-2}) Y_{N-2} + Y_{N-3} = h^2 r_{N-2}$$
$$\beta \quad - (2 + h^2 q_{N-1}) Y_{N-1} + Y_{N-2} = h^2 r_{N-1}$$

where, in forming the first and last equations, we have made use of the boundary conditions (7.59). Clearly we can express this system in matrix form as

$$A\mathbf{Y} = \mathbf{b}$$

where $\mathbf{Y}^T = (Y_1, Y_2, \ldots, Y_{N-1})$, $\mathbf{b}^T = (h^2 r_1 - \alpha, h^2 r_2, h^2 r_3, \ldots, h^2 r_{N-2}, h^2 r_{N-1} - \beta)$ and

Sec. 7.9] SECOND ORDER EQUATIONS 343

$$A = \begin{bmatrix} -(2+h^2q_1) & 1 & & & & \\ 1 & -(2+h^2q_2) & 1 & & 0 & \\ & \cdot & \cdot & \cdot & & \\ & \cdot & \cdot & \cdot & & \\ & & \cdot & \cdot & \cdot & \\ & 0 & & 1 & -(2+h^2q_{N-2}) & 1 \\ & & & & 1 & -(2+h^2q_{N-1}) \end{bmatrix}$$

We observe that the coefficient matrix A is tridiagonal.

EXAMPLE 7.11
Consider the problem

$$y'' = y + x \sinh(1)$$
$$y(0) = 0; \quad y(1) = 0$$

for which the solution is $y(x) = \sinh(x) - x \sinh(1)$. Choosing $h = 0.2$, we have the following system of equations defining approximations to $y(0.2)$, $y(0.4)$, $y(0.6)$ and $y(0.8)$

$$\begin{bmatrix} -2.04 & 1 & 0 & 0 \\ 1 & -2.04 & 1 & 0 \\ 0 & 1 & -2.04 & 1 \\ 0 & 0 & 1 & -2.04 \end{bmatrix} \begin{bmatrix} Y_1 \\ Y_2 \\ Y_3 \\ Y_4 \end{bmatrix} = \begin{bmatrix} 0.009\,402 \\ 0.018\,804 \\ 0.028\,206 \\ 0.037\,608 \end{bmatrix}.$$

We can now solve this system using the techniques of subsection 3.6.3. Using the notation introduced there we have $\alpha_1 = -2.04$, $\gamma_1 = 1/-2.04 = -0.490196$, $\alpha_2 = -2.04 + 0.490196 = -1.549804$, $\gamma_2 = 1/-1.549804 = -0.645243$, $\alpha_3 = -2.04 + 0.645243 = -1.394757$, $\gamma_3 = 1/-1.394757 = -0.716971$, $\alpha_4 = -2.04 + 0.716971 = -1.323029$. Hence, the coefficient matrix may be decomposed as

$$\begin{bmatrix} -2.04 & 0 & 0 & 0 \\ 1 & -1.549804 & 0 & 0 \\ 0 & 1 & -1.394757 & 0 \\ 0 & 0 & 1 & -1.323029 \end{bmatrix} \begin{bmatrix} 1 & -0.490196 & 0 & 0 \\ 0 & 1 & -0.645243 & 0 \\ 0 & 0 & 1 & -0.716971 \\ 0 & 0 & 0 & 1 \end{bmatrix}$$

Using forward substitution (and using z to denote the intermediate solution vector) we have

$$z_1 = 0.009402/-2.04 = -0.004609$$
$$z_2 = (0.018804 + 0.004609)/-1.549804 = -0.015107$$
$$z_3 = (0.028206 + 0.015107)/-1.394757 = -0.031054$$
$$z_4 = (0.037608 + 0.031054)/-1.323029 = -0.051898.$$

Finally, back substitution gives

$$Y_4 = -0.051898$$
$$Y_3 = -0.031054 - 0.051898(0.716971) = -0.068263$$
$$Y_2 = -0.015107 - 0.068263(0.645243) = -0.059153$$
$$Y_1 = -0.004609 - 0.059153(0.490196) = -0.033606.$$

In subsection 3.6.3 we noted that the decomposition employed in the above example is stable provided the initial coefficient matrix is diagonally dominant. Here, the coefficient matrix A defined by (7.79) will be diagonally dominant if $q(x)$ is everywhere positive.

In this subsection we have considered only a very restricted class of boundary value problems. Methods for solving the general problem (7.57) subject to the boundary conditions (7.59) are often based on the use of a *shooting* technique. Here some guess at a value for $y'(a)$ is made and a second order initial value problem solved. The value for Y_N so obtained is compared with the known value β. If $|Y_N - \beta|$ is, in some sense, large a new estimate of $y'(a)$ is made and the process repeated. Continuing in this way we stop when $|Y_N - \beta| < \varepsilon$, where ε is some chosen accuracy criterion. For further details of this technique see Conte and de Boor (1980).

7.10 AVAILABLE ALGORITHMS

A large amount of effort has been put into the development of efficient, general purpose software for the solution of ordinary differential equations. If we consider initial value problems only, the more sophisticated codes are of the variable step/variable order type; that is, the codes allow both the interval length and the numerical scheme employed to alter as progress is made along the domain of interest in order to ensure that approximate solutions are returned which are correct to the routine user's specified accuracy tolerance. Such codes will need to have some mechanism for deciding whether to decrease the interval length or to use a method which is potentially more accurate when difficulties are encountered. Routines of this form are available in the NAG library based on the use of Adams formulae. Other facilities available include routines based on Runge–Kutta formulae for integrating over a single step or an entire range, and routines specifically designed to deal with stiff problems. It is unlikely that a single call of any of these codes will be sufficient, particularly if the user is unaware of the stability and/or stiffness properties of his problem. The NAG d02 chapter introduction gives advice on how to use the routines to derive preliminary results indicating stability and stiffness characteristics. From this analysis, the appropriate routine may be selected, where by appropriate we mean that which is likely to give the most acceptable results in the most efficient manner over the entire range of interest.

EXERCISES

7.1 Consider the first order initial value problem

$$y' = y(x) + \cos(x) - \sin(x) \qquad x \geq 0$$
$$y(0) = 1 \ .$$

Use Euler's method with a step length of $\pi/6$ to obtain approximations to $y(\pi/6)$, $y(\pi/3)$ and $y(\pi/2)$. (The values of $y(x)$ at these points are 2.18809, 3.71568 and 5.81048.)

7.2 Using the first four terms in the Taylor series expansion for $y(x)$ about the point $x = 0$, obtain an approximation to $y(0.2)$, where $y(x)$ is the solution of the initial value problem given in Exercise 7.1.

7.3 Find approximations to $y(0.2)$ and $y(0.3)$, where $y(x)$ satisfies

$$y' = 1 - 2xy \qquad x \geq 0$$
$$y(0) = 0$$

using the midpoint rule–trapezium rule predictor-corrector pair. Choose a step length of $h = 0.1$ and use $Y_1 = 0.0993$ (obtained from the Taylor series expansion of $y(x)$) as a starting value. Correct until two successive iterates agree to four decimal places.

7.4 Derive a modified initial estimate for the Adams–Bashforth–Adams––Moulton predictor-corrector pair (7.51) and (7.52) of section 7.7.

7.5 Use the Runge–Kutta method

$$Y_j = Y_{j-1} + \frac{1}{4} k_1 + \frac{3}{4} k_2$$

$$k_1 = hf(x_{j-1}, Y_{j-1})$$

$$k_2 = hf\left(x_{j-1} + \frac{2h}{3}, Y_{j-1} + \frac{2}{3}k_1\right)$$

introduced in section 7.8 to obtain approximations to the solution of the problem

$$y' = x + y \quad x \geq 0$$
$$y(0) = 1$$

at $x = 0.25, 0.5, 0.75$ and 1.0 using a step length of $h = 0.25$.

7.6 Given the finite difference approximation

$$y'(x_j) \simeq \frac{y(x_{j+1}) - y(x_{j-1})}{2h}$$

derive a system of equations which defines solutions to the boundary value problem

$$y'' = p(x)y' + q(x)y + r(x) \quad x \in [a, b]$$
$$y(a) = \alpha; \quad y(b) = \beta$$

at $x = a + h, a + 2h, \ldots, a + (N-1)h$, where $h = (b-a)/N$. Given that

$$q(x) \geq 0$$
$$|p(x)| \leq p^* \quad x \in [a, b]$$

and that the step length is chosen so that $h < 2/p^*$, show that the coefficient matrix in your system of equations is strictly diagonally dominant.

7.7 Write a program to solve a system of first order initial value problems using Euler's method. Obtain results which can be directly compared with the sample output of Case Study 7.3.

7.8 Write a program to solve a first order initial value problem using the midpoint rule–trapezium rule predictor-corrector pair. Use the Runge–Kutta method (iii) of section 7.8 to provide a value for Y_1 and the modified initial estimate (7.40) at each subsequent step. Use your program to solve the problem

$$y' = -x/(3x + 2y) \quad x \in [1, 2]$$
$$y(1) = 1$$

with $h = 0.1$. (The solution is $y(x) = \dfrac{2}{x^2} - x$.)

7.9 Write a program to solve the second order linear boundary value problem (7.77) using the method based on the replacement of $y''(x_j)$ by a finite difference approximation. (Use the code of Case Study 3.4 to solve the system of equations.) Test your program using the problem

$$y'' = 2xy - 4x^3 + 4 \quad x \in [0, 1]$$
$$y(0) = 0; \quad y(1) = 2.$$

(The solution is $y(x) = 2x^2$.)

Solutions to Exercises

1.1 $I_n = \int_0^{\pi/2} \sin^n(x)\, dx = \int_0^{\pi/2} \sin(x) \sin^{n-1}(x)\, dx$

$= [-\cos(x) \sin^{n-1}(x)]_0^{\pi/2} + \int_0^{\pi/2} \cos(x)\, (n-1)\sin^{n-2}(x)\cos(x)\, dx$

$= (n-1) \int_0^{\pi/2} \cos^2(x) \sin^{n-2}(x)\, dx$

$= (n-1) \int_0^{\pi/2} (1 - \sin^2(x)) \sin^{n-2}(x)\, dx$

$= (n-1) \int_0^{\pi/2} \sin^{n-2}(x)\, dx - (n-1) \int_0^{\pi/2} \sin^n(x)\, dx$

$= (n-1) I_{n-2} - (n-1) I_n$.

Hence

$$I_n = \frac{n-1}{n} I_{n-2}.$$

Starting with $\pi \approx 3.1416$ we have $I_0 \approx 1.5708$. Then $I_2 = \frac{1}{2} I_0 = 0.7854$, $I_4 = \frac{3}{4} I_2 = 0.58905$, $I_6 = \frac{5}{6} I_4 = 0.49088$. Generating the odd terms we have $I_1 = 1$, $I_3 = \frac{2}{3} I_1 = 0.66667$, $I_5 = \frac{4}{5} I_3 = 0.53333$, $I_7 = \frac{6}{7} I_5 = 0.45714$.

1.2 Let $A = (n-1)x$ and $B = x$. Then

$$\cos(A+B) + \cos(A-B) = \cos((n-1)x + x) + \cos((n-1)x - x)$$
$$= \cos(nx) + \cos((n-2)x)$$
$$2\cos(A)\cos(B) = \cos((n-1)x)\cos(x) .$$

Hence
$$y_n + y_{n-2} = 2\cos(x) y_{n-1}$$
or
$$y_n = 2\cos(x) y_{n-1} - y_{n-2} .$$

Now $y_0 = 1$, $y_1 = \sqrt{3}/2 \approx 1.73205/2 = 0.866025$. Hence

$$y_2 = 2 \times 0.866025 \times 0.866025 - 1 = 0.5$$
$$y_3 = 2 \times 0.866025 \times 0.5 - 0.866025 = 0$$
$$y_4 = 2 \times 0.866025 \times 0 - 0.5 = -0.5$$
$$y_5 = 2 \times 0.866025 \times (-0.5) - 0 = -0.866025$$
$$y_6 = 2 \times 0.866025 \times (-0.866025) + 0.5 = -1.$$

1.3
$$(3001 x_{n-1} - 1000 x_{n-2})/3 = \left(3001 \frac{1}{3^{n-2}} - 1000 \frac{1}{3^{n-3}}\right)/3$$
$$= \left(3001 \frac{1}{3^{n-2}} - 3000 \frac{1}{3^{n-2}}\right)/3$$
$$= \left(\frac{1}{3^{n-2}}\right)/3 = \frac{1}{3^{n-1}} = x_n.$$

Starting with $x_0 = 3$, $x_1 = 1$ we have

$$x_2 = (3001 x_1 - 1000 x_0)/3 = (3001 - 3000)/3 = 0.333333$$
$$x_3 = (3001 \times 0.333333 - 1000 \times 1)/3 = 0.110778$$
$$x_4 = (3001 \times 0.110778 - 1000 \times 0.333333)/3 = -0.296074$$
$$x_5 = (3001 \times -0.296074 - 1000 \times 0.110778)/3 = -333.099.$$

In forming x_2 we incur a rounding error of $\frac{1}{3000000}$. This error is multiplied by 3001 in the calculation of x_3 and in the calculation of x_4 we multiply the error in x_3 by 3001 and the error in x_2 by 1000. The accumulation of error, therefore, completely swamps the values we are trying to estimate.

1.4 Using the iteration

$$x_n = x_{n-1} - \tfrac{1}{3} \frac{(x_{n-1} - 1)(x_{n-1} + 2)}{(x_{n-1} + 1)}$$

and the starting value $x_0 = 0.9$ we have $x_1 = 0.950877$; $x_2 = 0.975645$; $x_3 = 0.987872$; $x_4 = 0.993949$; $x_5 = 0.996971$; $x_6 = 0.998489$; $x_7 = 0.999245$.

Using the iteration

$$x_n = x_{n-1} - \tfrac{2}{3}\frac{(x_{n-1}-1)(x_{n-1}+2)}{(x_{n-1}+1)}$$

and the same starting value we have $x_1 = 1.001754$; $x_2 = 1.000001$.

1.5 $e^1 \simeq 1 + 1 + 0.5 + 0.16667 + 0.04167 + 0.00833 + 0.00139 + 0.00020 + 0.00002 = 2.71828$.

The errors introduced are

  (i) rounding errors from each term
  (ii) a truncation error from ignoring $1/9! \, e^\xi$.

That we have e^1 correct to five decimal places is due to the fact that not all errors are of the same sign and the net effect is that they tend to cancel each other out.

2.1 (i) $\phi'(x) = -1/x$ so that near the root $|\phi'(x)| > 1$.
  (ii) $\phi'(x) = -e^{-x}$ so that near the root $|\phi'(x)| < 1$.
  (iii) $\phi'(x) = (1 - e^{-x})/2$ so that near the root $|\phi'(x)| < 1$.

Hence (ii) and (iii) will converge but (i) will not. Now $e^{-0.55} \sim 0.58$ so that $(1 - e^{-0.55})/2 \simeq 0.21$ and we therefore would expect (iii) to give the fastest rate of convergence. For (ii), $\phi'(x) < 0$ and so we expect oscillatory convergence but for (iii), $\phi'(x) > 0$ and we expect monotonic convergence.

If $\phi(x) = (kx + e^{-x})/(k+1)$ then $\phi'(x) = (k - e^{-x})/(k+1)$ so for a fast rate of convergence set $k \simeq e^{-0.55}$.

See Table S1 for details of the iterates.

2.2 $\phi(x) = (1-k)x + k(x^3+1)/3$ so that $\phi'(x) = 1 - k + kx^2$.

  (i) If $x = 0.3$, $\phi'(x) = 1 - k + 0.09k = 1 - 0.91k$ and so $|\phi'(x)| < 1$ if $0 < k < 2.1978$. Choose $k = 1$.
  (ii) If $x = \tfrac{3}{2}$, $\phi'(x) = 1 - k + k\tfrac{9}{4} = 1 + \tfrac{5}{4}k$ and so $|\phi'(x)| < 1$ if $-\tfrac{8}{5} < k < 0$. Choose $k = -1$.
  (iii) If $x = -2$, $\phi'(x) = 1 - k + 4k = 1 + 3k$ and so $|\phi'(x)| < 1$ if $-\tfrac{2}{3} < k < 0$. Choose $k = -0.5$.

See Table S2.

Table S1

	$\phi(x) = -\ln(x)$	$\phi(x) = e^{-x}$	$\phi(x) = (x + e^{-x})/2$	$\phi(x) = (kx + e^{-x})/(k+1)$ $k = e^{-0.55}$
x_0	0.450 000	0.450 000	0.450 000	0.450 000
x_1	0.798 508	0.637 628	0.543 814	0.568 982
x_2	0.225 011	0.528 544	0.562 172	0.567 155
x_3	1.491 61	0.589 462	0.566 071	0.567 143
x_4	$-0.399\ 854$	0.554 625	0.566 911	0.567 143
x_5		0.574 287	0.567 093	
x_6		0.563 106	0.567 132	
x_7		0.569 438	0.567 141	
x_8		0.565 843	0.567 143	
x_9		0.567 881		
x_{10}		0.566 725		
x_{11}		0.567 380		
x_{12}		0.567 009		
x_{13}		0.567 219		
x_{14}		0.567 100		
x_{15}		0.567 168		
x_{16}		0.567 129		
x_{17}		0.567 151		
x_{18}		0.567 139		
x_{19}		0.567 146		
x_{20}		0.567 142		

Table S2

k	Start value	Final value	Number of iterations
1.0	0.300 000	0.347 296	6
-1.0	1.500 00	1.532 08	8
-0.5	$-2.000\ 00$	$-1.879\ 39$	8

2.3 See Table S3. For (i) $0.299306^2 = 0.089584$; $0.110490^2 = 0.012208$; $0.011356^2 = 0.000129$; $0.000128^2 = 0.000000$ and since the ratios $0.089584/0.110490$, $0.012208/0.011356$ and $0.000129/0.000128$ are all roughly the same we conclude that convergence is quadratic. For (ii)

Table S3

(i) $e^{2x} - 3 = 0$		(ii) $e^{-x} - \sinh(x) = 0$	
Iterate	|Error|	Iterate	|Error|
0.250 000	0.299 306	0.250 000	0.299 306
0.659 796	0.110 490	0.540 677	0.008 629
0.560 662	0.011 356	0.549 306	0.000 000
0.549 434	0.000 128	0.549 306	0.000 000
0.549 306	0.000 000		
0.549 306	0.000 000		

convergence is clearly much faster. In fact, Newton–Raphson iteration gives third order convergence for (ii) but in order to demonstrate this numerically we would have to use higher precision arithmetic.

2.4 For $f(x) = 1/x^2 - a$ we have $f'(x) = -2/x^3$ so that Newton–Raphson iteration is $x_n = x_{n-1} - \left(\dfrac{1}{x_{n-1}^2} - a\right)/(-2/x_{n-1}^3) = x_{n-1}(3 - ax_{n-1}^2)/2$.

For $f(x) = \dfrac{1}{x^m} - a$ we have $f'(x) = -m/x^{m+1}$ so that Newton–Raphson iteration is $x_n = x_{n-1} - \left(\dfrac{1}{x_{n-1}^m} - a\right)/(-m/x_{n-1}^{m+1}) = x_{n-1} + \dfrac{1}{m}(x_{n-1} - ax_{n-1}^{m+1}) = x_{n-1}(m+1 - ax_{n-1}^m)/m$. If $m = 3$, $x_n = x_{n-1}(4 - ax_{n-1}^3)/3$ and with $x_0 = 1.0$ we obtain $\sqrt[3]{\left(\dfrac{1}{0.5}\right)} = \sqrt[3]{2} \simeq 1.25992$ after 4 iterations.

2.5 In the notation of subsection 2.8.4 we have $a_5 = 1$, $a_4 = -5$, $a_3 = 11$, $a_2 = -15$, $a_1 = 7$, $a_0 = -1$, $m = 5$, $p = -3$ and $q = 1$. Hence, $f(x) = (x^2 - 3x + 1)(c_3 x^3 + c_2 x^2 + c_1 x + c_0) + Rx + S$ where $c_3 = 1$, $c_2 = -5 - (-3)1 = -2$, $c_1 = 11 - (-3)(-2) - 1(1) = 4$, $c_0 = -15 - (-3)(4) - 1(-2) = -1$, $R = 7 - (-3)(-1) - 1(4) = 0$ and $S = -1 - 1(-1) = 0$. Hence $x^2 - 3x + 1$ is a quadratic factor of $f(x)$. The roots of the quadratic, and hence of $f(x) = 0$, are $\tfrac{1}{2}(3 \pm \sqrt{5}) \simeq 2.618$ and 0.382.

2.6 $\dfrac{\partial f_1}{\partial x_1} = -e^{-x_1} - 6(x_1 - x_2)$; $\dfrac{\partial f_1}{\partial x_2} = 6(x_1 - x_2)$; $\dfrac{\partial f_2}{\partial x_1} = e^{x_1 - x_2}$; $\dfrac{\partial f_2}{\partial x_2} = -e^{x_1 - x_2} - 5$ and so $\left(\dfrac{\partial f_1}{\partial x_1}\right)_0 = -1$; $\left(\dfrac{\partial f_1}{\partial x_2}\right)_0 = 0$; $\left(\dfrac{\partial f_2}{\partial x_1}\right)_0 = 1$;

$$\left(\frac{\partial f_2}{\partial x_2}\right)_0 = -6.$$

Hence

$$-h_1^{[0]} = -f_1(0,0) = -1$$
$$h_1^{[0]} - 6h_2^{[0]} = -f_2(0,0) = -1$$

giving $h_1^{[0]} = 1$; $h_2^{[0]} = \frac{1}{3}$. Therefore $x_1^{[1]} = 1$; $x_2^{[1]} = \frac{1}{3}$.

3.1 We have

(i) $5x_1 + 3x_2 + 10x_3 = 5$
(ii) $2x_1 + 0.2x_2 + 3x_3 = -1$
(iii) $x_1 - 0.4x_2 + 3x_3 = 6$

so form (ii)$' = $ (ii) $- \frac{2}{5}$ (i) and (iii)$' = $ (iii) $- \frac{1}{5}$ (i) to give

(ii)$'$ $-x_2 - x_3 = -3$
(iii)$'$ $-x_2 + x_3 = 5$

and then form (iii)$'' = $ (iii)$'$ $-$ (ii)$'$

(iii)$''$ $2x_3 = 8.$

Back-substitution gives $x_3 = 4$, $x_2 = -1$ and $x_1 = -6.4$. Using full pivoting we first eliminate x_3 from (ii) and (iii) to give

(ii)$'$ $0.5x_1 - 0.7x_2 = -2.5$
(iii)$'$ $-0.5x_1 - 1.3x_2 = 4.5.$

Now eliminate x_2 from (ii)$'$ to give

(ii)$''$ $\dfrac{1}{1.3} x_1 = -\dfrac{6.4}{1.3}$

so that $x_1 = -6.4$, $x_2 = -1$ and $x_3 = 4$.

3.2 The components of the residual vector are

$$r_1 = 5 - 2 \times 0.9 + (-1) - 2.1 = 0.1$$
$$r_2 = 6 - 0.9 - (-1) - 3 \times 2.1 = -0.2$$
$$r_3 = 1 - 0.9 + 2(-1) + 2.1 = 0.2$$

so we need to solve

354 SOLUTIONS TO EXERCISES

(i) $2v_1 - v_2 + v_3 = 0.1$
(ii) $v_1 + v_2 + 3v_3 = -0.2$
(iii) $v_1 - 2v_2 - v_3 = 0.2$

for the corrections v_i to x_i. Using Gauss elimination without pivoting we form

(ii)' $1.5v_2 + 2.5v_3 = -0.25$
(iii)' $-1.5v_2 - 1.5v_3 = 0.15$

and

(iii)'' $v_3 = -0.1$

so that, from (ii)', $v_2 = 0$ and, from (i), $v_1 = 0.1$. Hence our improved solution is $x_1 = 0.9 + 0.1 = 1$, $x_2 = -1 + 0 = -1$, $x_3 = 2.1 - 0.1 = 2$.

3.3 Rearranging the equations we have

$$x_1 = 1 - 0.1x_2$$
$$x_2 = 2 - 0.1x_1 - 0.1x_3$$
$$x_3 = 3 - 0.1x_2 .$$

Starting with $x_1^{[0]} = x_2^{[0]} = x_3^{[0]} = 1$ we obtain

(i) from Jacobi iteration

$$\mathbf{x}^{[1]T} = (0.9, 1.8, 2.9); \quad \max_{1 \leq i \leq 3} |x_i^{[1]} - x_i^{[0]}| = \max (0.1, 0.8, 1.9)$$
$$= 1.9;$$

$$\mathbf{x}^{[2]T} = (0.82, 1.62, 2.82); \quad \max_{1 \leq i \leq 3} |x_i^{[2]} - x_i^{[1]}| = 0.18;$$

$$\mathbf{x}^{[3]T} = (0.838, 1.636, 2.838); \quad \max_{1 \leq i \leq 3} |x_i^{[3]} - x_i^{[2]}| = 0.018;$$

$$\mathbf{x}^{[4]T} = (0.8364, 1.6324, 2.8364); \quad \max_{1 \leq i \leq 3} |x_i^{[4]} - x_i^{[3]}| = 0.0036;$$

$$\mathbf{x}^{[5]T} = (0.83676, 1.63272, 2.83676); \quad \max_{1 \leq i \leq 3} |x_i^{[5]} - x_i^{[4]}| = 0.00036.$$

(ii) from Gauss–Seidel iteration

$\mathbf{x}^{[1]T} = (0.9, 1.81, 2.819); \quad \max_{1 \leq i \leq 3} |x_i^{[1]} - x_i^{[0]}| = 1.819;$

$\mathbf{x}^{[2]T} = (0.819, 1.6362, 2.83638); \quad \max_{1 \leq i \leq 3} |x_i^{[2]} - x_i^{[1]}| = 0.1738;$

$\mathbf{x}^{[3]T} = (0.83638, 1.632724, 2.8367276);$
$\quad \max_{1 \leq i \leq 3} |x_i^{[3]} - x_i^{[2]}| = 0.01738;$

$\mathbf{x}^{[4]T} = (0.8367276, 1.6326545, 2.8367346);$
$\quad \max_{1 \leq i \leq 3} |x_i^{[4]} - x_i^{[3]}| = 0.0003476.$

3.4 Reordering the equations we have the strictly diagonally dominant system

$$8x_1 - x_2 + x_3 = 1$$
$$-x_1 + 5x_2 + x_4 = 2$$
$$x_1 + 4x_3 - 2x_4 = 0$$
$$-x_2 - 2x_3 + 4x_4 = 1$$

which we rewrite as

$$x_1 = (1 + x_2 - x_3)/8$$
$$x_2 = (2 + x_1 - x_4)/5$$
$$x_3 = (-x_1 + 2x_4)/4$$
$$x_4 = (1 + x_2 + 2x_3)/4 \ .$$

Jacobi iteration gives

$\mathbf{x}^{[1]T} = (0.125, 0.4, 0.0, 0.25); \quad \sum_{i=1}^{4} |x_i^{[1]} - x_i^{[0]}| = 0.775;$

$\mathbf{x}^{[2]T} = (0.175, 0.375, 0.09375, 0.35); \quad \sum_{i=1}^{4} |x_i^{[2]} - x_i^{[1]}| = 0.26875;$

$\mathbf{x}^{[3]T} = (0.1601563, 0.365, 0.13125, 0.390625);$

$\sum_{i=1}^{4} |x_i^{[3]} - x_i^{[2]}| = 0.1029687;$

$\mathbf{x}^{[4]T} = (0.1542188, 0.3539063, 0.1552734, 0.406875);$

$\sum_{i=1}^{4} |x_i^{[4]} - x_i^{[3]}| = 0.0573046;$

$\mathbf{x}^{[5]T} = (0.1498291, 0.3494688, 0.1648828, 0.4161133);$

$$\sum_{i=1}^{4} |x_i^{[5]} - x_i^{[4]}| = 0.0276749 \ .$$

Gauss–Seidel iteration gives

$\mathbf{x}^{[1]T} = (0.125, 0.425, -0.03125, 0.340625);$

$$\sum_{i=1}^{4} |x_i^{[1]} - x_i^{[0]}| = 0.921875;$$

$\mathbf{x}^{[2]T} = (0.1820313, 0.3682813, 0.1248047, 0.4044727);$

$$\sum_{i=1}^{4} |x_i^{[2]} - x_i^{[1]}| = 0.3336524;$$

$\mathbf{x}^{[3]T} = (0.1554346, 0.3501924, 0.1633777, 0.4192370);$

$$\sum_{i=1}^{4} |x_i^{[3]} - x_i^{[2]}| = 0.0980229;$$

$\mathbf{x}^{[4]T} = (0.1483518, 0.3458230, 0.1725306, 0.4227210);$

$$\sum_{i=1}^{4} |x_i^{[4]} - x_i^{[3]}| = 0.0240890;$$

$\mathbf{x}^{[5]T} = (0.1466616, 0.3447881, 0.1746951, 0.4235446);$

$$\sum_{i=1}^{4} |x_i^{[5]} - x_i^{[4]}| = 0.0057132 \ .$$

Gauss–Seidel is converging at a faster rate than Jacobi.

3.5 Using the algorithm for Doolittle's method (subsection 3.6.2) we have $u_{11} = 1; u_{12} = 2; u_{13} = 3; l_{21} = 2/1 = 2; l_{31} = 2/1 = 2; u_{22} = 8 - 2 \times 2 = 4;$ $u_{23} = 7 - 2 \times 3 = 1; l_{32} = (16 - 2 \times 2)/4 = 3; u_{33} = 12 - 2 \times 3 - 3 \times 1 = 3$. Hence

$$\begin{bmatrix} 1 & 2 & 3 \\ 2 & 8 & 7 \\ 2 & 16 & 12 \end{bmatrix} = \begin{bmatrix} 1 & 0 & 0 \\ 2 & 1 & 0 \\ 2 & 3 & 1 \end{bmatrix} \begin{bmatrix} 1 & 2 & 3 \\ 0 & 4 & 1 \\ 0 & 0 & 3 \end{bmatrix}.$$

Now, using forward substitution we have $y_1 = 2$, $y_2 = 7 - 2 \times 2 = 3$, $y_3 = 10 - 2 \times 2 - 3 \times 3 = -3$ and, using back-substitution, we have $x_3 = -3/3 = -1$, $x_2 = (3 - 1(-1))/4 = 1$, $x_1 = (2 - 2 \times 1 - 3(-1))/1 = 3$.

3.6 Using the notation of subsection 3.6.3 we have $\alpha_1 = 2$; $\gamma_1 = 2/2 = 1$; $\alpha_2 = 4 - 2 \times 1 = 2$; $\gamma_2 = -2/2 = -1$; $\alpha_3 = 3 - (-2)(-1) = 1$. Hence

$$\begin{bmatrix} 2 & 2 & 0 \\ 2 & 4 & -2 \\ 0 & -2 & 3 \end{bmatrix} = \begin{bmatrix} 2 & 0 & 0 \\ 2 & 2 & 0 \\ 0 & -2 & 1 \end{bmatrix} \begin{bmatrix} 1 & 1 & 0 \\ 0 & 1 & -1 \\ 0 & 0 & 1 \end{bmatrix}.$$

If $\mathbf{b}^T = (1, 0, 0)$, forward substitution gives $\mathbf{y}^T = (\tfrac{1}{2}, -\tfrac{1}{2}, -1)$ and back-substitution gives $\mathbf{x}^T = (2, -\tfrac{3}{2}, -1)$. Similarly, if $\mathbf{b}^T = (0, 1, 0)$ then we have $\mathbf{y}^T = (0, \tfrac{1}{2}, 1)$ and $\mathbf{x}^T = (-\tfrac{3}{2}, \tfrac{3}{2}, 1)$ and if $\mathbf{b}^T = (0, 0, 1)$, $\mathbf{y}^T = (0, 0, 1)$, $\mathbf{x}^T = (-1, 1, 1)$. Hence

$$\begin{bmatrix} 2 & 2 & 0 \\ 2 & 4 & -2 \\ 0 & -2 & 3 \end{bmatrix}^{-1} = \tfrac{1}{2} \begin{bmatrix} 4 & -3 & -2 \\ -3 & 3 & 2 \\ -2 & 2 & 2 \end{bmatrix}.$$

4.1 (i) $\sin(x) = \sin(0) + x \cos(0) - \dfrac{x^2}{2} \sin(0) - \dfrac{x^3}{6} \cos(0) + \dfrac{x^4}{24} \sin(0)$

$+ \dfrac{x^5}{120} \cos(0) + \ldots = x - \dfrac{x^3}{6} + \dfrac{x^5}{120} \ldots$

(ii) $\cos(x) = \cos(0) - x \sin(0) - \dfrac{x^2}{2} \cos(0) + \dfrac{x^3}{6} \sin(0)$

$+ \dfrac{x^4}{24} \cos(0) - \dfrac{x^5}{120} \sin(0) \ldots = 1 - \dfrac{x^2}{2} + \dfrac{x^4}{24} \ldots$

(iii) $\ln(1+x) = \ln(1) + x \left(\dfrac{1}{1+x}\right)\Big|_{x=0} - \dfrac{x^2}{2} \left(\dfrac{1}{1+x}\right)^2\Big|_{x=0}$

$+ \dfrac{x^3}{3} \left(\dfrac{1}{1+x}\right)^3\Big|_{x=0} - \dfrac{x^4}{4} \left(\dfrac{1}{1+x}\right)^4\Big|_{x=0}$

$+ \dfrac{x^5}{5} \left(\dfrac{1}{1+x}\right)^5\Big|_{x=0} - \ldots = x - \dfrac{x^2}{2} + \dfrac{x^3}{3} - \dfrac{x^4}{4} + \dfrac{x^5}{5} - \ldots$.

Hence, truncating each series after the term involving x^5 we have $\sin(x) \simeq x - \dfrac{x^3}{6} + \dfrac{x^5}{120}$, $\cos(x) \simeq 1 - \dfrac{x^2}{2} + \dfrac{x^4}{24}$ and $\ln(1+x) \simeq x - \dfrac{x^2}{2} + \dfrac{x^3}{3} - \dfrac{x^4}{4} + \dfrac{x^5}{5}$, the truncation errors being $-\dfrac{x^6}{720} \sin(\xi)$, $-\dfrac{x^6}{720} \cos(\xi)$ and

$\left(-\frac{x^6}{6} \left(\frac{1}{1+x} \right)^6 \right) \Big|_{x=\xi}$ respectively where ξ is some point between 0 and x.

Hence for $x \in [-\pi/4, \pi/4]$,

$$\left| \sin(x) - \left(x - \frac{x^3}{6} + \frac{x^5}{120} \right) \right| < \frac{1}{720} \left(\frac{\pi}{4} \right)^6 \left(\frac{1}{\sqrt{2}} \right) \approx 0.00023;$$

for $x \in [-\pi/4, \pi/4]$, $\left| \cos(x) - \left(1 - \frac{x^2}{2} + \frac{x^4}{24} \right) \right| < \frac{1}{720} \left(\frac{\pi}{4} \right)^6 1 \approx 0.00033;$

for $x \in [-\frac{1}{2}, \frac{1}{2}]$, $\left| \ln(1+x) - \left(x - \frac{x^2}{2} + \frac{x^3}{3} - \frac{x^4}{4} + \frac{x^5}{5} \right) \right| < \frac{1}{6} \left(\frac{1}{2} \right)^6 (2)^6$
≈ 0.17.

4.2 (i) If we tackle this problem directly, the elements of the coefficient matrix are

$$A_{ij} = \left(\frac{\pi}{2} \right)^{i+j-1} / (i+j-1)$$

so that

$$A = \begin{bmatrix} \frac{\pi}{2} & \frac{\pi^2}{8} \\ \frac{\pi^2}{8} & \frac{\pi^3}{24} \end{bmatrix}.$$

The elements of the right-hand side vector are

$$b_1 = \int_0^{\pi/2} \sin(x) \, dx = [-\cos(x)]_0^{\pi/2} = 1$$

$$b_2 = \int_0^{\pi/2} x \sin(x) \, dx = [-x \cos(x)]_0^{\pi/2} + \int_0^{\pi/2} \cos(x) \, dx$$

$$= [\sin(x)]_0^{\pi/2} = 1$$

so that if $p_1(x) = \alpha_1 + \alpha_2 x$ is our straight line approximation

$$\begin{bmatrix} \frac{\pi}{2} & \frac{\pi^2}{8} \\ \frac{\pi^2}{8} & \frac{\pi^3}{24} \end{bmatrix} \begin{bmatrix} \alpha_1 \\ \alpha_2 \end{bmatrix} = \begin{bmatrix} 1 \\ 1 \end{bmatrix}.$$

The solution of this system is $\alpha_1 = \dfrac{8}{\pi}\left(1 - \dfrac{3}{\pi}\right)$, $\alpha_2 = \dfrac{96}{\pi^3}\left(1 - \dfrac{\pi}{4}\right)$.

(ii) Let $Q_0(x) = 1$. We want to find c_{11} such that $Q_1(x) = 1 + c_{11}x$ and $Q_0(x)$ are orthogonal on $[0, \pi/2]$.
Therefore we require

$$0 = \int_0^{\pi/2} Q_0(x) Q_1(x)\, dx = \int_0^{\pi/2} (1 + c_{11}x)\, dx = \dfrac{\pi}{2} + \dfrac{\pi^2}{8} c_{11}$$

so choose $c_{11} = -\dfrac{4}{\pi}$. Now

$$\int_0^{\pi/2} Q_0(x) \sin(x)\, dx = 1$$

$$\int_0^{\pi/2} Q_1(x) \sin(x)\, dx = \int_0^{\pi/2} \sin(x)\, dx - \dfrac{4}{\pi}\int_0^{\pi/2} x \sin(x)\, dx$$

$$= 1 - \dfrac{4}{\pi}$$

and

$$\int_0^{\pi/2} Q_0(x) Q_0(x)\, dx = \int_0^{\pi/2} dx = [x]_0^{\pi/2} = \dfrac{\pi}{2}$$

$$\int_0^{\pi/2} Q_1(x) Q_1(x)\, dx = \int_0^{\pi/2} \left(1 - \dfrac{4}{\pi}x\right)^2 dx$$

$$= \int_0^{\pi/2}\left(1 - \dfrac{8}{\pi}x + \dfrac{16}{\pi^2}x^2\right)dx$$

$$= \left[x - \dfrac{4}{\pi}x^2 + \dfrac{16}{3\pi^2}x^3\right]_0^{\pi/2}$$

$$= \dfrac{\pi}{2} - \pi + \dfrac{2\pi}{3} = \dfrac{\pi}{6}.$$

Hence if $p_1(x) = \beta_1 Q_0(x) + \beta_2 Q_1(x)$ we have $\beta_1 = 1 / \dfrac{\pi}{2} = \dfrac{2}{\pi}$,

$$\beta_2 = \left(1 - \dfrac{4}{\pi}\right) / \dfrac{\pi}{6} = \dfrac{6}{\pi}\left(1 - \dfrac{4}{\pi}\right);$$

that is, $p_1(x) = \dfrac{2}{\pi} + \dfrac{6}{\pi}\left(1 - \dfrac{4}{\pi}\right)\left(1 - \dfrac{4}{\pi}x\right).$

(iii) Use the change of variable $z = \left(2x - \dfrac{\pi}{2}\right)/\dfrac{\pi}{2} = \dfrac{4}{\pi}x - 1$ and

approximate $\sin\left((z+1)\dfrac{\pi}{4}\right)$ on $[-1, 1]$. Then

$$\int_{-1}^{1} P_0(z) \sin\left((z+1)\dfrac{\pi}{4}\right) dz = \int_{-1}^{1} \sin\left((z+1)\dfrac{\pi}{4}\right) dz$$

$$= \left[-\cos\left((z+1)\frac{\pi}{4}\right)\frac{4}{\pi}\right]_{-1}^{1} = \frac{4}{\pi}$$

$$\int_{-1}^{1} P_1(z) \sin\left((z+1)\frac{\pi}{4}\right) dz = \int_{-1}^{1} z \sin\left((z+1)\frac{\pi}{4}\right) dz =$$

$$\left[-z\cos\left((z+1)\frac{\pi}{4}\right)\frac{4}{\pi}\right]_{-1}^{1} + \frac{4}{\pi}\int_{-1}^{1} \cos\left((z+1)\frac{\pi}{4}\right) dz =$$

$$-\frac{4}{\pi} + \frac{16}{\pi^2}\left[\sin\left((z+1)\frac{\pi}{4}\right)\right]_{-1}^{1} = -\frac{4}{\pi} + \frac{16}{\pi^2}.$$

Hence, if $p_1(z) = \beta_1 P_0(z) + \beta_2 P_1(z)$ we have $\beta_1 =$

$$\frac{4}{\pi}/\int_{-1}^{1} P_0^2(z)\, dz = \frac{4}{\pi}/2 = \frac{2}{\pi}, \beta_2 = \left(-\frac{4}{\pi} + \frac{16}{\pi^2}\right)/\int_{-1}^{1} P_1^2(z)dz =$$

$$\left(-\frac{4}{\pi} + \frac{16}{\pi^2}\right)/\left(\frac{2}{3}\right) = -\frac{6}{\pi} + \frac{24}{\pi^2}.$$

Making the inverse change of variable we have

$$p_1(x) = \frac{2}{\pi} + \frac{6}{\pi}\left(\frac{4}{\pi} - 1\right)\left(\frac{4}{\pi}x - 1\right).$$

4.3 $e^x = 1 + x + \dfrac{x^2}{2} + \dfrac{x^3}{6} + \dfrac{x^4}{24} + \dfrac{x^5}{120} + \dfrac{x^6}{720}e^{\xi}$ $\xi \in]0,x[$, so if $p_5(x) = 1 + x +$
$\dfrac{x^2}{2} + \dfrac{x^3}{6} + \dfrac{x^4}{24} + \dfrac{x^5}{120}$ then $\max\limits_{x\in[0,1]} |e^x - p_5(x)| = \max\limits_{x\in[0,1]} |\dfrac{x^6}{720}e^x| = \dfrac{e}{720} \approx$
0.0038. Now, $T_0(x) = 1$, $T_1(x) = x$, $T_2(x) = 2x^2 - 1$, $T_3(x) = 4x^3 - 3x$, $T_4(x) = 8x^4 - 8x^2 + 1$, $T_5(x) = 16x^5 - 20x^3 + 5x$, so $1 = T_0(x)$, $x = T_1(x)$, $x^2 = \frac{1}{2}(T_2(x) + T_0(x))$, $x^3 = \frac{1}{4}(T_3(x) + 3T_1(x))$, $x^4 = \frac{1}{8}(T_4(x) + 4T_2(x) + 3T_0(x))$, $x^5 = \frac{1}{16}(T_5(x) + 5T_3(x) + 10T_1(x))$ and hence $p_5(x) = T_0(x) + T_1(x) + \frac{1}{4}(T_2(x) + T_0(x)) + \frac{1}{24}(T_3(x) + 3T_1(x)) + \frac{1}{192}(T_4(x) + 4T_2(x) + 3T_0(x)) + \frac{1}{1920}(T_5(x) + 5T_3(x) + 10T_1(x)) = T_0(x)(1 + \frac{1}{4} + \frac{1}{64})$
$+ T_1(x)\ (1 + \frac{1}{8} + \frac{1}{192}) + T_2(x)\ (\frac{1}{4} + \frac{1}{48}) + T_3(x)\ (\frac{1}{24} + \frac{1}{384}) + \frac{1}{192}\ T_4(x) +$
$\frac{1}{1920}\ T_5(x) = \frac{81}{64}\ T_0(x) + \frac{217}{192}\ T_1(x) + \frac{13}{48}\ T_2(x) + \frac{17}{384}\ T_3(x) + \frac{1}{192}\ T_4(x) +$
$\frac{1}{1920}\ T_5(x)$.

Now if $p_4(x) = p_5(x) - \frac{1}{1920} T_5(x)$ the additional error incurred is $\max_{x \in [0,1]} \frac{1}{1920} |T_5(x)| = \frac{1}{1920} = 0.000\,52$.

4.4 If $A\alpha = b$ we have $A_{11} = \int_{-\pi}^{\pi} dx = [x]_{-\pi}^{\pi} = 2\pi$, $A_{12} = \int_{-\pi}^{\pi} \sin(x)\, dx =$

$[-\cos(x)]_{-\pi}^{\pi} = 0 = A_{21}$, $A_{13} = \int_{-\pi}^{\pi} \cos(x)\, dx = [\sin(x)]_{-\pi}^{\pi} = 0 = A_{31}$,

$A_{22} = \int_{-\pi}^{\pi} \sin^2(x)\, dx = \int_{-\pi}^{\pi} \tfrac{1}{2}(1 - \cos(2x))\, dx = \tfrac{1}{2}\left[x - \frac{\sin(2x)}{2}\right]_{-\pi}^{\pi} = \pi$,

$A_{23} = \int_{-\pi}^{\pi} \sin(x) \cos(x)\, dx = \int_{-\pi}^{\pi} \tfrac{1}{2} \sin(2x)\, dx = \tfrac{1}{2}[-\tfrac{1}{2} \cos(2x)]_{-\pi}^{\pi} = 0$

$= A_{32}$, $A_{33} = \int_{-\pi}^{\pi} \cos^2(x)\, dx = \int_{-\pi}^{\pi} \tfrac{1}{2}(1 + \cos(2x))\, dx =$

$\tfrac{1}{2}\left[x + \frac{\sin(2x)}{2}\right]_{-\pi}^{\pi} = \pi$.

Now $b_1 = \int_{-\pi}^{\pi} e^x dx = [e^x]_{-\pi}^{\pi} = e^{\pi} - e^{-\pi}$. Also $b_2 = \int_{-\pi}^{\pi} e^x \sin(x)\, dx =$

$[-e^x \cos(x)]_{-\pi}^{\pi} + \int_{-\pi}^{\pi} e^x \cos(x)\, dx = e^{\pi} - e^{-\pi} + [e^x \sin(x)]_{-\pi}^{\pi} -$

$\int_{-\pi}^{\pi} e^x \sin(x)\, dx = e^{\pi} - e^{-\pi} - b_2$ and hence $b_2 = \tfrac{1}{2}(e^{\pi} - e^{-\pi})$. Finally,

$b_3 = \int_{-\pi}^{\pi} e^x \cos(x)\, dx = [e^x \sin(x)]_{-\pi}^{\pi} - \int_{-\pi}^{\pi} e^x \sin(x)\, dx = 0 - b_2 =$

$-b_2$. Hence, since the coefficient matrix is diagonal, we have $\alpha_1 = \frac{1}{2\pi}(e^{\pi} - e^{-\pi})$, $\alpha_2 = \frac{1}{2\pi}(e^{\pi} - e^{-\pi})$ and $\alpha_3 = -\frac{1}{2\pi}(e^{\pi} - e^{-\pi})$; that is, $\alpha_1 = \alpha_2 = -\alpha_3 = \tfrac{1}{2} \sinh(\pi)$.

4.5 From the three term recurrence relation for Chebyshev polynomials we have $T_6(x) = 2xT_5(x) - T_4(x) = 32x^6 - 40x^4 + 10x^2 - 8x^4 + 8x^2 - 1 = 32x^6 - 48x^4 + 18x^2 - 1$ and $T_7(x) = 2x T_6(x) - T_5(x) = 64x^7 - 96x^5 + 36x^3 - 2x - 16x^5 + 20x^3 - 5x = 64x^7 - 112x^5 + 56x^3 - 7x$. Hence, if $p_6(x)$ is the approximating polynomial $x^7 - p_6(x) = \tfrac{1}{64}(64x^7 - 112x^5 + 56x^3 - 7x) = x^7 - \tfrac{1}{64}(112x^5 - 56x^3 + 7x)$. Therefore $p_6(x) = \tfrac{1}{64}(112x^5 - 56x^3 + 7x)$.

4.6 $\int_{-1}^{1} (f(x) - p_{n-1}(x))^2\, dx = \int_{-1}^{1} f^2(x)\, dx - 2\int_{-1}^{1} p_{n-1}(x)f(x)\, dx$

$+ \int_{-1}^{1} p_{n-1}^2(x)\, dx$

$$= \int_{-1}^{1} f^2(x)\, dx - 2 \int_{-1}^{1} \sum_{i=1}^{n} \beta_i P_{i-1}(x) f(x)\, dx +$$

$$\int_{-1}^{1} \left(\sum_{i=1}^{n} \beta_i P_{i-1}(x)\right)\left(\sum_{j=1}^{n} \beta_j P_{j-1}(x)\right) dx$$

$$= \int_{-1}^{1} f^2(x)\, dx - 2 \sum_{i=1}^{n} \beta_i \int_{-1}^{1} P_{i-1}(x) f(x)\, dx +$$

$$\sum_{i=1}^{n} \sum_{j=1}^{n} \beta_i \beta_j \int_{-1}^{1} P_{i-1}(x) P_{j-1}(x)\, dx$$

$$= \int_{-1}^{1} f^2(x)\, dx - 2 \sum_{i=1}^{n} \beta_i \frac{2}{2i-1} \beta_i + \sum_{i=1}^{n} \beta_i^2 \int_{-1}^{1} P_{i-1}^2(x)\, dx$$

using the definition of the β_is from the solution of the least squares problem and the orthogonality of the Legendre polynomials. Hence

$$\int_{-1}^{1} (f(x) - p_{n-1}(x))^2\, dx = \int_{-1}^{1} f^2(x)\, dx - 2 \sum_{i=1}^{n} \frac{2}{2i-1} \beta_i^2$$

$$+ \sum_{i=1}^{n} \beta_i^2 \frac{2}{2i-1}$$

$$= \int_{-1}^{1} f^2(x)\, dx - 2 \sum_{i=1}^{n} \frac{\beta_i^2}{2i-1}.$$

5.1 $p_3(x) = \dfrac{(x-1)(x-2)(x-3)}{-1 \times -2 \times -3} \times (-1) + \dfrac{x(x-2)(x-3)}{1 \times -1 \times -2} \times 0$
$+ \dfrac{x(x-1)(x-3)}{2 \times 1 \times -1} \times 3 + \dfrac{x(x-1)(x-2)}{3 \times 2 \times 1} \times 14 = x^3 - 2x^2 + 2x - 1.$

5.2 $\phi_1(x) = \dfrac{(x-1)(x-3)}{-2 \times -4} = \tfrac{1}{8}(x^2 - 4x + 3);\ \phi_1'(x) = \tfrac{1}{4}(x-2);$
$\phi_1'(-1) = -\tfrac{3}{4}.$

$$\phi_2(x) = \frac{(x+1)(x-3)}{2 \times -2} = -\frac{1}{4}(x^2 - 2x - 3); \phi_2'(x) = -\frac{1}{2}(x-1);$$
$$\phi_2'(1) = 0.$$

$$\phi_3(x) = \frac{(x+1)(x-1)}{4 \times 2} = \frac{1}{8}(x^2 - 1); \phi_3'(x) = \frac{1}{4}x; \phi_3'(3) = \frac{3}{4}.$$

Hence

$$p_5(x) = (1 - 2(x+1) \times -\frac{3}{4}) \frac{1}{64} (x^2 - 4x + 3)^2 0$$
$$+ (1 - 2(x-1)0) \frac{1}{16} (x^2 - 2x - 3)^2 0 + (1 - 2(x-3)\frac{3}{4}) \frac{1}{64} (x^2 - 1)^2 208$$
$$+ (x+1) \frac{1}{64} (x^2 - 4x + 3)^2 12 + (x-1) \frac{1}{16} (x^2 - 2x - 3)^2 4$$
$$+ (x-3) \frac{1}{64} (x^2 - 1)^2 348$$
$$= x^5 - x^4 + 2x^3 - 3x + 1.$$

5.3 Choosing $\phi_1(x) = 1$ and $\phi_2(x) = x$ we have $<\phi_1, \phi_1> = 3$; $<\phi_1, \phi_2> = <\phi_2, \phi_1> = 6$; $<\phi_2, \phi_2> = 14$; $<f, \phi_1> = -2$; $<f, \phi_2> = 0$. Hence, if $p_1(x) = \alpha_1 + \alpha_2 x$

$$\begin{pmatrix} 3 & 6 \\ 6 & 14 \end{pmatrix} \begin{pmatrix} \alpha_1 \\ \alpha_2 \end{pmatrix} = \begin{pmatrix} -2 \\ 0 \end{pmatrix} \Rightarrow \begin{pmatrix} 3 & 6 \\ 0 & 2 \end{pmatrix} \begin{pmatrix} \alpha_1 \\ \alpha_2 \end{pmatrix} \begin{pmatrix} -2 \\ 4 \end{pmatrix}$$

so that $\alpha_2 = 2$ and $\alpha_1 = -\frac{14}{3}$, that is, $p_1(x) = -4\frac{2}{3} + 2x$. Using Gram-Schmidt we have

$$\psi_1(x) = \phi_1(x) = 1$$

$$\psi_2(x) = \phi_2(x) - \frac{<\psi_1, \phi_2>}{<\psi_1, \psi_1>} \psi_1(x) = x - \frac{6}{3} = x - 2.$$

Hence, if $p_1(x) = \alpha_1 + \alpha_2(x-2)$, $\alpha_1 = <f, \psi_1>/<\psi_1, \psi_1> = -\frac{2}{3}$ and $\alpha_2 = <f, \psi_2>/<\psi_2, \psi_2> = \frac{4}{2} = 2$ so that $p_1(x) = -\frac{2}{3} + 2(x-2) = -4\frac{2}{3} + 2x$. Now $p_1(1) = -2\frac{2}{3}; p_1(2) = -\frac{2}{3}; p_1(3) = 1\frac{1}{3}$ so that the sum of the squares of the residuals is

$$(-2 + 2\frac{2}{3})^2 + (-2 + \frac{2}{3})^2 + (2 - 1\frac{1}{3})^2 = \frac{4}{9} + \frac{16}{9} + \frac{4}{9} = \frac{24}{9} = 2\frac{2}{3}.$$

5.4 Using Gram-Schmidt, $\psi_1(x) = 1$, $\psi_2(x) = x - \frac{15}{6} = x - \frac{5}{2}$. Now $<f, \psi_1> = 49$; $<\psi_1, \psi_1> = 6$; $<f, \psi_2> = \frac{71}{2}$; $<\psi_2, \psi_2> = \frac{35}{2}$. Hence if $p_1(x) = \alpha_1 \psi_1(x) + \alpha_2 \psi_2(x)$ then $\alpha_1 = <f, \psi_1>/<\psi_1, \psi_1> = $

$\frac{49}{6}$ and $\alpha_2 = \frac{71}{2}/\frac{35}{2} = \frac{71}{35}$. Therefore $p_1(x) = \frac{49}{6} + \frac{71}{35}(x - \frac{5}{2}) = \frac{65}{21} + \frac{71}{35}x$.

For the l_1 approximation a sketch reveals that many of the possibilities may be ignored immediately. For example, the line passing through the first and second points gives an absolute residual sum of 49. Looking at each candidate in turn we find that the straight line which gives the minimum residual sum passes through the third and sixth coordinates and the equation of this line is $p_1(x) = 3 + 2x$. For this approximation the residual vector is $(2, -2, 0, -1, 2, 0)^T$ which equioscillates on three points with magnitude 2.

5.5 Let $\tilde{x} = \max_{1 \leq i \leq N} |x_i|$. Then $\left(\sum_{i=1}^{N} x_i^2\right)^{\frac{1}{2}} = (x_1^2 + x_2^2 + \ldots + x_N^2)^{\frac{1}{2}} \geq (\tilde{x}^2)^{\frac{1}{2}} = \tilde{x}$.

Also $\left(\sum_{i=1}^{N} x_i^2\right)^{\frac{1}{2}} \leq (\tilde{x}^2 + \tilde{x}^2 + \ldots + \tilde{x}^2)^{\frac{1}{2}} = (N\tilde{x}^2)^{\frac{1}{2}} = N^{\frac{1}{2}}\tilde{x}$.

For the given approximation the residual vector is $\mathbf{r}^T = (-\frac{2}{5}, -\frac{2}{5}, \frac{8}{5}, -\frac{2}{5}, -\frac{2}{5})$ and so $\max_{1 \leq i \leq 5} |r_i| = \frac{8}{5}$. Also, $\left(\sum_{i=1}^{5} r_i^2\right)^{\frac{1}{2}} = (\frac{4}{25} + \frac{4}{25} + \frac{64}{25} + \frac{4}{25} + \frac{4}{25})^{\frac{1}{2}} = \sqrt{\frac{80}{25}} = \frac{8}{5}\sqrt{\frac{5}{4}}$ and we have $\frac{8}{5} < \frac{8}{5}\sqrt{\frac{5}{4}} < \frac{8}{5}\sqrt{5}$.

6.1 $f(x) = 1 : \frac{2}{3}\{1 + 1 + 1\} = 2; \int_{-1}^{1} f(x)\,dx = [x]_{-1}^{1} = 2$.

$f(x) = x : \frac{2}{3}\left\{-\frac{1}{\sqrt{2}} + 0 + \frac{1}{\sqrt{2}}\right\} = 0; \int_{-1}^{1} x\,dx = \left[\frac{x^2}{2}\right]_{-1}^{1} = 0$.

$f(x) = x^2 : \frac{2}{3}\left\{\frac{1}{2} + 0 + \frac{1}{2}\right\} = \frac{2}{3}; \int_{-1}^{1} x^2\,dx = \left[\frac{x^3}{3}\right]_{-1}^{1} = \frac{2}{3}$.

$f(x) = x^3 : \frac{2}{3}\left\{-\frac{1}{2\sqrt{2}} + 0 + \frac{1}{2\sqrt{2}}\right\} = 0; \int_{-1}^{1} x^3\,dx = \left[\frac{x^4}{4}\right]_{-1}^{1} = 0$.

$f(x) = x^4 : \frac{2}{3}\left\{\frac{1}{4} + 0 + \frac{1}{4}\right\} = \frac{1}{3}; \int_{-1}^{1} x^4\,dx = \left[\frac{x^5}{5}\right]_{-1}^{1} = \frac{2}{5}$.

The rule integrates $1, x, x^2$ and x^3 exactly but not x^4 and hence has precision 3.

6.2 $R = (2-1)e^1 = 2.718282$. Now $\max_{x \in [1,2]} \left|\frac{d}{dx} e^x\right| = e^2$ so that $|E| < \frac{1}{2}e^2 = 3.694529$. But $I = \int_{1}^{2} e^x\,dx = [e^x]_1^2 = e^2 - e = 4.670775$ and so $I - R = 1.952493$ which is within the computed bound. Subdividing the interval we obtain $R = \frac{1}{2}e^1 + \frac{1}{2}e^{1.5} = 3.599986$ and the error in this estimate is 1.070288. Hence halving the interval length has resulted in a reduction in the error by a factor of $1.952493/1.070788 \simeq 2$, that is, the truncation error is proportional to h.

6.3 If $f(x) = 1/x$ then $f'''(x) = 2/x^3$ and $\max_{x \in [1,2]} |f''(x)| = 2$. Hence we must have $h \leq \sqrt{12 \times 0.01/2} = \sqrt{0.06}$. Choose $h = 0.2$. Then the composite trapezium rule approximation is $0.2 \left\{ \frac{1}{2} + \frac{1}{1.2} + \frac{1}{1.4} + \frac{1}{1.6} + \frac{1}{1.8} + \frac{1}{4} \right\} \approx 0.6956$. Now $\int_1^2 1/x \, dx = [\ln(x)]_1^2 = \ln(2) - \ln(1) = 0.6931$ and so the error in our approximation is, in absolute value, 0.0025, well within the bound 0.01.

6.4 Making use of the fact that the entries in the second column of the T-table are Simpson's rule approximations we have

$$T_{1,0} = \frac{b-a}{6} \left\{ f(a) + 4f\left(a + \frac{b-a}{2}\right) + f(b) \right\}$$

$$T_{1,1} = \frac{b-a}{12} \left\{ f(a) + 4f\left(a + \frac{b-a}{4}\right) + 2f\left(a + \frac{b-a}{2}\right) \right.$$
$$\left. + 4f\left(a + 3\frac{b-a}{4}\right) + f(b) \right\}$$

and so

$$T_{2,0} = \frac{16T_{1,1} - T_{1,0}}{15}$$

$$= \frac{1}{15} \left\{ 4\frac{b-a}{3} \left\{ f(a) + 4f\left(a + \frac{b-a}{4}\right) + 2f\left(a + \frac{b-a}{2}\right) \right. \right.$$
$$\left. + 4f\left(a + 3\frac{b-a}{4}\right) + f(b) \right\}$$
$$\left. - \frac{b-a}{6} \left\{ f(a) + 4f\left(a + \frac{b-a}{2}\right) + f(b) \right\} \right\}$$

$$= \frac{b-a}{90} \left\{ 7f(a) + 32f\left(a + \frac{b-a}{4}\right) + 12f\left(a + \frac{b-a}{2}\right) \right.$$
$$\left. + 32f\left(a + 3\frac{b-a}{4}\right) + 7f(b) \right\}.$$

6.5 $G_2 = e^{-1/\sqrt{3}} + e^{1/\sqrt{3}} = 0.561384 + 1.781312 = 2.342696$. Now $\int_{-1}^{1} e^x \, dx = e - e^{-1} = 2.718282 - 0.367879 = 2.350403$ so the error is 0.007707.

6.6 $x_1 = \cos\left(\frac{\pi}{6}\right) = \frac{\sqrt{3}}{2}$; $x_2 = \cos\left(\frac{\pi}{2}\right) = 0$; $x_3 = \cos\left(\frac{5\pi}{6}\right) = -\frac{\sqrt{3}}{2}$.

Hence $R_3 = \dfrac{\pi}{3}\left\{\left(\dfrac{\sqrt{3}}{2}\right)^4 + 0^4 + \left(-\dfrac{\sqrt{3}}{2}\right)^4\right\} = \dfrac{\pi}{3}\dfrac{9}{16}2 = \dfrac{3\pi}{8}$. Now

$$\int_{-1}^{1} \dfrac{x^4}{\sqrt{1-x^2}} dx = -\int_{\pi}^{0} \dfrac{\cos^4(\theta)}{\sqrt{1-\cos^2(\theta)}} \sin\theta\, d\theta = \int_{0}^{\pi} \cos^4(\theta)\, d\theta$$

$$= \int_{0}^{\pi} [\tfrac{1}{2}(1+\cos(2\theta))]^2\, d\theta$$

$$= \dfrac{1}{4}\int_{0}^{\pi} (1 + 2\cos(2\theta) + \cos^2(2\theta))\, d\theta$$

$$= \dfrac{1}{4}\int_{0}^{\pi} (1 + 2\cos(2\theta) + \tfrac{1}{2}(1+\cos(4\theta)))\, d\theta$$

$$= \dfrac{1}{4}\int_{0}^{\pi} \left(\dfrac{3}{2} + 2\cos(2\theta) + \dfrac{1}{2}\cos(4\theta)\right) d\theta$$

$$= \dfrac{1}{4}\left[\dfrac{3}{2}\theta + \sin(2\theta) + \dfrac{1}{8}\sin(4\theta)\right]_{0}^{\pi} = \dfrac{3\pi}{8}$$

and so the rule is exact for this integral.

7.1 $Y_0 = 1;\ Y_1 = 1 + \dfrac{\pi}{6}(1 + \cos(0) - \sin(0)) = 1 + \dfrac{2\pi}{6} \approx 2.04720;$

$Y_2 = 2.04720 + \dfrac{\pi}{6}\left(2.04720 + \cos\left(\dfrac{\pi}{6}\right) - \sin\left(\dfrac{\pi}{6}\right)\right)$

$= 2.04720 + \dfrac{\pi}{6}\left(2.04720 + \dfrac{\sqrt{3}}{2} - \dfrac{1}{2}\right) \approx 3.31076;$

$Y_3 = 3.31076 + \dfrac{\pi}{6}\left(3.31076 + \cos\left(\dfrac{\pi}{3}\right) - \sin\left(\dfrac{\pi}{3}\right)\right) \approx 4.85262$.

7.2 $y(0) = 1;\ y'(0) = y(0) + \cos(0) - \sin(0) = 2;\ y''(x) = y'(x) - \sin(x) - \cos(x) => y''(0) = 1;\ y'''(x) = y''(x) - \cos(x) + \sin(x) => y'''(0) = 0;\ y^{(iv)}(x) = y'''(x) + \sin(x) + \cos(x) => y^{(iv)}(0) = 1.$ Hence $y(0.2) \approx Y_1 = 1 + 0.2 \times 2 + \dfrac{0.2^2}{2} \times 1 + \dfrac{0.2^3}{6} \times 0 + \dfrac{0.2^4}{24} \times 1 \approx 1.42007.$

7.3 $Y_2^{[0]} = 0 + 0.2(1 - 2 \times 0.1 \times 0.0993) = 0.1960$;
$Y_2^{[1]} = 0.0993 + 0.05(0.9801 + 0.9216) = 0.1944$;
$Y_2^{[2]} = 0.0993 + 0.05(0.9801 + 0.9222) = 0.1944$;
$Y_3^{[0]} = 0.0993 + 0.2 \times 0.9222 = 0.2837$;
$Y_3^{[1]} = 0.1944 + 0.05(0.9222 + 0.8298) = 0.2820$;
$Y_3^{[2]} = 0.1944 + 0.05(0.9222 + 0.8308) = 0.2820$.

7.4 For the predictor we have $y(x_j) - Y_j^{[0]} = \frac{251}{720} h^5 y^{(v)} (\eta_{j-4})$ and for the corrector $y(x_j) - Y_j = -\frac{19}{720} h^5 y^{(v)} (\xi_{j-3})$. Assuming that the fifth derivative of $f(x)$ is constant, $Y_j - Y_j^{[0]} = \frac{3}{8} h^5 y^{(v)} (\xi)$ so that if we assume $Y_{j-1} - Y_{j-1}^{[0]} = \frac{3}{8} h^5 y^{(v)} (\xi)$ we have $y(x_j) - Y_j^{[0]} = \frac{251}{270} \left(Y_{j-1} - Y_{j-1}^{[0]} \right)$. Hence use as a modified initial estimate $\tilde{Y}_j = Y_j^{[0]} + \frac{251}{270} \left(Y_{j-1} - Y_{j-1}^{[0]} \right)$.

7.5 Step 1: $k_1 = 0.25(0 + 1) = 0.25$; $k_2 = 0.25 \left(0 + \frac{2}{3} 0.25 + 1 + \frac{2}{3} 0.25 \right) = 0.333333$; $Y_1 = 1 + \frac{1}{4} 0.25 + \frac{3}{4} 0.333333 = 1.31250$.

Step 2: $k_1 = 0.25(0.25 + 1.31250) = 0.390625$;
$k_2 = 0.25 \left(0.25 + \frac{2}{3} 0.25 + 1.31250 + \frac{2}{3} 0.390625 \right) = 0.497396$;
$Y_2 = 1.31250 + \frac{1}{4} 0.390625 + \frac{3}{4} 0.497396 = 1.78320$.

Step 3: $k_1 = 0.25(0.5 + 1.78320) = 0.570800$;
$k_2 = 0.25 \left(0.5 + \frac{2}{3} 0.25 + 1.78320 + \frac{2}{3} 0.570800 \right) = 0.707600$;
$Y_3 = 1.78320 + \frac{1}{4} 0.570800 + \frac{3}{4} 0.707600 = 2.45660$.

368 SOLUTIONS TO EXERCISES

Step 4: $k_1 = 0.25(0.75 + 2.45660) = 0.801650$;

$k_2 = 0.25\left(0.75 + \dfrac{2}{3}0.25 + 2.45660 + \dfrac{2}{3}0.801650\right) = 0.976925$;

$Y_4 = 2.45660 + \dfrac{1}{4}0.801650 + \dfrac{3}{4}0.976925 = 3.38971$.

7.6 We have

$$\dfrac{Y_{j+1} - 2Y_j + Y_{j-1}}{h^2} = p_j \dfrac{Y_{j+1} - Y_{j-1}}{2h} + q_j Y_j + r_j .$$

Multiply throughout by h^2 and rearrange to give

$$Y_{j+1}\left(1 - \dfrac{h}{2}p_j\right) - Y_j(2 + h^2 q_j) + Y_{j-1}\left(1 + \dfrac{h}{2}p_j\right) = h^2 r_j$$

or, in matrix form,

$$\begin{bmatrix} -(2+h^2q_1) & 1-\dfrac{h}{2}p_1 & & & \\ 1+\dfrac{h}{2}p_2 & -(2+h^2q_2) & 1-\dfrac{h}{2}p_2 & & \\ \cdot & \cdot & \cdot & & \\ \cdot & \cdot & \cdot & & \\ & & 1+\dfrac{h}{2}p_{N-2} & -(2+h^2q_{N-2}) & 1-\dfrac{h}{2}p_{N-2} \\ & & & 1+\dfrac{h}{2}p_{N-1} & -(2+h^2q_{N-1}) \end{bmatrix} \begin{bmatrix} Y_1 \\ Y_2 \\ \cdot \\ \cdot \\ Y_{N-2} \\ Y_{N-1} \end{bmatrix} = \begin{bmatrix} h^2 r_1 - \left(1+\dfrac{h}{2}p_1\right)\alpha \\ h^2 r_2 \\ \cdot \\ \cdot \\ h^2 r_{N-2} \\ h^2 r_{N-1} - \left(1-\dfrac{h}{2}p_{N-1}\right)\beta \end{bmatrix}$$

For diagonal dominance we require $|2+h^2q_j| > |1-(h/2)p_j| + |1+(h/2)p_j|$. Using the given conditions on $q(x)$, $p(x)$ and h we can remove the modulus signs which means that we require $2+h^2q_j > 1-(h/2)p_j + 1+(h/2)p_j = 2$ and hence the system is strictly diagonally dominant.

References

Atkinson, L. V. and Harley, P. J. (1983). *An Introduction to Numerical Methods with Pascal*. Addison-Wesley, London.

Barrodale, I. and Olesky, D. D. (1981). Exponential approximation using Prony's method. In *The Numerical Solution of Nonlinear Problems* (ed. C. T. H. Baker and C. Phillips), pp. 258–269. Clarendon Press, Oxford.

Barrodale, I. and Phillips, C. (1974). An improved algorithm for discrete Chebyshev linear approximation. *Proc. Fourth Manitoba Conf. on Numer. Math.*, pp. 177–190.

Barrodale, I. and Roberts, F. D. K. (1973). An improved algorithm for discrete L1 linear approximation. *S.I.A.M. J. Numer. Anal.* **10**, pp. 839–848.

Broyden, C. G. (1975). *Basic Matrices: An Introduction to Matrix Theory and Practice*. Macmillan, London.

BSI (1982). *BS6192:1982 — Specification for Computer Programming Language Pascal*. British Standards Institution, London.

Clenshaw, C. W. and Curtis, A. R. (1960). A method for numerical integration on an automatic computer. *Numer. Math.* **2**, pp. 197–205.

Colin, A. J. T. (1978). *Programming and Problem-solving in Algol 68*. Macmillan, London.

Conte, S. D. and De Boor, C. (1980). *Elementary Numerical Analysis. An Algorithmic Approach*. (3rd edition). McGraw-Hill, Kogakusha, Tokyo.

Dahlquist, G. and Björck, Å. (1974). *Numerical Methods*. (Translated by N. Anderson). Prentice-Hall, Englewood Cliffs, New Jersey.

Davies, D., Swann, W. H. and Campey, I. G. (1964). *Report on the Development of a New Direct Search Method of Optimization*. ICI Ltd, Central Instrument Laboratory Research Note 64/3. (Author W. H. Swann.)

Davis, P. J. and Rabinowitz, P. (1975). *Methods of Numerical Integration*. Academic Press, New York.

De Boor, C. (1978). *A Practical Guide to Splines*. Springer-Verlag, New York.

Findlay, W. and Watt, D. A. (1985). *Pascal: An Introduction to Methodical Programming*. (3rd edition). Pitman, London.

Fisher, A. J. (1983). *Algol 68 on the PRIME: A Manual for Users*, Department of Computer Studies Internal Report No. 83/1, University of Hull.

Flett, T. M. (1966). *Mathematical Analysis*. McGraw-Hill, London.

Gear, C. W. (1971). *Numerical Initial Value Problems in Ordinary Differential Equations*. Prentice-Hall, Englewood Cliffs, New Jersey.

Grant, J. A. and Hitchens, G. D. (1971). An always convergent minimization technique for the solution of polynomial equations. *J. Inst. Math. Appl.* **8**, pp. 122–129.

Henrici, P. (1962). *Discrete Variable Methods in Ordinary Differential Equations*. Wiley, New York.

Johnson, L. W. and Riess, R. D. (1982). *Numerical Analysis*. (2nd edition). Addison-Wesley, Reading, Massachusetts.

Keller, H. B. (1968). *Numerical Methods for Two-Point Boundary Value Problems*. Blaisdell, Waltham, Massachusetts.

Lambert, J. D. (1973). *Computational Methods in Ordinary Differential Equations*. Wiley, London.

O'Hara, H. and Smith, F. J. (1968). Error estimation in the Clenshaw–Curtis quadrature formula. *Comput. J.* **11**, pp. 213–219.

Oliver, J. (1972). A doubly-adaptive Clenshaw–Curtis quadrature method. *Comput. J.* **15**, pp. 141–147.

Ortega, J. M. and Poole, G. P. Jr. (1981). *An Introduction to Numerical Methods for Differential Equations*. Pitman, Marshfield, Massachusetts.

Ortega, J. M. and Rheinboldt, W. C. (1970). *Iterative Solution of Nonlinear Equations in Several Variables*. Academic Press, New York.

Patterson, T. N. L. (1968). The optimum addition of points to quadrature formulae. *Math. Comput.* **22**, pp. 847–856.

Phillips, G. M. and Taylor, P. J. (1973). *Theory and Applications of Numerical Analysis*. Academic Press, London.

Powell, M. J. D. (1981). *Approximation Theory and Methods*. Cambridge University Press, Cambridge.

Ralston, A. and Rabinowitz, P. (1978). *A First Course in Numerical Analysis*. (2nd edition). McGraw-Hill, Kogakusha, Tokyo.

Rice, J. R. (1964). *The Approximation of Functions. Vol. 1, Linear Theory*. Addison-Wesley, Reading, Massachusetts.

Roberts, F. D. K. (1976). An algorithm for minimal degree linear Chebyshev approximation on a discrete set. *Int. J. Numer. Meth. Engng.* **10**, pp. 619–635.

Schumaker, L. L. (1981). *Spline Functions: Basic Theory*. Wiley, New York.

Smith, F. J. (1965). Quadrature methods based on the Euler–MacLaurin formula and on the Clenshaw–Curtis method of integration. *Numer. Math.* **7**, pp. 406–411.

Smith, G. D. (1978). *Numerical Solution of Partial Differential Equations: Finite Difference Methods*. (2nd edition). Clarendon Press, Oxford.

Stroud, A. H. and Secrest, D. (1966). *Gaussian Quadrature Formulas*. Prentice-Hall, Englewood Cliffs, New Jersey.

Stummel, F. and Hainer, K. (1980). *Introduction to Numerical Analysis*. (Translated by E. R. Dawson and W. N. Everitt). Scottish Academic Press, Edinburgh.

Van Wijngaarden, A., Mailloux, B. J., Peck, J. E. L., Koster, C. H. A., Sintzoff, M., Lindsey, C. H., Meertens, L. G. L. T. and Fisker, R. G. (1976). *Revised Report on the Algorithmic Language Algol 68*. Springer-Verlag, Berlin.

Wilkinson, J. H. (1959). The evaluation of the zeros of ill-conditioned polynomials: part 1. *Numer. Math.* **1**, pp. 150–166.

Wilkinson, J. H. (1963). *Rounding Errors in Algebraic Processes*. HMSO, London.

Wolfe, M. A. (1978). *Numerical Methods for Unconstrained Optimization*. Van Nostrand Reinhold, New York.

Index

absolute error, 26
Adams-Bashforth method, 328
Adams-Moulton method, 328
adaptive quadrature, 284
Aitken acceleration, 49, 118
Algol 68, 22, 29
asymptotic error constant
   see functional iteration

back-substitution, 90
backward error analysis, 107
Bairstow iteration, 80
band matrix, 131, 138
band width, 131
basis functions, 169, 171, 173, 178, 227
best approximation, 148, 229
binary numbers, 23, 26
bisection method, 38
bits, 24
boundary value problem, 332, 341

Chebyshev approximation
   see minimax approximation
Chebyshev polynomials, 176, 181, 184, 234, 281
   of the second kind, 193
Chebyshev series approximation, 184
Choleski factorisation, 128, 139
Clenshaw-Curtis quadrature, 283, 290
closed interval, 31
computational error, 250, 296, 302
computing resources, 22
conditioning
   see ill-conditioning
conformant array parameter, 100
constant identifier, 159
contraction mapping theorem, 76
Crout factorisation, 128, 138

damped iterates
   see Newton's method
decimal numbers, 23, 25
   fractional part, 24
   integer part, 24
decimal place accuracy, 25
determinant of a matrix, 92
diagonal dominance, 87
   strict diagonal dominance, 87, 111
differences, 212, 214, 341
   backward, 212
   central, 213
   forward, 212
Doolittle factorisation, 125
dynamic array, 29, 70, 100, 114

economisation, 195
eigenvalues, 340
eigenvectors, 341
Euler's method, 303, 330, 333
   error in, 308
Euler's rule, 299, 314, 315, 316, 321

false position, 63
fixed point format, 24
floating point format, 24
   exponent, 24
   mantissa, 24
   normalised, 24
function minimisation, 205, 218
functional iteration, 40, 314
   asymptotic error constant, 59
   convergence theorem, 42
   for a system of equations, 75
   order of convergence, 59
   termination criteria, 46

Gauss elimination, 88, 127
  full pivoting, 98
  operations count, 92
  partial pivoting, 98
  pivot, 91
  pivotal equation, 90
  scaling, 99
  several right hand sides, 96
  updating formulae, 91
Gauss quadrature, 273
  Gauss Chebyshev, 281
  Gauss Hermite, 281
  Gauss Laguerre, 281
  Gauss Legendre, 273, 290
  Gauss rational, 281
Gauss-Seidel iteration, 109
  convergence theorem, 110
  termination criteria, 112
Gram-Schmidt orthogonalisation, 174, 233

Hamming's method, 328
Hermite interpolating polynomial, 215, 224, 275, 280
  error term, 217
Hermite polynomials, 179, 281
Hilbert matrix, 159

identity declaration, 159
ill-conditioning, 28
  in least squares approximation, 159, 233
  of a polynomial equation, 73
  of a system of equations, 105
Illinois method, 64
infinity norm, 112, 148
initial value problem, 301, 331
  explicit method, 305
  implicit method, 305
  k-step, 305
  one-step, 305
  second order, 331
  simultaneous first order equations, 332
  test problem, 302, 305, 324, 333
inner product, 161, 230
input data, 27
intermediate value theorem, 34
interpolation, 192, 197
inverting a matrix
  see matrix inversion
iteration, 19
iterative refinement, 106

Jacobi iteration, 108
  convergence theorem, 110
  termination criteria, 112

knot, 225
Kronecker delta function, 200

Lagrange interpolating polynomial, 198, 211, 251
  error term, 203
  uniqueness theorem, 202
Laguerre polynomials, 180, 281
least first power approximation, 180, 191, 236, 238
least squares approximation, 149, 230
  normal equations, 151, 173
  weighted least squares, 172, 232
Legendre polynomials, 160, 275
linear convergence, 50
linear independence, 172
linear multistep method, 326
local truncation error, 322
L_p approximation, 180
l_p approximation, 236
  non-uniqueness, 239

machine numbers, 26
machine round-off unit, 26
mapping a range, 164, 188
matrix
  full, 88
  identity, 88, 97
  Jacobian, 77
  lower triangular, 87, 129
  positive definite, 124, 138
  sparse, 88
  symmetric, 87, 123, 130
  transpose, 123
  tridiagonal, 87, 128, 131
  unit, 88
  upper triangular, 87, 129
matrix factorisation, 120
  for tridiagonal matrices, 128
matrix inversion, 97, 137, 141
mean value theorem, 33
midpoint method, 316, 324, 334
midpoint rule, 255, 320
Milne's device, 323
Milne's method, 327
Milne's rule, 256
minimal degree approximation, 244
minimax approximation, 180, 181, 194, 236
  characterisation theorem, 184, 236
  equioscillation property, 182, 236
minimum deviation theorem, 181
MODE declarations, 277, 298
modified initial estimate, 323
monotonic convergence, 46
multiple length precision, 31, 107, 158

NAG, 23, 81, 138, 193, 208, 277, 298, 345
Newton's forward difference formula, 212
Newton's method, 76
  damped iterates, 79
  quasi-Newton method, 79

374 INDEX

Newton-Cotes quadrature rules, 251
  composite rules, 257
Newton-Raphson iteration, 51
  convergence theorem, 52
  for mth root, 20, 45, 60
  for multiple roots, 60
  for square roots, 19, 45, 60
  termination criteria, 54
norm, 107, 112
numerical differentiation, 213
numerical instability, 27
numerical integration
  see quadrature

open interval, 31
operators in Algol 68, 158, 169, 187, 335
ordinary differential equation, 301
orthogonal basis, 169
orthogonal polynomial, 160, 233
orthonormal polynomial, 168
oscillatory convergence, 46
output data, 27
overdetermined system of equations, 229
overflow, 31

Padé approximation, 194
Pascal, 22, 29, 31
Patterson quadrature, 282
Pegasus method, 64
piecewise polynomial approximation, 223
  cubic, 224
  linear, 223
polynomial, 37, 66
  deflated equation, 68
  evaluation of, 67
  evaluation of derivative of, 69
  nested form, 67
  polynomial equation, 37, 66
  synthetic division, 68, 72
polynomial orthogonalisation, 164
  orthonormalisation, 167
PR1ME 550 computer, 26, 29
precision of a quadrature rule, 249
predictor-corrector method, 321, 326, 334
  starting value, 321

quadratic convergence, 60
quadrature, 247
quadrature rule, 248
  closed, 248
  half-open, 248
  open, 248

rational approximation, 194
rectangle rule
  see midpoint rule
recurrence relation, 18, 160, 177, 179
recursive procedure, 290

relative error, 26
residual function, 148
residual vector, 93, 106, 113, 158, 229, 236
Richmond's iteration, 61
Rolle's theorem, 32
Romberg quadrature, 266
  Romberg T-table, 268
roots of an equation, 37
  isolated, 37
  polynomial equations, 66
  roots of a system, 74
  simple, 37
rounding error, 25
Runge-Kutta method, 328, 334

secant method, 65
Sherman-Morrison formula, 138
shooting technique, 344
significant figure accuracy, 25
Simpson's method, 327
Simpson's rule, 253, 285, 327
  adaptive, 284
  composite, 260
  error bound, 255
  error in, 254
spline approximation, 223
  B-splines, 227
  cardinal splines, 229
  end conditions, 225
  hat function, 227
  piecewise cubic approximation, 224
  piecewise linear approximation, 223, 227
stability, 324
Steffensen iteration, 61
step-by-step method, 302
stiffness, 340
  stiffness ratio, 341
Stirling's formula, 36
subrange type, 71
successive over-relaxation, 118

Taylor series expansion, 143, 311
Taylor's theorem, 32
  in two dimensions, 32
trapezium rule, 15, 252, 257, 266, 313
  composite, 257
  error bound, 254
  error in, 16, 254
trapezoidal method, 313
  error in, 313
trim, 188
truncation error, 25, 249, 302, 308

underflow, 31
UPB operator, 101, 139

vector norm, 112

weighting function, 172
wordlength, 29

zeros of an equation
  see roots of an equation